BIOS INSTANT NOTES

Biochemistry

FOURTH EDITION

BIOS INSTANT NOTES

Biochemistry

FOURTH EDITION

David Hames and Nigel Hooper
Faculty of Biological Sciences, University of Leeds

Garland Science
Taylor & Francis Group
NEW YORK AND LONDON

Garland Science
Vice President: Denise Schanck
Editor: Elizabeth Owen
Editorial Assistant: Louise Dawnay
Production Editor: Georgina Lucas
Copyeditor: Alison Gibbs
Typesetting and illustrations: Phoenix Photosetting, Chatham, Kent
Proofreader: Jo Clayton
Printed by: MPG Books Limited

ISBN 978-0-4156-0845-9

Library of Congress Cataloging-in-Publication Data

Hames, B. D.
   Biochemistry / David Hames and Nigel Hooper. — 4th ed.
      p. ; cm. — (BIOS instant notes)
   Includes bibliographical references and index.
   ISBN 978-0-415-60845-9
   1. Biochemistry—Outlines, syllabi, etc. I. Hooper, N. M. II. Title. III.
   Series: BIOS instant notes.
   [DNLM: 1. Biochemical Phenomena—Outlines. 2. Biochemistry—Outlines.
   QU 18.2]
   QP518.3.H355 2011
   612'.015—dc22   2011004854

Published by Garland Science, Taylor & Francis Group, LLC, an informa business,
270 Madison Avenue, New York NY 10016, USA, and 2 Park Square, Milton Park, Abingdon,
OX14 4RN, UK.

15 14 13 12 11 10 9 8 7 6 5 4 3 2 1

Taylor & Francis Group

Visit our web site at http://www.garlandscience.com

Contents

Preface

It is 5 years since we wrote the third edition of *Instant Notes in Biochemistry*. Much has happened in the subject since then; it has truly been an exciting period for biochemical discovery in so many areas. Not surprisingly therefore we faced the same desire experienced by textbook authors the world over – to tell student readers all about the new progress that has been made. However, the reason we wrote the first edition many years ago was to highlight the core information that a first-year student needs to know as a firm foundation for future years of study. This new edition remains true to our original goal but throughout we have taken the opportunity to significantly update the content.

Our firm belief is that, as with earlier editions, this book will provide a fast way for student readers to recognize and understand the main facts and concepts – the bedrock of the subject. For those who wish to know more, in Further Reading we have provided details of some of the excellent mainstream Biochemistry textbooks that take the reader to the cutting edge of the subject, as well as more detailed reviews of specific topic areas.

As we remarked in the Preface of the third edition, although this book is aimed at supporting students primarily in the first and second years of their degree, we know from personal experience of teaching students that it can rapidly become a valuable and long-term companion that can be checked whenever there is a need to refresh one's memory. Therefore, for some students it becomes a useful revision aid.

In writing the fourth edition, we have updated all of the core topic areas. The overall aim was to produce a book that was about the same size as the third edition, since student (and academic staff) feedback was that its relatively small size and clear focus is a major strength compared with the very large textbooks that are generally available.

Finally, a few words directly to the student reader. Like any subject studied at advanced level, Biochemistry can initially seem a daunting (although intellectually highly rewarding) subject. This book is organized in such a way that you do not need to read it from cover to cover. You can dip into it in line with your university coursework, seeking understanding and guidance here and there as your studies progress. Keep it by you as the dreaded examinations approach; it will prove to be a quick way to revise – and to reassure yourself that you do know the fundamental aspects. We hope you will find it as useful as many earlier students told us they found the earlier editions.

David Hames

Nigel Hooper

A1 Prokaryotic cells

Key Notes

The origin and evolution of cells	All organisms on Earth evolved from a common ancestor. In the prebiotic era, simple organic molecules formed and spontaneously polymerized into macromolecules. These macromolecules then acquired the ability to self-replicate and catalyze other reactions, before being enclosed within a membrane to form the first cell. Present-day cells are divided into three main groups: bacteria, archaea and eukaryotes.
Prokaryotes	Prokaryotes are the most abundant organisms on Earth and fall into two distinct groups, the bacteria (or eubacteria) and the archaea (or archaebacteria). A prokaryotic cell does not contain a membrane-bound nucleus.
Prokaryote cell structure	Each prokaryotic cell is surrounded by a plasma membrane. The cell has no subcellular organelles. The deoxyribonucleic acid (DNA) is condensed within the cytosol to form the nucleoid.
Bacterial cell walls	The peptidoglycan (protein and oligosaccharide) cell wall protects the prokaryotic cell from mechanical and osmotic pressure. Some antibiotics, such as penicillin, target enzymes involved in the synthesis of the cell wall. Gram-positive bacteria have a thick cell wall surrounding the plasma membrane, whereas Gram-negative bacteria have a thinner cell wall and an outer membrane, between which is the periplasmic space.
Bacterial flagella	Some prokaryotes have tail-like flagella. By rotation of their flagella, bacteria can move through their surrounding media in response to chemicals (chemotaxis). Bacterial flagella are made of the protein flagellin that forms a long filament whose rotation is driven by a flow of protons through the flagellar motor proteins.
Related topics	(A2) Eukaryotic cells (E2) Membrane structure (B1) Amino acid structure (F2) Genes and chromosomes (B5) Molecular motors (L2) Electron transport and (E1) Membrane lipids oxidative phosphorylation

The origin and evolution of cells

Despite the huge variety of living systems, all organisms on Earth are remarkably uniform at the molecular level, indicating that they have evolved from a **common ancestor**. Life first emerged at least 3.8 billion years ago, although how life originated and how the first cell came into existence are matters of speculation. Experiments have shown that simple organic molecules can form and spontaneously polymerize into macromolecules under the conditions thought to exist in primitive Earth's atmosphere, the so-called **pre-biotic era**. The next critical step was the ability of the macromolecules to **self-replicate**, as seen with the present-day nucleic acids, and to catalyze other reactions, as demonstrated for ribonucleic acid (**RNA**) (Section G1). The first cell is presumed to have arisen by the enclosure of the self-replicating RNA in a **membrane** composed of **phospholipids** (Section E1), thus separating the interior of the cell from its external environment. The encapsulated macromolecules would thus have been maintained as a unit, capable of self-reproduction and further evolution to give rise to the variety of life forms found on Earth today. The analysis of the deoxyribonucleic acid (**DNA**) sequences (Section F1) of organisms has allowed a possible evolutionary path from a common ancestral cell to the present-day cells and organisms to be deduced (Figure 1).

The living world therefore has three major divisions or domains: **bacteria**, **archaea**, and **eukaryotes** (Section A2). The bacteria are the commonly encountered prokaryotes in soil, water and living in or on larger organisms, and include *Escherichia coli* (*E. coli*) and the *Bacillus* species, as well as the cyanobacteria (photosynthetic blue-green algae). The archaea mainly inhabit unusual environments such as salt brines, hot acid springs, and the ocean depths, and include the sulfur bacteria and the methanogens, although some are found in less hostile environments.

Prokaryotes

Prokaryotes are the most numerous and widespread organisms on Earth, and are so classified because they have no defined membrane-bound nucleus. Prokaryotes comprise two separate but related groups: the bacteria (or eubacteria) and the archaea (or archaebacteria). These two distinct groups of prokaryotes diverged early in the history of life on Earth (Figure 1).

Prokaryote cell structure

Prokaryotes generally range in size from 0.1 to 10 µm, and have one of three basic shapes: spherical (cocci), rod-like (bacilli) or helically coiled (spirilla). Like all cells, a prokaryotic

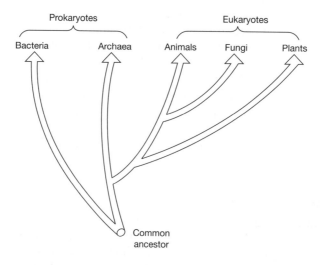

Figure 1. Evolution of cells.

cell is bounded by a **plasma membrane** that completely encloses the cytosol and separates the cell from the external environment (Figure 2). The plasma membrane, which is about 8 nm thick, consists of a lipid bilayer containing proteins (Sections E1 and E2). Prokaryotes lack the membranous subcellular organelles characteristic of eukaryotes (Section A2). The aqueous cytosol contains the macromolecules [enzymes, messenger ribonucleic acid (mRNA), transfer RNA (tRNA) and ribosomes], organic compounds and ions needed for cellular metabolism. Also within the cytosol is the prokaryotic 'chromosome' consisting of a single circular molecule of DNA, which is condensed to form a body known as the **nucleoid** (Figure 2) (Section F2).

Bacterial cell walls

To protect the cell from mechanical injury and osmotic pressure, most prokaryotes are surrounded by a rigid 3–25 nm-thick **cell wall** (Figure 2). The cell wall is composed of **peptidoglycan**, a complex of **oligosaccharides** and **proteins**. The oligosaccharide component consists of linear chains of alternating *N*-acetylglucosamine (GlcNAc) and *N*-acetylmuramic acid (NAM) linked β1–4 (Section J1). Attached via an amide bond to the lactic acid group on NAM is a D-**amino acid**-containing tetrapeptide. Adjacent parallel peptidoglycan chains are covalently cross-linked through the tetrapeptide side-chains by other short peptides. The extensive cross-linking in the peptidoglycan cell wall gives it its strength and rigidity. The presence of D-amino acids in the peptidoglycan renders the cell wall resistant to the action of **proteases**, which act on the more commonly occurring L-amino acids (Section B1), but provides a unique target for the action of certain **antibiotics** such as **penicillin**. Penicillin acts by inhibiting the enzyme that forms the covalent cross-links in the peptidoglycan, thereby weakening the cell wall. The β1–4 glycosidic linkage between NAM and GlcNAc is susceptible to hydrolysis by the enzyme **lysozyme**, which is present in tears, mucus and other body secretions.

Bacteria can be classified as either **Gram-positive** or **Gram-negative** depending on whether they take up the **Gram stain**. Gram-positive bacteria (e.g. *Bacillus polymyxa*) have a thick (25 nm) cell wall surrounding their plasma membrane, whereas Gram-negative bacteria (e.g. *E. coli*) have a thinner (3 nm) cell wall and a second **outer membrane** (Figure 3). In contrast with the plasma membrane, this outer membrane is very

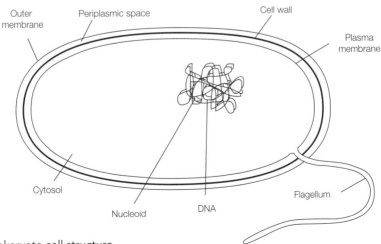

Figure 2. Prokaryote cell structure.

Figure 3. Cell wall structure of (a) Gram-positive and (b) Gram-negative bacteria.

permeable to the passage of relatively large molecules (molecular weight >1000 Da) due to **porin proteins**, which form pores in the lipid bilayer (Section E3). Between the outer membrane and the cell wall is the **periplasm**, a space occupied by proteins secreted from the cell.

Bacterial flagella

Many bacterial cells have one or more tail-like appendages known as **flagella**. By **rotating** their flagella, bacteria can move through the extracellular medium towards attractants and away from repellents, so called **chemotaxis**. Bacterial flagella are different from eukaryotic cilia and flagella in two ways: 1. each bacterial flagellum is made of the protein **flagellin** (53 kDa subunit) as opposed to tubulin (Section B5); and 2. it **rotates** rather than bends. An *E. coli* bacterium has about six flagella that emerge from random positions on the surface of the cell. Flagella are thin helical filaments, 15 nm in diameter and 10 μm long. Electron microscopy has revealed that the **flagellar filament** contains 11 subunits in two helical turns, which, when viewed end-on, has the appearance of an 11-bladed propeller with a hollow central core. Flagella grow by the addition of new flagellin subunits to the end away from the cell, with the new subunits diffusing through the central core. At its base, situated in the plasma membrane, is the **flagellar motor**, an intricate assembly of 40 or so proteins. The rotation of the flagella is driven by a flow of protons through certain proteins of the flagellar motor. A similar proton-driven motor is found in the F_0F_1-adenosine triphosphatase (ATPase) that synthesizes adenosine triphosphate (ATP) (Section L2).

A2 Eukaryotic cells

Key Notes

Eukaryotes
Eukaryotic cells have a membrane-bound nucleus and a number of other membrane-bound subcellular (internal) organelles, each of which has a specific function.

Plasma membrane
The plasma membrane surrounds the cell, separating it from the external environment. The plasma membrane is a selectively permeable barrier due to the presence of specific transport proteins and has receptor proteins that bind specific ligands. It is also involved in the processes of exocytosis and endocytosis.

Nucleus
The nucleus stores the cell's genetic information as DNA in chromosomes. It is bounded by a double membrane but pores in this membrane allow molecules to move in and out of the nucleus. The nucleolus within the nucleus is the site of ribosomal RNA synthesis.

Endoplasmic reticulum
This interconnected network of membrane vesicles is divided into two distinct parts. The rough endoplasmic reticulum (RER), which is studded with ribosomes, is the site of membrane and secretory protein biosynthesis and their post-translational modification. The smooth endoplasmic reticulum (SER) is involved in phospholipid biosynthesis and in the detoxification of toxic compounds.

Golgi apparatus
The Golgi apparatus is a system of flattened membrane-bound sacs. It receives membrane vesicles from the RER, further modifies the proteins within them, and then packages the modified proteins in other vesicles, which eventually fuse with the plasma membrane or other subcellular organelles.

Mitochondria
Mitochondria have an inner and an outer membrane separated by the intermembrane space. The outer membrane is more permeable than the inner membrane due to the presence of porin proteins. The inner membrane, which is folded to form cristae, is the site of oxidative phosphorylation, which produces ATP. The central matrix is the site of fatty acid degradation and the citric acid cycle.

Chloroplasts
Chloroplasts in plant cells are surrounded by a double membrane and have an internal membrane system of thylakoid vesicles that are stacked up to form grana. The thylakoid vesicles contain chlorophyll and are the site of photosynthesis. Carbon dioxide (CO_2) fixation takes place in the stroma, the soluble matter around the thylakoid vesicles.

Lysosomes	Lysosomes in animal cells are bounded by a single membrane. They have an acidic internal pH (pH 4–5), maintained by proteins in the membrane that pump in H^+ ions. Within the lysosomes are acid hydrolases – enzymes involved in the degradation of macromolecules, including those internalized by endocytosis.
Peroxisomes	Peroxisomes contain enzymes involved in the breakdown of amino acids and fatty acids, a byproduct of which is hydrogen peroxide. This toxic compound is rapidly degraded by the enzyme catalase, also found within the peroxisomes.
Cytosol	The cytosol is the soluble part of the cytoplasm where a large number of metabolic reactions take place, e.g. glycolysis, gluconeogenesis, fatty acid synthesis. Within the cytosol is the cytoskeleton.
Cytoskeleton	The cytoskeleton is an internal scaffold that controls the shape and movement of the cell and the organelles within it. The cytoskeleton consists of microfilaments, intermediate filaments and microtubules. Microfilaments are 5–9 nm diameter helical polymers of the protein actin that have a mechanically supportive function in the cell. Intermediate filaments are 7–11 nm diameter rope-like fibers made from a family of intermediate filament proteins that provide mechanical strength and resistance to shear stress. Microtubules are hollow cylinders of 25 nm diameter made of the protein tubulin. The mitotic spindle involved in separating the chromosomes during cell division is made of microtubules. Colchicine and vinblastine inhibit microtubule formation, whereas taxol stabilizes microtubules. Through interfering with mitosis, some of these compounds are used as anticancer drugs.
Plant cell wall	The cell wall surrounding a plant cell is made up of the polysaccharide cellulose. In wood, the phenolic polymer called lignin gives the cell wall additional strength and rigidity.
Plant cell vacuole	The membrane-bound vacuole is used to store nutrients and waste products, has an acidic pH and, due to the influx of water, creates turgor pressure inside the cell as it pushes out against the cell wall.
Related topics	(A1) Prokaryotic cells (F2) Genes and chromosomes (E4) Membrane transport: (H4) Protein targeting macromolecules

Eukaryotes

Eukaryotes include species as diverse as animals, plants, fungi and slime molds. In all eukaryotes the cell is surrounded by a **plasma membrane**, has a **membrane-bound nucleus** and contains a number of other distinct **subcellular organelles** (Figure 1). These organelles are membrane-bounded structures, each having a unique role and each containing a specific complement of proteins and other molecules. The key differences between eukaryotic and prokaryotic cells (Section A1) are listed in Table 1. Animal and plant cells have the same basic structure, although some organelles and structures are found in one and not the other (e.g. chloroplasts, vacuoles and cell wall in plant cells, lysosomes in animal cells).

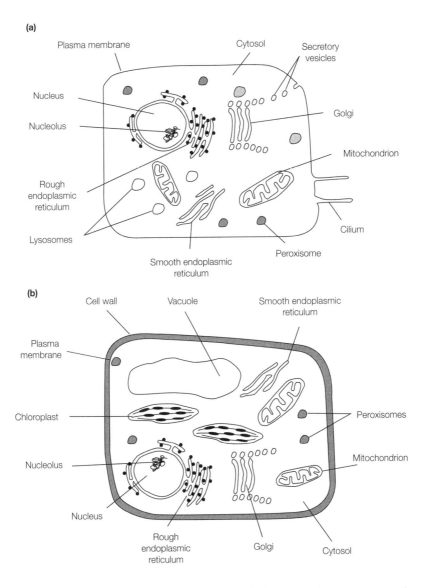

Figure 1. Eukaryote cell structure: (a) structure of a typical animal cell; (b) structure of a typical plant cell.

Table 1. The key differences between eukaryotic and prokaryotic cells

Characteristic	Prokaryote	Eukaryote
Diameter	Approx. 1 μm	10–100 μm
Nucleus	Absent	Present
Cytoplasmic organelles	Absent	Present
DNA content (base pairs)	1×10^6 to 5×10^6	1.5×10^7 to 5×10^9
Chromosomes	Single circular DNA molecule	Multiple linear DNA molecules

Plasma membrane

The plasma membrane envelops the cell, separating it from the external environment and maintaining the correct ionic composition and osmotic pressure of the cytosol. The plasma membrane, like all membranes, is **selectively permeable** to certain molecules due to the presence of specific proteins in the membrane (Setion E3). The plasma membrane is also involved in communicating with other cells, in particular through the binding of ligands (small molecules such as hormones, neurotransmitters, etc.) to **receptor proteins** on its surface (Section E5). The plasma membrane is also involved in the **exocytosis** (secretion) and **endocytosis** (internalization) of proteins and other macromolecules (Section E4).

Nucleus

The nucleus is bounded by two membranes, the **inner and outer nuclear membranes**. These two membranes fuse together at the **nuclear pores** through which molecules (mRNA, proteins, ribosomes, etc.) can move between the nucleus and the cytosol. Other proteins, for example those involved in regulating gene expression, can pass through the pores from the cytosol to the nucleus. The outer nuclear membrane is often continuous with the rough endoplasmic reticulum (RER). Within the nucleus, the **DNA** is tightly coiled around **histone proteins** and organized into complexes called **chromosomes** (Section F2). Visible under the light microscope (Section A4) is the **nucleolus**, a subregion of the nucleus, which is the site of ribosomal ribonucleic acid (rRNA) synthesis.

Endoplasmic reticulum

The endoplasmic reticulum (ER) is an interconnected network of membrane vesicles. The **rough endoplasmic reticulum (RER)** is studded on the cytosolic face with **ribosomes**, the sites of **membrane and secretory protein biosynthesis** (Section H4). Within the lumen of the RER are enzymes involved in the **post-translational modification** (glycosylation, proteolysis, etc.) of membrane and secretory proteins (Section H5). The **smooth endoplasmic reticulum (SER)**, which is not studded with ribosomes, is the site of **phospholipid biosynthesis**, and is where a number of **detoxification reactions** take place.

Golgi apparatus

The Golgi apparatus is a system of flattened membrane-bound sacs. Membrane vesicles from the RER, containing membrane and secretory proteins, fuse with the *cis* face of the Golgi apparatus and release their contents into it. On transit through the Golgi apparatus, further **post-translational modifications** to these proteins take place and they are then sorted and packaged into different vesicles (Section H5). These vesicles bud off from the

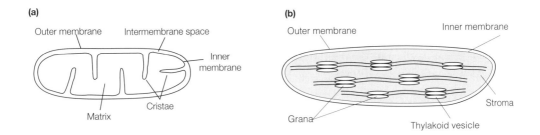

Figure 2. Structure of (a) a mitochondrion and (b) a chloroplast.

trans face of the Golgi apparatus (also called the trans-Golgi network) and are transported through the cytosol, eventually fusing either with the plasma membrane to release their contents into the extracellular space (a process known as **exocytosis**; Section E4) or with other internal organelles (e.g. lysosomes) (for details Section H4).

Mitochondria

A mitochondrion has an **inner and an outer membrane** between which is the **intermembrane space** (Figure 2a). The outer membrane contains **porin proteins**, which make it permeable to molecules of up to 10 kDa. The inner membrane, which is considerably less permeable, has large infoldings called **cristae**, which protrude into the **central matrix**. The inner membrane is the site of oxidative phosphorylation and electron transport involved in ATP production (Section L2). The central matrix is the site of numerous metabolic reactions including the citric acid cycle (Section L1) and fatty acid breakdown (Section K2). Also within the matrix is found the mitochondrial DNA, which encodes some of the mitochondrial proteins.

Chloroplasts

Chloroplasts, present exclusively in plant cells, also have **inner and outer membranes**. In addition, there is an extensive internal membrane system made up of **thylakoid vesicles** (interconnected vesicles flattened to form disks) stacked upon each other to form **grana** (Figure 2b). Within the thylakoid vesicles is the green pigment **chlorophyll** (Section M4), along with the enzymes that trap light energy and convert it into chemical energy in the form of ATP (Section L3). The **stroma**, the space surrounding the thylakoid vesicles, is the site of carbon dioxide (CO_2) fixation – the conversion of CO_2 into organic compounds. Chloroplasts, like mitochondria, contain DNA, which encodes some of the chloroplast proteins.

Lysosomes

Lysosomes, which are found only in animal cells, have a single boundary membrane. The internal pH of these organelles is **mildly acidic** (pH 4–5), and is maintained by integral membrane proteins that pump H^+ ions into them (Section E3). The lysosomes contain a range of hydrolases that are optimally active at this acidic pH (and hence are termed **acid hydrolases**) but which are inactive at the neutral pH of the cytosol and extracellular fluid. These enzymes are involved in the degradation of host and foreign macromolecules into their monomeric subunits; **proteases** degrade proteins, **lipases** degrade lipids, **phosphatases** remove phosphate groups from nucleotides and phospholipids, and **nucleases** degrade DNA and RNA. Lysosomes are involved in the degradation of extracellular macromolecules that have been brought into the cell by **endocytosis** (Section E4), as well as in the degradation and recycling of normal aged cellular organelles, a process called **autophagy**.

Peroxisomes

These organelles have a single boundary membrane and contain enzymes that degrade fatty acids and amino acids. A byproduct of these reactions is **hydrogen peroxide**, which is toxic to the cell. The presence of large amounts of the enzyme **catalase** in the peroxisomes rapidly converts the toxic hydrogen peroxide into harmless H_2O and O_2:

$$2H_2O_2 \xrightarrow{\text{Catalase}} 2H_2O + O_2$$

Cytosol

The cytosol is that part of the **cytoplasm** not included within any of the subcellular organelles, and is a major site of cellular metabolism, containing a large number of different enzymes and other proteins. For example, glycolysis (Section J3), gluconeogenesis (Section J4), the pentose phosphate pathway (Section J5) and fatty acid synthesis (Section K3) all take place in the cytosol. The cytosol is not a homogeneous 'soup' but has within it the **cytoskeleton**. Also found within the cytosol of many cells are **inclusion bodies** (granules of material that are not membrane-bounded) such as glycogen granules in liver and muscle cells, and droplets of triacylglycerol in the fat cells of adipose tissue.

Cytoskeleton

The cytoskeleton is an **internal scaffold** important in maintaining and altering the shape of the cell, in enabling cells such as sperm and white blood cells to move from one place to another, in transporting intracellular vesicles, and in pulling the chromosomes apart at mitosis and then dividing the cell in two. Three types of **filaments** make up the cytoskeleton: microfilaments, intermediate filaments and microtubules, each with distinct mechanical properties and dynamics.

Microfilaments

The **microfilaments** (also known as **actin filaments**), diameter 5–9 nm, have a mechanically supportive function, determining the shape of the cell's surface and they are involved in whole cell movement. Microfilaments are two-stranded helical polymers of the protein **actin**, which appear as flexible structures organized into a variety of linear bundles and more extensive networks. Through their interaction with myosin, the microfilaments form contractile assemblies that are involved in various intracellular movements such as cytoplasmic streaming and the formation of membrane invaginations (Section B5).

Intermediate filaments

The **intermediate filaments** (7–11 nm in diameter) provide mechanical strength and resistance to shear stress. They are made of **intermediate filament proteins**, which constitute a large and heterogeneous family, that form rope-like fibers. The skin in higher animals contains an extensive network of intermediate filaments made up of the protein **keratin** that has a two-stranded α-helical coiled-coil structure, while the nuclear lamina, a meshwork just beneath the inner nuclear membrane, is formed from another type of intermediate filament.

Microtubules

The third type of cytoskeletal filaments, the **microtubules**, determines the position of membrane-bound organelles and directs their intracellular transport. For example, the **mitotic spindle** involved in separating the replicated chromosomes during mitosis is an assembly of microtubules. Microtubules are hollow cylindrical structures with an outer diameter of 25 nm that are built from the protein **tubulin** (Figure 3). The rigid wall of a microtubule is made up of a helical array of alternating α- and β-tubulin subunits, each 50 kDa in size. A cross-section through a microtubule reveals that there are 13 tubulin subunits per turn of the filament. Microtubules in cells are formed by the addition of α- and β-tubulin molecules to pre-existing filaments or nucleation centers. One end of the microtubule is usually attached to a **microtubule-organizing center** called a **centrosome**. The drugs **colchicine** and **vinblastine** inhibit the polymerization of microtubules, thus blocking cell processes such as cell division that depend on functioning microtubules. Another compound, **taxol**, stabilizes tubulin in microtubules and promotes polymerization. Some of these compounds, such as vinblastine and taxol, are being used as **anticancer drugs**, since they block the proliferation of rapidly dividing cells by interfering with the mitotic spindle.

Plant cell wall

Surrounding the plasma membrane of a plant cell is the cell wall, which imparts strength and rigidity to the cell. This is built primarily of **cellulose**, a rod-like **polysaccharide** of repeating glucose units linked β1–4 (Section J2). These cellulose molecules are aggregated together by hydrogen bonding into bundles of fibers, and the fibers in turn

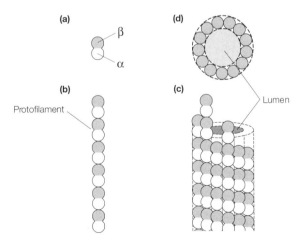

Figure 3. The structure of a microtubule: (a) tubulin consists of α- and β-subunits; (b) a tubulin protofilament consisting of many adjacent subunits; (c) the microtubule is formed from 13 protofilaments aligned in parallel; (d) cross-section of the hollow microtubule.

are cross-linked together by other polysaccharides. In wood another compound, **lignin**, imparts added strength and rigidity to the cell wall. Lignin is a complex water-insoluble phenolic polymer.

Plant cell vacuole

Plant cells usually contain one or more **membrane-bounded vacuoles**. These are used to store nutrients (e.g. sucrose), water, ions and waste products (especially excess nitrogen-containing compounds). Like lysosomes in animal cells, vacuoles have an **acidic pH** maintained by H^+ pumps in the membrane and contain a variety of **degradative enzymes**. Entry of water into the vacuole causes it to expand, creating hydrostatic pressure (**turgor**) inside the cell, which is balanced by the mechanical resistance of the cell wall.

A3 Cell growth

Key Notes

Cell culture: overview	Cells can be grown in culture under appropriate conditions, allowing cell biological studies and genetic manipulation to be undertaken.
Prokaryotic cell culture and division	Prokaryotic cells can be grown in simple media containing inorganic salts and a source of energy. In the prokaryotic cell cycle, the replication of DNA occurs as the cell enlarges, prior to division into two daughter cells by binary fission.
Eukaryotic cell culture	Animal and plant cells can be grown in culture but, in addition to inorganic salts and glucose, require various amino acids, vitamins and growth factors to survive. Primary cultures are prepared directly from tissues and have a finite life-span. Immortal cell lines derived from stem cells or tumor cells can proliferate indefinitely in culture.
Eukaryotic cell cycle	In eukaryotes, the cell cycle consists of M, G_1, S and G_2 phases. In the M phase, which lasts about 1 h, mitosis and cell division occur, while in the S phase the chromosomal DNA is replicated. Most of the cell cycle (95%) is spent in interphase (G_1, S and G_2 phases). Quiescent cells are said to be in the G_0 phase.
Related topics	(A1) Prokaryotic cells (A2) Eukaryotic cells (F2) Genes and chromosomes (F3) DNA replication in bacteria (F4) DNA replication in eukaryotes

Cell culture: overview

Both prokaryotic and eukaryotic cells can be grown and manipulated in **culture**. Cell culture provides a large number of identical cells that can be used for a variety of cell biological and biochemical studies. Such *in vitro* cell culture systems allow the growth of the cells to be studied, and genetic manipulations to understand gene structure and function to be performed.

Prokaryotic cell culture and division

Prokaryotes such as *E. coli* can be grown in simple media solutions containing inorganic salts and a source of energy, for example, glucose. Some prokaryotes can replicate in approximately 20 min under ideal conditions (i.e. presence of an appropriate energy source, growth at the optimal temperature), which means that a single cell can multiply to many millions over a few hours.

The **prokaryote cell cycle** is relatively simple since the replication of the **DNA** in the **single circular chromosome** (Section A1) occurs continuously as the cell enlarges. The cell then divides by **binary fission**. In this process, a new cell membrane and cell wall forms around the middle of the cell, which ultimately divides the parent cell into two daughter cells. Each of the daughter cells receives a copy of the chromosomal DNA.

Eukaryotic cell culture

Although **yeast** can be grown readily in culture using minimal media as for bacterial cells, other eukaryotic cells (**animal and plant cells**) require much more complex culture media that contains in addition to salts and glucose, various amino acids and vitamins, which the cells cannot make themselves. Animal cell growth media also contains serum, which provides the various protein growth factors required to support cell division.

Animal cell cultures are initiated by the dispersion (for example by gentle homogenization, Section A5) of a piece of tissue into a suspension of the component cells. The isolated cells are then grown in a plastic **culture dish** under appropriate conditions with **defined growth medium**. Cultures prepared directly from the tissues of an organism are referred to as **primary cultures**. Such primary cultures can usually be cultured for 50–100 population doublings, after which they stop growing and die. In contrast, cells derived from tumors and embryonic stem cells frequently proliferate indefinitely in culture, and are referred to as permanent or immortal **cell lines**. Under the appropriate conditions, many cells in culture retain the differentiated properties appropriate to their origin. For example, fibroblasts continue to secrete collagen, and nerve cells extend axons. **Stem cells**, often cultured from early embryos, maintain their ability to differentiate into all of the cell types of the adult organism.

Eukaryotic cell cycle

The division of eukaryotic cells must be carefully regulated with both cell growth and DNA replication being carefully coordinated in order to ensure the formation of daughter cells containing the complete complement of intact chromosomes (that is the complete genome, Section F2). The life of a eukaryotic cell can be defined as a **cell cycle**, which consists of four coordinated processes: cell growth, DNA replication, distribution of the duplicated chromosomes to daughter cells, and cell division.

Mitosis (nuclear division) in which the daughter chromosomes are separated and **cell division** (**cytokinesis**) occur in the **M phase**, which lasts for only about 1 h (Figure 1).

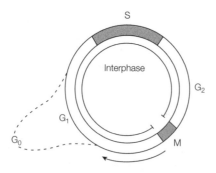

Figure 1. The eukaryotic cell cycle. The S phase is typically 6–8 h long, G_2 is a phase in which the cell prepares for mitosis and lasts for 2–6 h, mitosis (M) itself is short and takes only about 1 h. The length of G_1 is very variable and depends on the cell type. Cells can enter G_0, a quiescent phase, instead of continuing with the cell cycle.

The remainder of the cell cycle (approximately 95%) is spent in **interphase** (Figure 1), the period between mitoses. During interphase, the chromosomes are decondensed and distributed throughout the nucleus; at the molecular level, this is the time when both cell growth and DNA replication occur.

Mitosis is followed by the **G_1 phase** (G for gap) (Figure 1), which corresponds to the interval between mitosis and the initiation of DNA replication when the cell grows continuously. Then the cell enters the **S phase** (S for synthesis), during which time the chromosomal DNA is replicated, and finally the **G_2 phase** when cell growth continues and proteins are synthesized in preparation for mitosis. Eukaryotic cells in culture typically have cell cycle times of 16–24 h but the cell cycle time can be much longer (>100 days) for some cells in a multicellular organism. Most of the variation in cell cycle times occurs by differences in the length of the G_1 phase. Some cells *in vivo*, such as neurons, stop dividing completely and are said to be quiescent, locked in a **G_0 phase** (Figure 1).

A4 Cell imaging

Key Notes

Light microscopy

In light microscopy, a beam of light is focused using glass lenses to produce an enlarged image of the specimen. In bright-field microscopy, the specimen is illuminated from below with the beam of light being focused on to it by the condenser lens. The incident light that passes through the specimen is then focused by the objective lens on to its focal plane, creating a magnified image. The specimen to be viewed is first fixed with alcohol or formaldehyde, and then stained with a chemical, such as hematoxylin or eosin. Phase-contrast microscopy and the more complex differential interference contrast microscopy can be used to visualize living cells.

Fluorescence microscopy

In fluorescence microscopy, fluorescent compounds (which absorb light at the exciting wavelength and then emit it at the emission wavelength) are used to visualize a particular component in the cell. In confocal microscopy, a laser focuses light of the exciting wavelength on to the specimen so that only a thin section of it is illuminated. The laser beam is moved through the sample, producing a series of images, which are then reassembled by a computer to produce a three-dimensional picture of the specimen. Fluorescent compounds such as rhodamine and fluorescein can be coupled to antibodies that recognize the proteins of interest. The naturally occurring green fluorescent protein (GFP) from a jellyfish can be tagged on to other proteins and used to visualize the location and movement of proteins in living cells. Interactions between one protein and another can be monitored by fluorescence resonance energy transfer (FRET) by labeling the two proteins of interest with different fluorochromes. The lateral movement of a fluorescently tagged protein can be monitored by fluorescence recovery after photobleaching (FRAP).

Electron microscopy

In electron microscopy, a beam of electrons is focused using electromagnetic lenses. The specimen is mounted within a vacuum so that the electrons are not absorbed by atoms in the air. In transmission electron microscopy, the beam of electrons is passed through a thin section of the specimen that has been stained with heavy metals. The electron-dense metals scatter the incident electrons, thereby producing an image of the specimen. In scanning electron microscopy, the surface of a whole specimen is coated with a layer of heavy metal and then scanned with an electron beam to produce a three-dimensional image of the specimen.

Related topics	(A2) Eukaryotic cells	(C4) Immunodetection
	(B6) Antibodies	(E2) Membrane structure

Light microscopy

In light microscopy, **glass lenses** are used to focus a beam of light on to the **specimen** under investigation. The light passing through the specimen is then focused by other lenses to produce a **magnified image**.

Several different types of light microscopy are routinely used to study various aspects of cell structure. The simplest is **bright-field microscopy** in which the specimen is illuminated from underneath by a lamp in the base of the microscope (Figure 1), with the light being focused on to the plane of the specimen by a **condenser lens**. Incident light coming through the specimen is picked up by the **objective lens** and focused on to its focal plane, creating a magnified image. This image is further magnified by the eyepiece, with the total magnification achieved being the sum of the magnifications of the individual lenses. In order to increase the resolution achieved by a compound microscope, the specimen is often overlaid with **immersion oil** into which the objective lens is placed. The limit of resolution of the light microscope using visible light is approximately 0.2 μm.

Fixing and staining specimens

In bright-field microscopy the specimen to be examined is usually first fixed with a solution containing alcohol or formaldehyde. These compounds denature proteins and, in the case of formaldehyde, introduce covalent cross-links between amino groups on adjacent molecules, which stabilize protein–protein and protein–nucleic acid interactions. The **fixed specimen** may then be embedded in paraffin wax or a resin and cut into **thin sections** (0.5–10 μm thick) using a **microtome**. Each section is mounted on a glass slide and then positioned on the movable specimen stage of the microscope (Figure 1). The various subcellular constituents (nucleus, mitochondria, cytosol, etc.) absorb about the same degree of visible light, making it difficult to distinguish them under the light

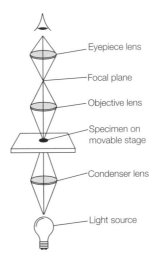

Figure 1. Optical pathway of a compound microscope.

microscope without first staining the specimen. Many **chemical stains** bind to biological molecules; for example, **hematoxylin** binds to the basic amino acids Arg and Lys in proteins, and **eosin** binds to acidic molecules (such as DNA and the side-chains of the amino acids Asp and Glu). Another way of visualizing specific structures within cells is **cytochemical staining** in which an enzyme catalyzes the production of many molecules of a localized, colored reaction product from a colorless precursor. The colored product can then be seen in the light microscope wherever the enzyme is present. For example, peroxisomes can be visualized by using a cytochemical stain for catalase (Section A2).

Phase-contrast microscopy

When light passes through a living cell, the phase of the light wave is changed according to the **refractive index** of the cell: light passing through a relatively thick or dense part of the cell, such as the nucleus, is retarded; consequently, its **phase** is shifted relative to light that has passed through an adjacent thinner region of the cytoplasm. Both **phase-contrast microscopy** and, in a more complex way, **differential interference contrast microscopy** (or Nomarski interference microscopy), exploit the interference effects produced when the two sets of light waves recombine, thereby creating an image of the cell's structure. As these types of microscopy do not require specimens to be fixed or stained they are useful for examining the structure of larger organelles (nucleus, mitochondria, etc.) in living cells. **Video-enhanced differential interference contrast microscopy** can be used to visualize the movement of organelles within cells, for example along microtubules (Section A2).

Fluorescence microscopy

In fluorescence microscopy, the light microscope is adapted to detect the light emitted by a **fluorescent compound** that is used to stain selectively components within the cell. A chemical is said to be fluorescent if it absorbs light at one wavelength (the **excitation wavelength**) and then emits light at a longer wavelength (the **emission wavelength**). Two commonly used compounds in fluorescent microscopy are **rhodamine** and **Texas red**, which both emit red light, and **fluorescein**, which emits green light. First, an antibody against the antigen of interest (so-called **primary antibody**; Section B6) is added to the specimen. A fluorescent compound is chemically coupled to a **secondary antibody** that recognizes the primary antibody. Then the fluorescently tagged secondary antibody is added to the tissue section or permeabilized cell, and the specimen is illuminated with light at the exciting wavelength (Figure 2). The structures in the specimen to which the antibody has bound can then be visualized.

Figure 2. Labeling of protein with a fluorescently tagged antibody for fluorescent microscopy. The primary antibody recognizes the antigen of interest and binds to it in the specimen. Several molecules of the secondary antibody bind to the primary antibody providing amplification of the signal. The secondary antibody is covalently coupled to a fluorescent dye that emits light when illuminated at its excitation wavelength.

Confocal microscopy is a refinement of normal fluorescence microscopy, which produces clearer images of whole cells or larger specimens. In normal fluorescence microscopy, the fluorescent light emitted by the compound comes from molecules above and below the **plane of focus**, blurring the image and making it difficult to determine the actual three-dimensional molecular arrangement. With the confocal microscope, only molecules in the plane of focus fluoresce due to the use of a focused **laser beam** at the exciting wavelength. The laser beam is moved to different parts of the specimen, allowing a series of images to be taken at different depths through the sample. The images are then combined by a computer to provide the complete three-dimensional image. **Deconvolution microscopy** achieves the same image-sharpening effect as confocal microscopy but through a different process.

Green fluorescent protein

Visualization of proteins in living cells has been revolutionized by the discovery of a naturally fluorescent protein found in the **jellyfish** *Aequorea victoria*. In this 238 amino acid protein, called **green fluorescent protein** (GFP), certain amino acid side-chains have spontaneously cyclized to form a green-fluorescing chromophore. Using recombinant DNA techniques (Section I1), the DNA encoding GFP can be tagged on to the DNA sequences encoding other proteins, and then introduced into living cells in culture or into specific cells of a whole animal. Cells containing the introduced gene will then produce the **protein tagged with GFP**, which will fluoresce green under the fluorescence microscope. The localization and movement of the GFP-tagged protein can then be studied in living cells in real time. Multiple variations of GFP have been engineered, which emit light at different wavelengths, e.g. cyan fluorescent protein (CFP) and yellow fluorescent protein (YFP), allowing several proteins to be visualized simultaneously in the same cell.

Fluorescence resonance energy transfer (FRET)

Interactions between one protein and another can be monitored by **fluorescence resonance energy transfer** (FRET). The two proteins of interest are each labeled with a **different fluorochrome** (tagged with different variants of GFP, see above), chosen so that the emission spectrum of one fluorochrome overlaps with the excitation spectrum of the other (Figure 3a). If the two proteins come into very **close proximity** (closer than 2 nm), the energy of the absorbed light can be transferred directly from one fluorochrome to the other (Figure 3c). Thus, when the sample is illuminated at the excitation wavelength of the first fluorochrome, light is emitted at the emission wavelength of the second. If the two proteins fail to come into close proximity then no transfer of fluorescence occurs (Figure 3b).

Fluorescence recovery after photobleaching (FRAP)

The **lateral movement** of proteins in a membrane can be visualized through the use of the technique **fluorescence recovery after photobleaching** (FRAP). In this technique, a small region of interest in the cell expressing a GFP-labeled protein is exposed to a very intense light pulse from a laser, which destroys (**bleaches**) the fluorescent molecules in that area. The fluorescence of this bleached region is subsequently monitored as a function of time, as other fluorescently labeled proteins move into the bleached region of the membrane. The rate of recovery of the fluorescence depends on the lateral mobility of the fluorescently labeled protein.

Figure 3. FRET. To determine whether two proteins interact inside the cell, the proteins are first tagged with two different variants of GFP. (a) In this example, protein X is coupled to CFP, which is excited at 440 nm and emits blue light at 490 nm, while protein Y is coupled to YFP, which is excited at 490 nm and emits yellow light at 527 nm. (b) If protein X and Y do not interact, illuminating the sample at 440 nm yields fluorescence at 490 nm from CFP only. (c) When protein X and Y interact, FRET now occurs. Illuminating the sample at 440 nm excites CFP, whose emission in turn excites YFP, resulting in the emission of yellow light at 527 nm.

Electron microscopy

In contrast with light microscopy where optical lenses focus a beam of light, in electron microscopy **electromagnetic lenses** focus a beam of **electrons**. Because electrons are absorbed by atoms in the air, the specimen has to be mounted in a **vacuum** within an evacuated tube. The resolution of the electron microscope with biological materials is at best 0.10 nm. In **transmission electron microscopy**, a beam of electrons is directed through the specimen and electromagnetic lenses are used to focus the **transmitted electrons** to produce an image either on a viewing screen or on photographic film (Figure 4a). As in standard light microscopy, thin sections of the specimen are viewed. However, for transmission electron microscopy the sections must be much thinner (50–100 nm thick). Since electrons pass uniformly through biological material, unstained specimens give very poor images. Therefore, the specimen must routinely be stained in order to scatter some of the incident electrons, which are then not focused by the electromagnetic lenses and so do not form the image. **Heavy metals** such as gold and osmium are often used to stain biological materials. In particular, **osmium tetroxide** preferentially stains certain cellular components, such as membranes, which appear black in the image. The transmission electron microscope has sufficiently high resolution that it can be used to obtain information about the shapes of purified proteins, viruses and subcellular organelles. Three-dimensional views of structures with resolutions of 2–10 nm can be obtained using the technique of **electron tomography**, which generates three-dimensional images by computer analysis of multiple two-dimensional images obtained over a range of viewing directions. **Cryoelectron microscopy**, in which the sample is rapidly frozen at very low temperatures, is used to determine the three-dimensional structures of proteins (Section B2).

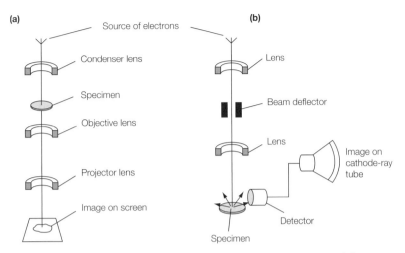

Figure 4. Principal features of (a) a transmission electron microscope and (b) a scanning electron microscope.

Antibodies can be tagged with electron-dense gold particles in a similar way to being tagged with a fluorescent compound in fluorescence microscopy, and then bound to specific target proteins in the thin sections of the specimen. When viewed in the electron microscope, small dark spots due to the gold particles are seen in the image wherever an antibody molecule has bound to its antigen (Section C4) and so the technique can be used to localize specific proteins.

In **scanning electron microscopy**, an (unsectioned) specimen is fixed and then coated with a thin layer of a **heavy metal** such as **platinum**. An electron beam then **scans** over the specimen, exciting molecules within it that release secondary electrons. These secondary electrons are focused on to a detector and the resulting image displayed (Figure 4b). The scanning electron microscope produces a **three-dimensional image** because the number of secondary electrons produced by any one point on the specimen depends on the angle of the electron beam in relation to the surface of the specimen. The resolution of the scanning electron microscope is only about 10 nm.

A5 Cell fractionation

Isolating cells and their parts: overview

Most animal and plant tissues contain a mixture of cell types, and most cells contain multiple **subcellular organelles** (Section A2). Although microscopy techniques (Section A4) can be used to visualize organelles and large molecules inside cells, many studies on cell structure and function require samples of a particular type of cell, subcellular organelle or components within them. Most biochemical procedures require obtaining large numbers of cells and then physically disrupting them to isolate their components. **Tissue samples** will often provide large quantities of material but will contain a heterogeneous mix of cells. Techniques have been developed whereby **homogeneous populations of**

cells can be isolated, grown in culture to amplify them, and subsequently studied or fractionated into their component parts.

Flow cytometry

Different cells can be identified by measuring the light they scatter, or the fluorescence they emit, as they pass a laser beam in a **flow cytometer**. In a **fluorescence-activated cell sorter** or **FACS** (Figure 1), an instrument based on flow cytometry, cells can be identified and separated from each other. The cells of interest are first labeled with an **antibody**, which is specific for a particular cell surface molecule. The antibody is coupled to a fluorescent dye (Section A4), such that when the individual cells pass a laser beam in single file in a narrow stream, the fluorescence of each cell is measured. A vibrating nozzle then

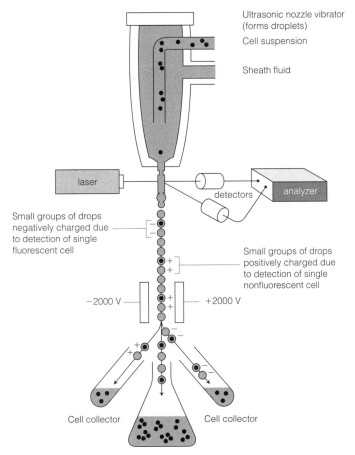

Figure 1. A fluorescence-activated cell sorter. An antibody specific for a particular cell surface protein is linked to a fluorescent molecule and then added to a mixture of cells. When the individual cells pass through a laser beam they are monitored for fluorescence. Droplets containing single cells are given a positive or negative charge, depending on whether the cell has bound the fluorescently tagged antibody or not. The droplets containing a single cell are then deflected by an electric field into collection tubes according to their charge. The cell concentration is such that most droplets contain no cells and flow through undeflected to a waste container together with any clumps of cells.

forms tiny droplets, each containing a single cell, which are given a positive or negative charge depending on whether the cell they contain is fluorescing. A strong electric field deflects the different charged droplets into separate containers so that each container eventually has a **homogeneous population of cells** with respect to the cell surface molecule tagged with fluorescent antibody. These homogeneous populations can then be used for biochemical analysis or grown in culture. The DNA and RNA content of a cell can also be measured by flow cytometry.

Subcellular fractionation

In order to study macromolecules and metabolic processes within cells, it is often helpful to isolate one type of **subcellular organelle** (Section A2) from the rest of the cell contents by subcellular fractionation. Initially, the plasma membrane (and cell wall if present) has to be ruptured. To do this, the tissue or cell sample is suspended in an isotonic sucrose solution (0.25–0.32 M) buffered at the appropriate pH, and the cells are then broken open by **homogenization** in a blender or homogenizer, by **sonication**, or by subjecting them to high pressures (**French press** or **nitrogen bomb**). The initial homogenization, and the following subcellular fractionation, are usually carried out at 4°C in order to minimize enzymic degradation of the cell's constituents. The sample of broken cells is often strained through muslin or other fine gauze to remove larger lumps of material before proceeding further.

Differential centrifugation

In differential centrifugation, the various subcellular organelles are separated from one another on the basis of their **size** and **density**, with the largest and heaviest structures sedimenting most rapidly. A **centrifuge** is used to generate powerful forces; up to 100 000 times the force of gravity (g). The homogenized sample is placed in an appropriate centrifuge tube, which is then loaded in the **rotor** of the centrifuge and subjected to centrifugation (Figure 2a). At first relatively low g forces are used for short periods of time but then increasingly higher g forces are used for longer time periods. For example, centrifugation at $600g$ for 3 min would pellet the **nuclei**, the largest organelles (Figure 2b). The supernatant from this step is removed to a fresh tube and then centrifuged at $6000g$ for 8 min to pellet out **mitochondria**, **peroxisomes** and, if present, **lysosomes** or **chloroplasts**. Centrifugation of this next supernatant at $40 000g$ for 30 min will pellet out the **plasma membrane**, and fragments of the **endoplasmic reticulum** and **Golgi apparatus**. A final centrifugation at $100 000g$ for 90 min would result in a **ribosomal pellet** and a supernatant that is essentially free of particulate matter and is considered to be the true soluble **cytosolic fraction**. However, the fractions isolated by differential centrifugation are not usually entirely free of other subcellular organelles and so may need to be purified further. For separations at low g forces, a preparative centrifuge is used, which has a rotor spinning in air at ambient pressure. However, an ultracentrifuge is required for separations at higher g forces in which the chamber is kept in a high vacuum to reduce friction, and subsequent heating, which would otherwise occur between the spinning rotor and air.

Equilibrium density-gradient centrifugation

Equilibrium density-gradient centrifugation is often used to purify further organelles following their partial separation by differential centrifugation. In this procedure, the organelles are separated on the basis of their **density**. The impure organelle fraction is loaded at the top of a centrifuge tube that contains a gradient of a **dense solution**

Figure 2. Cell fractionation by differential centrifugation: (a) scheme for subcellular fractionation of a tissue sample; (b) appearance of a sample in the centrifuge tube before and after centrifugation.

(e.g. a sucrose solution; Figure 3). The sucrose solution is most concentrated (dense) at the bottom of the tube, and decreases in concentration (and density) towards the top of the tube. During centrifugation (e.g. 160 000g for 3 h) the various organelles move down the tube to an **equilibrium position**, where their density is equal to that of the sucrose at that position. The forces of sedimentation tend to make the organelles move further down the tube but, if they do so, they enter a region of higher density than the organelle density and so they float back to their previous position. Mitochondria, lysosomes and peroxisomes all differ in density and so can be effectively separated from one another by density-gradient centrifugation (Figure 3). Similarly, the rough endoplasmic reticulum, Golgi apparatus and plasma membrane can be separated using a gradient of lower density. The more dense **cesium chloride** is used to make the density gradient for the separation of denser particles such as DNA, RNA and proteins by equilibrium density-gradient centrifugation.

Figure 3. Separation of organelles by equilibrium density-gradient centrifugation.

Marker proteins

When the cell sample has been fractionated, the purity of the different organelle preparations needs to be assessed. One way in which this can be done is by assessing **morphology** in the electron microscope (Section A4). A more readily available alternative though is to measure the activity of (to assay for) a particular **enzyme** (Section D1), which is characteristic of that organelle and is not found elsewhere in the cell. For example, **catalase** is a good marker enzyme for peroxisomes, **succinate dehydrogenase** for mitochondria, **cathepsin C** or **acid phosphatase** for lysosomes, and **alkaline phosphatase** for the plasma membrane. Thus, the presence of catalase in a fraction of lysosomes would indicate its contamination by peroxisomes. A good indication of the **purity**/degree of contamination of an organelle preparation can be ascertained by measuring the activity of such enzymes in the various isolated fractions. Alternatively, a marker protein can be detected following SDS-PAGE (Section C2) and Western blotting with a specific antibody (Section C4).

B1 Amino acid structure

Key Notes

Amino acids

All proteins are made up from the same set of 20 standard amino acids. A typical amino acid has a primary amino group, a carboxyl group, a hydrogen atom and a side-chain (R group) attached to a central α-carbon atom (C_α).

Enantiomers

Amino acids with four different groups arranged tetrahedrally around the C_α atom can exist in either the D or L configuration. These two enantiomers are nonsuperimposable mirror images that can be distinguished on the basis of their different rotation of plane-polarized light. Only the L isomer is found in proteins.

The 20 standard amino acids

The different side-chains or R groups display different physicochemical properties. They may be polar, acidic, basic, aromatic, bulky, conformationally inflexible, able to form hydrogen bonds, able to cross-link and be chemically reactive. Glycine (Gly, G) has a hydrogen atom as its R group. Alanine (Ala, A), valine (Val, V), leucine (Leu, L), isoleucine (Ile, I) and methionine (Met, M) have aliphatic side-chains of differing structures that are hydrophobic and chemically inert. The aromatic side-chains of phenylalanine (Phe, F), tyrosine (Tyr, Y) and tryptophan (Trp, W) are also hydrophobic in nature. The conformationally rigid proline (Pro, P) has its aliphatic side-chain bonded back on to the amino group and thus is really an imino acid. The hydrophobic, sulfur-containing side-chain of cysteine (Cys, C) is highly reactive and can form a disulfide bond with another cysteine residue. The basic amino acids arginine (Arg, R) and lysine (Lys, K) have positively charged side-chains, whilst the side-chain of histidine (His, H) can be either positively charged or uncharged at neutral pH. The side-chains of the acidic amino acids aspartic acid (Asp, D) and glutamic acid (Glu, E) are negatively charged at neutral pH. The amide side-chains of asparagine (Asn, N) and glutamine (Gln, Q), and the hydroxyl side-chains of serine (Ser, S) and threonine (Thr, T) are uncharged and polar, and can form hydrogen bonds.

Acids, bases and pH

pH is a measure of the concentration of H^+ in a solution. An acid is a proton donor; a base is a proton acceptor. Ionization of an acid yields its conjugate base, and the two are termed a conjugate acid–base pair, for example acetic acid (CH_3COOH) and acetate (CH_3COO^-). The pK of an acid is the pH at which it is half dissociated. The

	Henderson–Hasselbalch equation expresses the relationship between pH, pK and the ratio of acid to base.
Buffers	An acid–base conjugate pair can act as a buffer, resisting changes in pH. From a titration curve of an acid, the inflection point indicates the pK value. The buffering capacity of the acid–base pair is the p$K \pm 1$ pH unit. In biological fluids, the phosphate and carbonate ions act as buffers. Amino acids, proteins, nucleic acids and lipids also have some buffering capacity.
Ionization of amino acids	The α-amino and α-carboxyl groups on amino acids act as acid–base groups, donating or accepting a proton as the pH is altered. At low pH, both groups are fully protonated, but as the pH is increased first the carboxyl group and then the amino group loses a hydrogen ion. For the standard 20 amino acids, the pK is in the range 1.8–2.9 for the α-carboxyl group and 8.8–10.8 for the α-amino group. Those amino acids with an ionizable side-chain have an additional acid–base group with a distinctive pK.
Related topics	(B2) Protein structure and function (B4) Collagen (B3) Myoglobin and hemoglobin

Amino acids

Amino acids are the building blocks of **proteins** (Section B2). Proteins of all species, from bacteria to humans, are made up from the same set of **20 standard amino acids**. Nineteen of these are α-amino acids with a **primary amino group** ($-NH_3^+$) and a **carboxylic acid** (carboxyl; $-COOH$) group attached to a central carbon atom, which is called the α-**carbon atom** (C_α) because it is adjacent to the carboxyl group (Figure 1a). Also attached to the C_α atom is a hydrogen atom and a **variable side-chain** or 'R' group. The one exception to this general structure is proline, which has a secondary amino group and is really an α-**imino acid**. The names of the amino acids are often abbreviated, either to three letters or to a single letter. Thus, for example, proline is abbreviated to Pro or P (Figure 2).

Enantiomers

All of the amino acids, except for glycine (Gly or G; Figure 2), have four different groups arranged **tetrahedrally** around the central C_α atom, which is thus known as an **asymmetric center** or **chiral center** and has the property of **chirality** (Greek; *cheir*, hand) (Figure 1b). The two **nonsuperimposable mirror images** are termed **enantiomers**. Enantiomers are physically and chemically indistinguishable by most techniques, but can be distinguished on the basis of their different optical rotation of plane-polarized light. Molecules are classified as dextrorotatory (D; Greek '*dextro*'=right) or levorotatory (L; Greek '*levo*'=left) depending on whether they rotate the plane of plane-polarized light clockwise or anticlockwise. D- and L-amino acids can also be distinguished by enzymes that usually only recognize one or other enantiomer. Only the L-**amino acids** are found in proteins. D-**Amino acids** rarely occur in nature, but are found in bacterial cell walls (Section A1) and certain antibiotics.

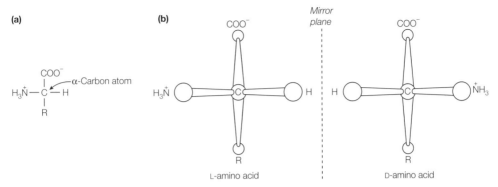

Figure 1. (a) Basic structure of an amino acid showing the four different groups around the central α-carbon atom; (b) the two enantiomers of an amino acid.

Figure 2. The standard amino acids: (a) hydrophobic, aliphatic R groups; (b) hydrophobic, aromatic R groups. The molecular weights of the amino acids are given in Table 1.

The 20 standard amino acids

The standard 20 amino acids differ only in the structure of the **side-chain** or 'R' group (Figures 2 and 3). They can be subdivided into smaller groupings on the basis of similarities in the properties of their side-chains. They display different **physicochemical properties** depending on the nature of their side-chain. Some are acidic, others are basic. Some have small side-chains, others large, bulky side-chains. Some have hydrophobic (water hating) side-chains, others are polar or hydrophilic (water loving). Some confer conformational inflexibility, others can participate either in hydrogen bonding or covalent bonding. Some are chemically reactive.

Hydrophobic, aliphatic amino acids

Glycine (Gly, G) (Figure 2a), the smallest amino acid with the simplest structure, has a hydrogen atom in the side-chain position, and thus does not exist as a pair of stereoisomers, since there are two identical groups (hydrogen atoms) attached to the C_α atom. The aliphatic side-chains of **alanine** (Ala, A), **valine** (Val, V), **leucine** (Leu, L), **isoleucine** (Ile, I) and **methionine** (Met, M) (Figure 2a) are chemically unreactive, but hydrophobic in nature. **Proline** (Pro, P) (Figure 2a) is also hydrophobic but, with its aliphatic side-chain bonded back on to the amino group, it is conformationally rigid. The sulfur-containing side-chain of **cysteine** (Cys, C) (Figure 2a) is also hydrophobic and is highly reactive, capable of reacting with another cysteine to form a disulfide bond (Section B2).

Figure 3. The standard amino acids: (a) polar, charged R groups; (b) polar, uncharged R groups. The molecular weights of the amino acids are given in Table 1.

Hydrophobic, aromatic amino acids

Phenylalanine (Phe, F), **tyrosine** (Tyr, Y) and **tryptophan** (Trp, W) (Figure 2b) are hydrophobic by virtue of their aromatic rings.

Polar, charged amino acids

The remaining amino acids all have polar, hydrophilic side-chains, some of which are charged at neutral pH. The amino groups on the side-chains of the basic amino acids **arginine** (Arg, R) and **lysine** (Lys, K) (Figure 3a) are protonated and thus positively charged at neutral pH. The side-chain of **histidine** (His, H) (Figure 3a) can be either positively charged or uncharged at neutral pH. In contrast, at neutral pH the carboxyl groups on the side-chains of the acidic amino acids **aspartic acid** (aspartate; Asp, D) and **glutamic acid** (glutamate; Glu, E) (Figure 3a) are de-protonated and possess a negative charge.

Polar, uncharged amino acids

The side-chains of **asparagine** (Asn, N) and **glutamine** (Gln, Q) (Figure 3b), the amide derivatives of Asp and Glu, respectively, are uncharged but can participate in hydrogen bonding. **Serine** (Ser, S) and **threonine** (Thr, T) (Figure 3b) are polar amino acids due to the reactive hydroxyl group in the side-chain, and can also participate in hydrogen bonding (as can the hydroxyl group of the aromatic amino acid Tyr).

Acids, bases and pH

The pH of a solution is a measure of its concentration of protons (H^+), and pH is defined as:

$$pH = \log_{10} \frac{1}{H^+} = -\log_{10} [H^+]$$

in which the square brackets denote a **molar concentration**.

An **acid** can be defined as a proton donor and a **base** as a proton acceptor:

$$Acid \rightleftharpoons H^+ + base$$

For example;

$$CH_3COOH \rightleftharpoons H^+ + CH_3COO^-$$
Acetic acid   Acetate

$$NH_4^+ \rightleftharpoons H^+ + NH_3$$
Ammonium ion   Ammonia

The species formed by the **ionization** of an acid is its conjugate base. Conversely, protonation of a base yields its conjugate acid. So, for example, acetic acid and acetate are a **conjugate acid–base pair**.

The ionization of a weak acid is given by:

$$HA \rightleftharpoons H^+ + A^-$$

The apparent **equilibrium constant** (K) for this ionization is defined as:

$$K = \frac{[H^+][A^-]}{[HA]}$$

(Equation 1)

The **pK** of an acid is defined as:

$$pK = -\log K = \log \frac{1}{K}$$

The pK of an acid is the pH at which it is half dissociated, i.e. when $[A^-]=[HA]$.

The **Henderson–Hasselbalch equation** expresses the relationship between pH and the ratio of acid to base. It is derived as follows. Rearrangement of Equation 1 gives:

$$\frac{1}{[H^+]} = \frac{1}{K} \times \frac{[A^-]}{[HA]}$$

Taking the logarithm of both sides of this equation gives:

$$\log \frac{1}{[H^+]} = \log \frac{1}{K} + \log \frac{[A^-]}{[HA]}$$

Substituting pH for $\log 1/[H^+]$ and pK for $\log 1/K$ gives:

$$pH = pK + \log \frac{[A^-]}{[HA]}$$

which is the Henderson–Hasselbalch equation. This equation indicates that the pK of an acid is numerically equal to the pH of the solution when the molar concentration of the acid is equal to that of its conjugate base. The pH of a solution can be calculated from the Henderson–Hasselbalch equation if the molar concentrations of A^- and HA, and the pK of HA are known. Similarly, the pK of an acid can be calculated if the molar concentrations of A^- and HA, and the pH of the solution are known.

Buffers

An acid–base conjugate pair, such as acetic acid and acetate, is able to resist changes in the pH of a solution. That is, it can act as a **buffer**. On addition of hydroxide (OH^-) to a solution of acetic acid the following happens:

$$CH_3COOH + OH^- \rightleftharpoons CH_3COO^- + H_2O$$

A plot of the dependence of the pH of this solution on the amount of OH^- added is called a **titration curve** (Figure 4). There is an inflection point in the curve at pH 4.8, which is the pK of acetic acid. In the vicinity of this pH, a relatively large amount of OH^- (or H^+) produces

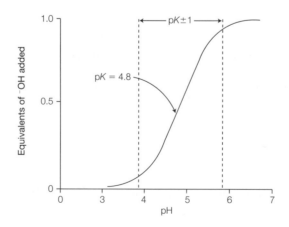

Figure 4. Titration curve of acetic acid.

little change in pH as the added OH^- (or H^+) reacts with CH_3COOH (or CH_3COO^-), respectively. Weak acids are most effective in buffering against changes in pH within 1 pH unit of the pK (Figure 4), often referred to as $pK \pm 1$, the **buffering capacity**.

Biological fluids, including the cytosol and extracellular fluids such as blood, are buffered. For example, in healthy individuals the pH of the blood is carefully controlled at pH 7.4. The major buffering components in most biological fluids are the phosphate ion ($H_2PO_4^-$, pK 6.82) and the carbonate ion (HCO_3^-, pK 6.35) because they have pK values in this range. However, many biological molecules, including amino acids, proteins, nucleic acids and lipids, have multiple acid–base groups that are effective at buffering in the physiological pH range (pH 6–8).

When working with enzymes, proteins and other biological molecules it is often crucial to buffer the pH of the solution in order to avoid **denaturation** (loss of activity) of the component of interest (Section D3). Numerous buffers are used in laboratories for this purpose. One of the commonest is tris(hydroxy methyl)aminomethane or **Tris**, which has a pK of 8.08.

Ionization of amino acids

The 20 standard amino acids have two **acid–base groups**: the α-amino and α-carboxyl groups attached to the C_α atom. Those amino acids with an **ionizable side-chain** (Asp, Glu, Arg, Lys, His, Cys, Tyr) have an additional acid–base group. The **titration curve** of Gly is shown in Figure 5a. At low pH (i.e. high hydrogen ion concentration) both the amino group and the carboxyl group are fully protonated so that the amino acid is in the cationic form $H_3N^+CH_2COOH$ (Figure 5b). As the amino acid in solution is titrated with increasing amounts of a strong base (e.g. NaOH), it loses two protons, first from the carboxyl group, which has the lower **pK** value (pK=2.3), and then from the amino group, which has the higher pK value (pK=9.78). The pH at which Gly has no net charge is termed its **isoelectric point, pI**. The α-carboxyl groups of all the 20 standard amino acids have pK values in the range 1.8–2.5, whilst their α-amino groups have pK values in the range 8.7–10.7

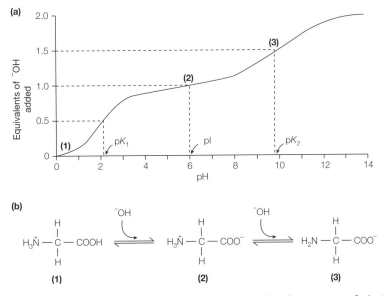

Figure 5. Ionization of glycine: (a) titration curve of glycine; (b) dissociation of glycine. Numbers in bold in parentheses in (a) correspond to the structures in (b).

(Table 1). The side-chains of the acidic amino acids Asp and Glu have pK values of 3.9 and 4.1, respectively, whereas those of the basic amino acids Arg and Lys have pK values of 12.5 and 10.5, respectively. Only the side-chain of His, with a pK value of 6.0, is ionized within the physiological pH range (pH 6–8). It should be borne in mind that when the amino acids are linked together in proteins, only the side-chain groups and the terminal α-amino and α-carboxyl groups of the polypeptide chain are free to ionize (Section B2).

Table 1. pK values and molecular weights of the 20 standard amino acids

Amino acid	Mol. Wt	pK α-COOH	pK α-NH$_3^+$	pK side-chain
Alanine	89.1	2.35	9.87	
Arginine	174.2	1.82	8.99	12.48
Asparagine	132.1	2.14	8.72	
Aspartic acid	133.1	1.99	9.90	3.90
Cysteine	121.2	1.92	10.70	8.37
Glutamic acid	147.1	2.10	9.47	4.07
Glutamine	146.2	2.17	9.13	
Glycine	75.1	2.35	9.78	
Histidine	155.2	1.80	9.33	6.04
Isoleucine	131.2	2.32	9.76	
Leucine	131.2	2.33	9.74	
Lysine	146.2	2.16	9.06	10.54
Methionine	149.2	2.13	9.28	
Phenylalanine	165.2	2.20	9.31	
Proline	115.1	1.95	10.64	
Serine	105.1	2.19	9.21	
Threonine	119.1	2.09	9.10	
Tryptophan	204.2	2.46	9.41	
Tyrosine	181.2	2.20	9.21	10.46
Valine	117.1	2.29	9.74	

B2 Protein structure and function

Key Notes

Proteins: an overview	Proteins perform a variety of functions, including the transport and storage of other molecules, mechanical support, generation of movement, immune protection, acting as catalysts, involvement in cellular signaling, and transmission of nerve impulses. Each protein has a unique amino acid sequence that is genetically determined.
Peptide bond	A protein is a linear sequence of amino acids linked together by peptide bonds. The peptide bond is a covalent bond between the α-amino group of one amino acid and the α-carboxyl group of another. The backbone conformation of a polypeptide is specified by the rotation angles about the C_α–N bond (*phi*, ϕ) and C_α–C bond (*psi*, ψ) of each of its amino acid residues. The sterically allowed values of ϕ and ψ are visualized in a Ramachandran plot.
Primary structure	The linear sequence of amino acids joined together by peptide bonds is termed the primary structure of the protein. The position of covalent disulfide bonds between cysteine residues is also included in the primary structure.
Secondary structure	Secondary structure in a protein refers to the regular folding of regions of the polypeptide chain. The two most common types of secondary structure are the α-helix and the β-pleated sheet. The α-helix is a cylindrical, rod-like helical arrangement of the amino acids in the polypeptide chain, which is maintained by hydrogen bonds parallel to the helix axis. In a β-pleated sheet, hydrogen bonds form between adjacent sections of polypeptides that are running either in the same direction (parallel) or in the opposite direction (antiparallel).
Tertiary structure	Tertiary structure in a protein refers to the three-dimensional arrangement of all the amino acids in the polypeptide chain. This biologically active, native conformation is maintained by multiple noncovalent bonds.
Quaternary structure	If a protein is made up of more than one polypeptide chain, it is said to have quaternary structure. This refers to the spatial arrangement of the polypeptide subunits and the nature of the interactions between them.
Protein stability	In addition to the peptide bonds between individual amino acid residues, the three-dimensional structure of a protein

	is maintained by a combination of noncovalent interactions (electrostatic forces, van der Waals forces, hydrogen bonds, hydrophobic forces) and covalent interactions (disulfide bonds).
Protein structure determination	The three-dimensional structure of a protein can be determined using complex physical techniques such as X-ray crystallography, nuclear magnetic resonance (NMR) spectroscopy and cryoelectron microscopy.
Protein folding	Proteins spontaneously fold into their native conformation, with the primary structure of the protein dictating its three-dimensional structure. Protein folding is driven primarily by hydrophobic forces and proceeds through an ordered set of pathways. Accessory proteins, including molecular chaperones, assist proteins to fold correctly in the cell. Diseases, such as Alzheimer's, are due to proteins misfolding.
Related topics	(B1) Amino acid structure (B4) Collagen (B3) Myoglobin and hemoglobin (B6) Antibodies

Proteins: an overview

Proteins are the most versatile macromolecules in living organisms and carry out a diverse array of functions. They **transport** and **store** other molecules (e.g. hemoglobin and myoglobin; Section B3), they provide **mechanical support** (e.g. collagen; Section B4), they **generate movement** (e.g. actin and myosin; Section B5), they provide **immune protection** (e.g. antibodies; Section B6), they act as **catalysts** (enzymes; Sections D1–D5), they are involved in **cellular signaling** (Section E5), and they **transmit nerve impulses** (Section E6). Each protein has a **unique amino acid sequence** that is genetically determined (Section H1). Most proteins contain between 50 and 2000 amino acids linked together via **peptide bonds**. With the average molecular weight of an amino acid being about 110 (Table 1 in Section B1), the **molecular weights** of most proteins are between 5500 and 220 000. We can also refer to the mass of a protein, which is expressed in units of **Daltons**, where one Dalton is equal to one atomic mass unit. Thus, a protein with a molecular weight of 25 000 has a mass of 25 000 daltons or 25 kDa.

Peptide bond

Proteins are linear sequences of amino acids linked together by peptide bonds. The peptide bond is a chemical, covalent bond formed between the α-amino group of one amino acid and the α-carboxyl group of another (Figure 1a) (Section B1). Once two amino acids are joined together via a peptide bond to form a dipeptide, there is still a free amino group at one end and a free carboxyl group at the other, each of which can in turn be linked to further amino acids. Thus, long, unbranched chains of amino acids can be linked together by peptide bonds to form oligopeptides (up to 25 amino acid residues) and polypeptides (>25 amino acid residues). Note that the polypeptide still has a free α-amino group and a free α-carboxyl group. Convention has it that peptide chains are written down with the free α-amino group on the left, the free α-carboxyl group on the right and a hyphen between the amino acids to indicate the peptide bonds. Thus, the tripeptide ^+H_3N-serine–leucine–phenylalanine-COO$^-$ would be written simply as Ser-Leu-Phe (in three-letter code) or S-L-F (in single-letter code).

(a)

Peptide bond

(b)

(c)

Rigid, planar peptide units

Free rotation about C_α–N and C_α–C bonds

Figure 1. (a) Formation of a peptide bond; (b) resonance structures of the peptide bond; (c) peptide units within a polypeptide.

The peptide bond between the carbon and nitrogen exhibits **partial double-bond character** due to the closeness of the carbonyl carbon–oxygen double bond allowing the **resonance structures** in Figure 1b to exist. Because of this, the C–N bond length is also shorter than normal C–N single bonds. The **peptide unit**, which is made up of the CO–NH atoms is thus relatively rigid and planar, although free rotation can take place about the C_α–N and C_α–C bonds (the bonds either side of the peptide bond), permitting adjacent peptide units to be at different angles (Figure 1c). The hydrogen of the amino group is nearly always on the opposite side (*trans*) of the double bond to the oxygen of the carbonyl group, rather than on the same side (*cis*).

The backbone of a protein is a linked sequence of rigid planar peptide groups. The backbone conformation of a polypeptide is specified by the **rotation angles** or **torsion angles** about the C_α–N bond (*phi*, ϕ) and C_α–C bond (*psi*, ψ) of each of its amino acid residues. When the polypeptide chain is in its planar, fully extended (all-*trans*) conformation the ϕ and ψ angles are both defined as 180°, and increase for a clockwise rotation when viewed from C_α (Figure 2). The **conformational range** of the torsion angles, ϕ and ψ, in a polypeptide backbone is restricted by steric hindrance. The sterically allowed values of ϕ and ψ can be determined by calculating the distances between the atoms of a tripeptide at all values of ϕ and ψ for the central peptide unit. These values are visualized in a steric contour diagram, otherwise known as a conformation map or **Ramachandran plot** (Figure 3). From Figure 3, it can be seen that most areas of the Ramachandran plot (most combinations of ϕ and ψ) are conformationally inaccessible to a polypeptide chain. Only three small regions of the conformation map are physically accessible to a polypeptide chain, and within these regions are the ϕ–ψ values that produce the right-handed α-helix, the parallel and antiparallel β-pleated sheets and the collagen helix (see below and Section B4).

The polypeptide chain folds up to form a specific shape (**conformation**) in the protein. This conformation is the **three-dimensional arrangement** of atoms in the structure and is determined by the amino acid sequence. There are four levels of structure in proteins: **primary**, **secondary**, **tertiary** and, sometimes but not always, **quaternary**.

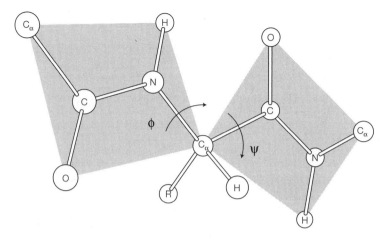

Figure 2. A segment of a polypeptide chain showing the torsion angles about the C_α–N bond (ϕ) and C_α–C bond (ψ).

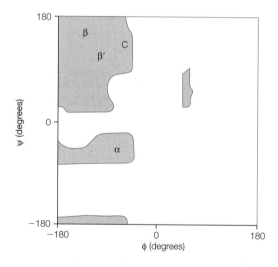

Figure 3. Ramachandran plot showing the allowed angles for poly-L-alanine (gray regions). α, ϕ–ψ values that produce the right-handed α-helix; β, the antiparallel β-pleated sheet; β', the parallel β-pleated sheet; C, the collagen helix.

Primary structure

The primary level of structure in a protein is the **linear sequence of amino acids** as joined together by peptide bonds. This sequence is determined by the sequence of nucleotide bases in the gene encoding the protein (Section H1). Also included under primary structure is the location of any other **covalent bonds**. These are primarily **disulfide bonds** between cysteine residues that are adjacent in space but not in the linear amino acid sequence. These covalent cross-links between separate polypeptide chains or between different parts of the same chain are formed by the oxidation of the SH groups on cysteine residues that are juxtaposed in space (Figure 4). The resulting disulfide is called a **cystine** residue. Disulfide bonds are often present in extracellular proteins, but are rarely found in

Figure 4. Formation of a disulfide bond between two cysteine residues, generating a cystine residue.

intracellular proteins. Some proteins, such as collagen, have covalent cross-links formed between the side-chains of Lys residues (Section B4).

Secondary structure

The secondary level of structure in a protein is the regular folding of regions of the polypeptide chain. The two most common types of protein fold are the **α-helix** and the **β-pleated sheet**. In the rod-like α-helix, the amino acids arrange themselves in a regular helical conformation (Figure 5a). The carbonyl oxygen of each peptide bond is **hydrogen bonded** to the hydrogen on the amino group of the fourth amino acid away (Figure 5b), with the hydrogen bonds running nearly parallel to the axis of the helix. In an α-helix there are 3.6 amino acids per turn of the helix covering a distance of 0.54 nm, and each amino acid residue represents an advance of 0.15 nm along the axis of the helix (Figure 5a). The side-chains of the amino acids are all positioned along the outside of the cylindrical helix (Figure 5c). Certain amino acids are more often found in α-helices than others. In particular, Pro is rarely found in α-helical regions, as it cannot form the correct pattern of hydrogen bonds due to the lack of a hydrogen atom on its nitrogen atom. For this reason, Pro is often found at the end of an α-helix, where it alters the direction of the polypeptide chain and terminates the helix. Different proteins have a different amount of the polypeptide chain folded up into α-helices. For example, the single polypeptide chain of myoglobin has eight α-helices (Section B3).

In the **β-pleated sheet**, hydrogen bonds form between the peptide bonds either in different polypeptide chains or in different sections of the same polypeptide chain (Figure 6a). The planarity of the peptide bond forces the polypeptide to be pleated with the side-chains of the amino acids protruding above and below the sheet (Figure 6b). Adjacent polypeptide chains in β-pleated sheets can be either **parallel** or **antiparallel** depending on whether they run in the same direction or in opposite directions, respectively (Figure 6c). The polypeptide chain within a β-pleated sheet is fully extended, such that there is a distance of 0.35 nm from one C_α atom to the next. β-Pleated sheets are always slightly curved and, if several polypeptides are involved, the sheet can close up to form a **β-barrel**. Multiple β-pleated sheets provide strength and rigidity in many structural proteins, such as silk fibroin, which consists almost entirely of stacks of antiparallel β-pleated sheets.

In order to fold tightly into the compact shape of a globular protein, the polypeptide chain often reverses direction, making a **β-turn** (hairpin or reverse turn). In these β-turns,

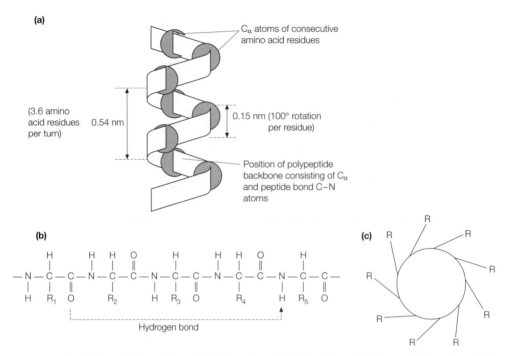

Figure 5. The folding of the polypeptide chain into an α-helix: (a) model of an α-helix with only the C_α atoms along the backbone shown; (b) in the α-helix the CO group of residue *n* is hydrogen bonded to the NH group on residue (*n*+4); (c) cross-sectional view of an α-helix showing the positions of the side-chains (R groups) of the amino acids on the outside of the helix.

Figure 6. The folding of the polypeptide chain in a β-pleated sheet: (a) hydrogen bonding between two sections of a polypeptide chain forming a β-pleated sheet; (b) a side-view of one of the polypeptide chains in a β-pleated sheet showing the side-chains (R groups) attached to the C_α atoms protruding above and below the sheet; (c) because the polypeptide chain has polarity, either parallel or antiparallel β-pleated sheets can form.

Figure 7. The folding of the polypeptide chain in a β-turn.

the carbonyl oxygen of one amino acid is hydrogen bonded to the hydrogen on the amino group of the fourth amino acid along (Figure 7). β-Turns are often found connecting the ends of antiparallel β-pleated sheets. Regions of the polypeptide chain that are not in a regular secondary structure are said to have a **coil** or **loop conformation**. About half the polypeptide chain of a typical globular protein will be in such a conformation.

Tertiary structure

The third level of structure found in proteins, tertiary structure, refers to the spatial arrangement of amino acids that are far apart in the linear sequence as well as those residues that are adjacent. Again, it is the sequence of amino acids that specifies this final **three-dimensional structure** (Figures 8 and 9). In water-soluble globular proteins such as myoglobin (Section B3), the main driving force behind the folding of the polypeptide chain is the energetic requirement to bury the nonpolar amino acids in the hydrophobic interior away from the surrounding aqueous, hydrophilic medium. The polypeptide chain folds spontaneously so that the majority of its hydrophobic side-chains are buried

(a)   – Arg–Val–Glu–Lys–Met–Val–Leu–Ala–Gly –

(b)

(c)

(d)

Figure 8. The four levels of structure in proteins: (a) primary structure (amino acid sequence); (b) secondary structure (α-helix); (c) tertiary structure; (d) quaternary structure.

Figure 9. Various graphic representations of the structure of RND3/RHOE a small GTP-binding protein complexed with GTP (guanosine triphosphate in space-filling representation): (a) the ball-and-stick representation reveals the location of all the atoms in the protein; (b) C_α backbone trace shows how the polypeptide chain is folded; (c) the ribbon representation emphasizes how α-helices and β-strands are organized in the protein; (d) a model of the water-accessible surface reveals the numerous bumps and crevices on the surface of the protein.

in the interior, and the majority of its polar, charged side-chains are on the surface. Once folded, the **three-dimensional biologically active (native) conformation** of the protein is maintained not only by hydrophobic interactions, but also by electrostatic forces, hydrogen bonding and, if present, the covalent disulfide bonds. The electrostatic forces include salt bridges between oppositely charged groups and the multiple weak van der Waals interactions between the tightly packed aliphatic side-chains in the interior of the protein.

In many larger proteins, the polypeptide chain folds into two or more compact, globular regions, called **domains**. Most domains consist of 30–400 amino acids in which the polypeptide chain folds into a specific structure with regions of relatively unfolded peptide linking domains together. Consequently, many domains are structurally independent units that have the characteristics of small globular proteins. Somewhat diverse proteins can have domains in common even if their overall tertiary structures are different. Examples of domains in proteins include the **immunoglobulin fold** that is found in antibodies and other proteins of the immune system (Section B6), and the **helix-turn-helix**, **zinc finger** and **basic domains** found in DNA-binding proteins (Section G6).

Quaternary structure

Proteins containing more than one polypeptide chain, such as hemoglobin (Section B3), exhibit a fourth level of protein structure called **quaternary structure** (Figure 8). This level of structure refers to the spatial arrangement of the polypeptide **subunits** and the nature of the interactions between them. These interactions may be covalent links (e.g. disulfide bonds) or noncovalent interactions (electrostatic forces, hydrogen bonding, hydrophobic interactions).

Protein stability

The native three-dimensional conformation of a protein is maintained by a range of noncovalent interactions (electrostatic forces, hydrogen bonds, hydrophobic forces) and covalent interactions (disulfide bonds) in addition to the peptide bonds between individual amino acids.

- **Electrostatic forces**: these include the interactions between two ionic groups of opposite charge, for example the ammonium group of Lys and the carboxyl group of Asp, often referred to as an **ion pair** or **salt bridge**. In addition, the noncovalent associations between electrically neutral molecules, collectively referred to as **van der Waals forces**, arise from electrostatic interactions between permanent and/or induced dipoles, such as the carbonyl group in peptide bonds.

- **Hydrogen bonds**: these are predominantly electrostatic interactions between a weakly acidic donor group and an acceptor atom that bears a lone pair of electrons, which thus has a partial negative charge that attracts the hydrogen atom. In biological systems, the donor group is an oxygen or nitrogen atom that has a covalently attached hydrogen atom, and the acceptor is either oxygen or nitrogen (Figure 10). Hydrogen bonds are normally in the range 0.27–0.31 nm and are highly directional, i.e. the donor, hydrogen and acceptor atoms are collinear. Hydrogen bonds are stronger than van der Waals forces but much weaker than covalent bonds. Hydrogen bonds not only play an important role in protein structure, but also in the structure of other biological macromolecules such as the DNA double helix (Section F1) and lipid bilayers (Section E1). In addition, hydrogen bonds are critical to both the properties of water and to its role as a biochemical solvent.

H-bond   H-bond
donor   acceptor

O — H · · · · N

O — H · · · · O

N — H · · · · N

N — H · · · · O

Figure 10.   Examples of hydrogen bonds (shown as dotted lines).

- **Hydrophobic forces**: the **hydrophobic effect** is the name given to those forces that cause nonpolar molecules to minimize their contact with water. This is clearly seen with amphipathic molecules such as lipids and detergents that form micelles in aqueous solution (Section E1). Proteins, too, find a conformation in which their nonpolar side-chains are largely out of contact with the aqueous solvent, and thus hydrophobic forces are an important determinant of protein structure, folding and stability. In proteins, the effects of hydrophobic forces are often termed **hydrophobic bonding**, to indicate the specific nature of protein folding under the influence of the hydrophobic effect.

- **Disulfide bonds**: these covalent bonds form between Cys residues that are close together in the final conformation of the protein (Figure 4) and function to stabilize its three-dimensional structure. Disulfide bonds are really only formed in the oxidizing environment of the endoplasmic reticulum (Section A2), and thus are found primarily in extracellular and secreted proteins.

Protein structure determination

The presence of α-helices and β-pleated sheets in proteins can often be predicted from the primary amino acid sequence. However, it is not possible to predict the precise three-dimensional structure of a protein from its amino acid sequence, unless its sequence is very similar to that of a protein whose three-dimensional structure is already known. Sophisticated physical methods and complex analyses of the experimental data are required to determine the conformation of a protein. The three-dimensional structure of a protein can be determined to the atomic level by the techniques of X-ray crystallography, nuclear magnetic resonance (NMR) spectroscopy and cryoelectron microscopy.

In **X-ray crystallography** the first requirement is **crystals** of highly purified protein. In the crystal, millions of protein molecules are precisely aligned with one another in a rigid array characteristic of that particular protein. **Beams of X-rays** are then passed through the crystal (Figure 11). The wavelengths of X-rays are 0.1–0.2 nm, short enough to resolve the atoms in the protein crystal. The atoms in the crystal scatter the X-rays, producing a **diffraction pattern** of discrete spots on photographic film. The intensities of the diffraction maxima (the darkness of the spots on the film) are then used mathematically to construct the three-dimensional image of the protein crystal.

Nuclear magnetic resonance (NMR) spectroscopy can be used to determine the three-dimensional structures of small (up to approximately 30 kDa) proteins in **aqueous solution**. It does not require the crystallization of the protein. In this technique, a concentrated protein solution is placed in a **magnetic field** and the effects of different radio frequencies on the spin of different atoms in the protein measured. The behavior of any particular atom is influenced by neighboring atoms in adjacent residues, with closer

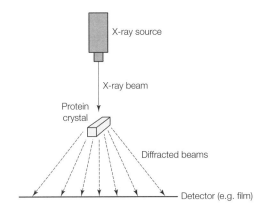

Figure 11. X-Ray crystallography. When a narrow beam of X-rays strikes a crystal, part of it passes straight through and the rest is scattered (diffracted) in various directions. The intensity of the diffracted waves is recorded on photographic film or with a solid-state electronic detector. From the diffraction data, the three-dimensional structure of the protein can be determined.

residues causing more perturbation than distant ones. From the magnitude of the effect, the distances between residues can be calculated and then used to generate the three-dimensional structure of the protein.

Cryoelectron microscopy is often used to determine the three-dimensional structures of proteins, particularly multisubunit proteins, which are difficult to crystallize. In this technique, the protein sample is **rapidly frozen** in liquid helium to preserve its structure. The frozen, hydrated protein is then examined in a cryoelectron microscope using a low dose of **electrons** to prevent radiation-induced damage to the structure. The resulting images are analyzed by complex computer programs and the three-dimensional structure of the protein reconstructed.

Protein folding

Under appropriate physiological conditions, proteins **spontaneously fold** into their native conformation. As there is no need for external templates, this implies that the primary structure of the protein dictates its three-dimensional structure. From experiments with the protein **RNase A**, it has been observed that it is mainly the internal residues of a protein that direct its folding to the native conformation. Alteration of surface residues by mutation is less likely to affect the folding than changes to internal residues. It has also been observed that protein folding is driven primarily by **hydrophobic forces**. Proteins fold into their native conformation through an **ordered set of pathways** rather than by a random exploration of all the possible conformations until the correct one is stumbled upon.

Although proteins can fold *in vitro* (in the laboratory) without the presence of accessory proteins, this process can take minutes to days. *In vivo* (in the cell) this process requires only a few minutes because the cells contain **accessory proteins**, which assist the polypeptides to fold to their native conformation. There are three main classes of protein folding accessory proteins:

- **Protein disulfide isomerases** catalyze disulfide interchange reactions, thereby facilitating the shuffling of the disulfide bonds in a protein until they achieve their correct pairing.

- **Peptidyl prolyl *cis–trans* isomerases** catalyze the otherwise slow interconversion of X–Pro peptide bonds between their *cis* and *trans* conformations, thereby accelerating the folding of Pro-containing polypeptides. One of the classes of peptidyl prolyl *cis–trans* isomerases is inhibited by the **immunosuppressive drug** cyclosporin A.

- **Molecular chaperones**, which include proteins such as the **heat shock proteins** 70 (hsc70), the **chaperonins**, and the **lectins calnexin** and **calreticulin**. These prevent the improper folding and aggregation of proteins that may otherwise occur as internal hydrophobic regions are exposed to one another.

Some diseases are caused by **protein misfolding**. For example in the fatal neurodegenerative **Alzheimer's disease** the small **amyloid-β peptide** misfolds and aggregates in the brain, killing the surrounding neurons. While in the infectious **prion diseases**, such as Creutzfeldt–Jakob disease (CJD) in humans and bovine spongiform encephalopathy (BSE or mad cow disease), the normal cellular form of the **prion protein**, which has mainly α-helical secondary structure, undergoes a conformational transition to the infectious form of the protein that has predominantly β-sheet structure.

B3 Myoglobin and hemoglobin

Key Notes

Oxygen-binding proteins

Hemoglobin and myoglobin are the two oxygen-binding proteins present in large multicellular organisms. Hemoglobin transports oxygen in the blood and is located in the erythrocytes; myoglobin stores the oxygen in the muscles.

Myoglobin

Myoglobin, whose three-dimensional structure was solved by X-ray crystallography, is a globular protein made up of a single polypeptide chain of 153 amino acid residues that is folded into eight α-helices. The heme prosthetic group is located within a hydrophobic cleft of the folded polypeptide chain.

Hemoglobin

Adult hemoglobin (HbA) has quaternary structure as it is made up of four polypeptide chains: two α-chains and two β-chains ($\alpha_2\beta_2$), each with a heme prosthetic group. The individual polypeptides of hemoglobin have a three-dimensional structure almost identical to the polypeptide chain of myoglobin.

Binding of oxygen to heme

The heme prosthetic group consists of a protoporphyrin IX ring and a central Fe^{2+} atom, which forms four bonds with the porphyrin ring. In addition, on one side of the porphyrin ring the Fe^{2+} forms a bond with the proximal histidine (His F8); a residue eight amino acids along the F-helix of hemoglobin. The sixth bond from the Fe^{2+} is to a molecule of O_2. Close to where the O_2 binds is the distal histidine (His E7), which prevents carbon monoxide binding most efficiently.

Allostery

Hemoglobin is an allosteric protein. The binding of O_2 is cooperative; the binding of O_2 to one subunit increases the ease of binding of further O_2 molecules to the other subunits. The oxygen dissociation curve for hemoglobin is sigmoidal, whereas that for myoglobin is hyperbolic. Myoglobin has a greater affinity for O_2 than does hemoglobin. Oxyhemoglobin has a different quaternary structure from deoxyhemoglobin. As O_2 binds to the Fe^{2+} it distorts the heme group and moves the proximal histidine. This in turn moves helix F and alters the interactions between the four subunits. H^+, CO_2 and 2,3-bisphosphoglycerate (BPG) are allosteric effectors, promoting the release of O_2 from hemoglobin. BPG binds in the central cavity between the four subunits.

Fetal hemoglobin	Hemoglobin F (HbF), which consists of two α-chains and two γ-chains ($\alpha_2\gamma_2$), is present in the fetus. HbF binds BPG less strongly than HbA and thus has a higher affinity for O_2, which promotes the transfer of O_2 from the maternal to the fetal circulation.
Hemoglobinopathies	Hemoglobinopathies are diseases caused by abnormal hemoglobins. The best characterized of these is the genetically transmitted, hemolytic disease sickle-cell anemia. This is caused by the nonconservative substitution of a Glu by a Val, resulting in the appearance of a hydrophobic sticky patch on the surface of the protein. This allows long aggregated fibers of hemoglobin molecules to form, which distort the shape of the red blood cells. Heterozygotes carrying only one copy of the sickle-cell gene are more resistant to malaria than those homozygous for the normal gene.
Related topics	(B2) Protein structure and function (D5) Regulation of enzyme activity (M4) Hemes and chlorophylls

Oxygen-binding proteins

Hemoglobin is one of two **oxygen-binding proteins** found in vertebrates. The function of hemoglobin is to carry O_2 in the blood from the lungs to the other tissues in the body, in order to supply the cells with the O_2 required by them for the oxidative phosphorylation of foodstuffs (Section L2). The hemoglobin is found in the blood within the **erythrocytes** (red blood cells). These cells essentially act, amongst other things, as a sac for carrying hemoglobin, since mature erythrocytes lack any internal organelles (nucleus, mitochondria, etc.). The other O_2-binding protein is **myoglobin**, which stores the oxygen in the tissues of the body ready for when the cells require it. The highest concentrations of myoglobin are found in skeletal and cardiac **muscle**, which require large amounts of O_2 because of their need for large amounts of energy during contraction (Section B5).

Myoglobin

Myoglobin is a relatively small protein of mass 17.8 kDa made up of 153 amino acids in a single polypeptide chain. It was the first protein to have its **three-dimensional structure** determined by **X-ray crystallography** (Section B2) by John Kendrew in 1957. Myoglobin is a typical **globular protein** in that it is a highly folded compact structure with most of the hydrophobic amino acid residues buried in the interior and many of the polar residues on the surface. X-ray crystallography revealed that the single polypeptide chain of myoglobin consists entirely of α-**helical secondary structure** (Section B2). In fact, there are eight α-helices (labeled A–H) in myoglobin (Figure 1a). Within a hydrophobic crevice formed by the folding of the polypeptide chain is the **heme prosthetic group** (Figure 1a). This nonpolypeptide unit is noncovalently bound to myoglobin and is essential for the biological activity of the protein (i.e. the binding of O_2).

Figure 1. Structure of (a) myoglobin and (b) hemoglobin, showing the α- and β-polypeptide chains. C_α-backbone traces of (c) human myoglobin and (d) human hemoglobin, showing the α-helices, the heme prosthetic group in space-filling representation and how the monomer of myoglobin maps onto the structure of hemoglobin (circle).

Hemoglobin

The three-dimensional structure of hemoglobin was solved using **X-ray crystallography** (Section B2) in 1959 by Max Perutz. This revealed that hemoglobin is made up of four polypeptide chains, each of which has a very similar three-dimensional structure to the single polypeptide chain in myoglobin (Figure 1b) despite the fact that their amino acid sequences differ at 83% of the residues. This highlights a relatively common theme in protein structure: that very different primary sequences can specify very similar three-dimensional structures. The major type of hemoglobin found in adults (HbA) is made up of two different polypeptide chains: the **α-chain** that consists of 141 amino acid residues, and the **β-chain** of 146 residues ($\alpha_2\beta_2$; Figure 1b). Each chain, like that in myoglobin, consists of eight α-helices and each contains a heme prosthetic group (Figure 1b). Therefore, hemoglobin can bind four molecules of O_2. The four polypeptide chains are packed tightly together in a tetrahedral array to form an overall spherically shaped molecule that is held together by multiple noncovalent interactions (Section B2).

Binding of oxygen to heme

The **heme prosthetic group** in myoglobin and hemoglobin is made up of a **protoporphyrin IX** ring structure with an **iron atom** in the ferrous (Fe^{2+}) oxidation state (Section M4; Figure 2). This Fe^{2+} bonds with four nitrogen atoms in the center of the protoporphyrin ring and forms two additional bonds on either side of the plane of the protoporphyrin ring. One of these is to a histidine residue, which lies eight residues along helix F of hemoglobin, the **proximal histidine** (His F8) (Figure 2). The sixth bond is to one of the oxygen atoms in a molecule of **O_2** (Figure 2). Near to where the O_2 binds to the heme group is another histidine residue, the **distal histidine** (His E7) (Figure 2). This serves two very important functions. First, it prevents heme groups on neighboring hemoglobin molecules coming into contact with one another and oxidizing to the Fe^{3+} state in which they can no longer bind O_2. Second, it prevents **carbon monoxide** (CO) binding with the most favorable configuration to the Fe^{2+}, thereby lowering the affinity of heme for CO. This is important because once CO has bound irreversibly to the heme, the protein can no longer bind O_2. Thus, although the oxygen binding site in hemoglobin and myoglobin is only a small part of the whole protein, the polypeptide chain modulates the function of the heme prosthetic group.

Allostery

Hemoglobin is an **allosteric protein** (Section D5 for a fuller discussion of allostery). This means that the binding of O_2 to one of the subunits is affected by its interactions with the other subunits. In fact, the binding of O_2 to one hemoglobin subunit induces conformational changes (see below and Figure 2) that are relayed to the other subunits, making them more able also to bind O_2 by raising their affinity for this molecule. Thus binding of O_2 to hemoglobin is said to be **cooperative**. In contrast, the binding of O_2 to the single polypeptide unit of myoglobin is **noncooperative**. This is clearly apparent from

Figure 2. Binding of O_2 to heme. The Fe^{2+} of the protoporphyrin ring is bonded to His F8 but not to His E7, which is located nearby. As the heme Fe^{2+} binds O_2, helix F moves closer to helix E (see the text for details).

Figure 3. Oxygen dissociation curves for hemoglobin and myoglobin.

the **oxygen dissociation curves** for the two proteins: that for hemoglobin is **sigmoidal**, reflecting this cooperative binding, whereas that for myoglobin is **hyperbolic** (Figure 3). From the O_2 dissociation curve, it can also be seen that for any particular oxygen pressure the degree of saturation of myoglobin is higher than that for hemoglobin. In other words, myoglobin has a higher affinity for O_2 than does hemoglobin. This means that in the blood capillaries in the muscle, for example, hemoglobin will release its O_2 to myoglobin for storage there.

Mechanism of the allosteric change

X-Ray crystallography revealed that **oxyhemoglobin**, the form that has four O_2 molecules bound, differs markedly in its **quaternary structure** from **deoxyhemoglobin**, the form with no O_2 bound. In the absence of bound O_2, the Fe^{2+} lies slightly to one side of the porphyrin ring, which itself is slightly curved (Figure 2). As a molecule of O_2 binds to the heme prosthetic group it pulls the Fe^{2+} into the plane of the porphyrin ring (Figure 2), flattening out the ring in the process. Movement of the Fe^{2+} causes the **proximal histidine** to move also. This, in turn, shifts the position of helix F and regions of the polypeptide chain at either end of the helix. Thus, movement in the center of the subunit is transmitted to the surfaces, where it causes the ionic interactions holding the four subunits together to be broken and to re-form in a different position, thereby altering the quaternary structure, leading to the cooperative binding of O_2 to Hb.

The Bohr effect

The binding of O_2 to hemoglobin is affected by the concentration of **H^+ ions** and **CO_2** in the surrounding tissue; the Bohr effect. In actively metabolizing tissue, such as muscle, the concentrations of these two substances are relatively high. This effectively causes a shift of the O_2 dissociation curve for hemoglobin to the right, promoting the release of O_2. This comes about because there are H^+ binding sites, primarily His146 in the β-chain, which have a higher affinity for binding H^+ in deoxyhemoglobin than in oxyhemoglobin. An increase in CO_2 also causes an increase in H^+ due to the action of the enzyme **carbonic anhydrase**, which catalyzes the reaction:

$$CO_2 + H_2O \rightleftharpoons HCO_3^- + H^+$$

In addition, CO_2 can react with the primary amino groups in the polypeptide chain to form a negatively charged carbamate. Again, this change from a positive to a negative charge favors the conformation of deoxyhemoglobin. On returning in the blood to the lungs, the concentrations of H^+ and CO_2 are relatively lower and that of O_2 higher, so that the process is reversed and O_2 binds to hemoglobin. Thus, it can be seen that not only does hemoglobin carry O_2 but it also carries CO_2 back to the lungs where it is expelled.

2,3-Bisphosphoglycerate (**BPG**) is a highly anionic organic phosphate molecule (Figure 4) that is present in erythrocytes along with the hemoglobin. This molecule promotes the release of O_2 from hemoglobin by lowering the affinity of the protein for O_2. BPG binds in the small cavity in the center of the four subunits. In oxyhemoglobin this cavity is too small for it, whereas in deoxyhemoglobin it is large enough to accommodate a single molecule of BPG. On binding in the central cavity of deoxyhemoglobin it forms ionic bonds with the positively charged amino acid side-chains in the β-subunits, stabilizing the quaternary structure. H^+, CO_2 and BPG are all **allosteric effectors** (Section D5) as they favor the conformation of deoxyhemoglobin and therefore promote the release of O_2. Because these three molecules act at different sites, their effects are additive.

Fetal hemoglobin

In the fetus there is a different kind of hemoglobin, **hemoglobin F** (HbF), which consists of two α-chains and two γ-chains ($\alpha_2\gamma_2$), in contrast to adult hemoglobin (HbA, $\alpha_2\beta_2$). HbF has a **higher affinity** for O_2 under physiological conditions than HbA, which optimizes the transfer of oxygen from the maternal to the fetal circulation across the placenta. The molecular basis for this difference in O_2 affinity is that HbF binds BPG less strongly than does HbA. Near birth, the synthesis of the γ-chain is switched off, and that of the β-chain (which is present in HbA) is switched on.

Hemoglobinopathies

Comparison of the primary sequences of hemoglobin chains from more than 60 different species reveals that only nine residues in the polypeptide chain are **invariant** (i.e. the same) between all of the species. These nine residues include the **proximal and distal histidines**, which are essential for the correct functioning of the protein. Many of the other residues are replaced from one species to another by residues with similar properties (e.g. the hydrophobic Val is replaced with the hydrophobic Ile, or the polar Ser is replaced with the polar Asn), so-called **conservative substitutions**. In contrast, only a few residues have changed between species to a completely different residue (e.g. a hydrophobic Leu to a positively charged Lys or a negatively charged Glu to a positively charged Arg), so-called **nonconservative substitutions**, since this type of change could have a major effect on the structure and function of the protein.

Several hundred **abnormal hemoglobins** have been characterized, giving rise to the so-called **hemoglobinopathies**. Probably the best characterized hemoglobinopathy is

Figure 4. 2,3-Bisphosphoglycerate.

sickle-cell anemia (sickle-cell hemoglobin; HbS). This disease is characterized by the patient's erythrocytes having a characteristic sickle or crescent shape. The molecular basis for this disease is the change of a Glu for a Val at position 6 of the β-chain, resulting in the substitution of a polar residue by a hydrophobic one. This **nonconservative substitution** of Val for Glu gives HbS a **sticky hydrophobic patch** on the outside of each of its β-chains. In the corner between helices E and F of the β-chain of deoxy-HbS is a hydrophobic site that is complementary to the sticky patch (Figure 5). Thus the complementary site on one deoxy-HbS molecule can bind to the sticky patch on another deoxy-HbS molecule, resulting in the formation of **long fibers** of hemoglobin molecules that distort the erythrocyte. Electron microscopy (Section A4) has revealed that the fibers have a diameter of 21.5 nm and consist of a 14-stranded helix. Multiple polar interactions, in addition to the critical interaction between the sticky patches, stabilize the fiber. In oxy-HbS the complementary site is masked, so the formation of the long fibers occurs only when there is a high concentration of the deoxygenated form of HbS.

Sickle-cell anemia is a **genetically transmitted**, hemolytic disease. The sickled cells are more fragile than normal erythrocytes, lysing more easily and having a shorter half-life, which leads to severe anemia. As sickle-cell anemia is genetically transmitted, **homozygotes** have two copies of the abnormal gene whereas **heterozygotes** have one abnormal and one normal copy. Homozygotes often have a reduced life-span as a result of infection, renal failure, cardiac failure or thrombosis, due to the sickled cells becoming trapped in small blood vessels leading to tissue damage. In contrast, heterozygotes are usually not symptomatic as only approximately 1% of their erythrocytes are sickled, compared with approximately 50% in a homozygote. The frequency of the sickle gene is relatively high in certain parts of Africa and correlates with the incidence of **malaria**. The reason for this is that heterozygotes are protected against the most lethal form of malaria, whereas normal homozygotes are more vulnerable to the disease. Inheritance of the abnormal hemoglobin gene can now be monitored by recombinant DNA techniques (Section I1).

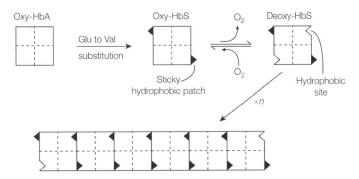

Figure 5. Molecular basis for the aggregation of deoxyhemoglobin molecules in sickle-cell anemia.

B4 Collagen

Key Notes

Function and diversity

Collagen is the name given to a family of structurally related proteins that form strong insoluble fibers. Collagens consist of three polypeptide chains, the identity and distribution of which vary between collagen types. The different types of collagen are found in different locations in the body.

Biosynthesis: overview

The collagen polypeptides are post-translationally modified by hydroxylation and glycosylation on transport through the rough endoplasmic reticulum and Golgi. The three polypeptides form the triple-helical procollagen, which is secreted out of the cell. The extension peptides are removed to form tropocollagen, which then aggregates into a microfibril and is covalently cross-linked to form the mature collagen fiber.

Composition and post-translational modifications

One-third of the amino acid residues in collagen are Gly, while another quarter are Pro. The hydroxylated amino acids 4-hydroxyproline (Hyp) and 5-hydroxylysine (Hyl) are formed post-translationally by the action of proline hydroxylase and lysine hydroxylase. These Fe^{2+}-containing enzymes require ascorbic acid (vitamin C) for activity. In the vitamin C deficiency disease scurvy, collagen does not form correctly due to the inability to hydroxylate Pro and Lys. Hyl residues are often post-translationally modified with carbohydrate.

Structure

Collagen contains a repeating tripeptide sequence of Gly–X–Y, where X is often Pro and Y is often Hyp. Each polypeptide in collagen folds into a helix with 3.3 residues per turn. Three polypeptide chains then come together to form a triple-helical cable that is held together by hydrogen bonds between the chains. Every third residue passes through the center of the triple helix, which is so crowded that only Gly is small enough to fit.

Secretion and aggregation

The extension peptides on both the N and C termini of the polypeptide chains direct the formation of the triple-helical cable and prevent the premature aggregation of the procollagen molecules within the cell. Following secretion out of the cell, the extension peptides are cleaved off by peptidases, and the resulting tropocollagen molecules aggregate together in a staggered array.

Cross-links

Covalent cross-links both between and within the tropocollagen molecules confer strength and rigidity on the collagen fiber. These cross-links are formed between Lys and

	its aldehyde derivative allysine. Allysine is derived from Lys by the action of the copper-containing lysyl oxidase, which requires pyridoxal phosphate for activity.
Bone formation	Hydroxyapatite (calcium phosphate) is deposited in nucleation sites between the ends of tropocollagen molecules as the first step in bone formation.
Related topics	(B1) Amino acid structure (H4) Protein targeting (B2 Protein structure and function)

Function and diversity

Collagen, which is present in all multicellular organisms, is not one protein but a family of structurally related proteins. It is the most abundant protein in mammals and is present in most organs of the body, where it serves to hold cells together in discrete units. It is also the major **fibrous element** of skin, bones, tendons, cartilage, blood vessels and teeth. The different collagen proteins have very diverse functions. The extremely hard structures of bones and teeth contain collagen and a calcium phosphate polymer. In tendons, collagen forms rope-like fibers of high tensile strength, while in the skin collagen forms loosely woven fibers that can expand in all directions. The different types of collagen are characterized by **different polypeptide compositions** (Table 1). Each collagen is composed of three polypeptide chains, which may be all identical (as in types II and III) or may be of two different chains (types I, IV and V). A single molecule of type I collagen has a molecular mass of 285 kDa, a width of 1.5 nm and a length of 300 nm.

Table 1. Types of collagen

Type	Polypeptide composition	Distribution
I	$[\alpha1(I)]_2\,\alpha2(I)$	Skin, bone, tendon, cornea, blood vessels
II	$[\alpha1(II)]_3$	Cartilage, intervertebral disk
III	$[\alpha1(III)]_3$	Fetal skin, blood vessels
IV	$[\alpha1(IV)]_2\,\alpha2(IV)$	Basement membrane
V	$[\alpha1(V)]_2\,\alpha2(V)$	Placenta, skin

Biosynthesis: overview

Like other secreted proteins, collagen polypeptides are synthesized by ribosomes on the rough endoplasmic reticulum (RER; Section H4). The polypeptide chain then passes through the RER and Golgi apparatus before being secreted. Along the way it is **post-translationally modified**: Pro and Lys residues are hydroxylated and carbohydrate is added (Figure 1). Before secretion, three polypeptide chains come together to form a triple-helical structure known as **procollagen**. The procollagen is then secreted into the extracellular spaces of the connective tissue where extensions of the polypeptide chains at both the N and C termini (**extension peptides**) are removed by peptidases to form **tropocollagen** (Figure 1). The tropocollagen molecules aggregate and are extensively **cross-linked** to produce the mature **collagen fiber** (Figure 1).

Figure 1. Overview of the biosynthesis of collagen.

Composition and post-translational modifications

The amino acid composition of collagen is quite distinctive. Nearly *one-third* of its residues are **Gly**, while another *one-quarter* are **Pro**, significantly higher proportions than are found in other proteins. The hydroxylated amino acids **4-hydroxyproline** (**Hyp**) and **5-hydroxylysine** (**Hyl**) (Figure 2) are found exclusively in collagen. These hydroxylated amino acids are formed from the parent amino acid by the action of **proline hydroxylase** and **lysine hydroxylase**, respectively (Figure 2). These enzymes have an Fe^{2+} ion at their active site and require **ascorbic acid** (**vitamin C**) for activity. The ascorbic acid acts as an antioxidant, keeping the Fe^{2+} ion in its reduced state. Proline hydroxylase and lysine hydroxylase are dioxygenases, using a molecule of O_2. α-Ketoglutarate, the citric acid cycle intermediate (Section L1), is an obligatory substrate and is converted into succinate during the reaction (Figure 2). Both enzymes will hydroxylate only Pro and Lys residues that are incorporated in a polypeptide chain, and then only when the residue is on the N-terminal side of Gly. Hyp is important in stabilizing the structure of collagen through hydrogen bond formation (see below). In **vitamin C deficiency**, Hyp (and Hyl) are not synthesized, resulting in the weakening of the collagen fibers. This leads to the skin lesions, fragile blood vessels and poor wound healing that are characteristic of the disease **scurvy**.

The other post-translational modification that occurs to collagen is **glycosylation**. In this case, the sugar residues, usually only glucose, galactose and their disaccharides, are attached to the hydroxyl group in the newly formed Hyl residues, rather than to Asn or

Figure 2. Formation of hydroxyproline and hydroxylysine.

Ser/Thr residues as occurs in the more widespread N- and O-linked glycosylation (Section H5). The amount of attached carbohydrate in collagen varies from 0.4% to 12% by weight depending on the tissue in which it is synthesized.

Structure

The **primary structure** of each polypeptide in collagen is characterized by a repeating **tripeptide** sequence of **Gly–X–Y** where X is often, but not exclusively, Pro and Y is often Hyp. Each of the three polypeptide chains in collagen is some 1000 residues long and they each fold up into a **helix** that has only 3.3 residues per turn, rather than the 3.6 residues per turn of an α-helix (Section B2). This **secondary structure** is unique to collagen and is often called the **collagen helix**. The three polypeptide chains lie parallel and wind round one another with a slight right-handed, rope-like twist to form a **triple-helical cable** (Figure 3). Every third residue of each polypeptide passes through the center of the triple helix, which is so crowded that only the small side-chain of Gly can fit in. This explains the absolute requirement for Gly at every third residue. The residues in the X and Y positions are located on the outside of the triple-helical cable, where there is room for the bulky side-chains of Pro and other residues. The three polypeptide chains are also staggered so that the Gly residue in one chain is aligned with the X residue in the second and the Y residue in the third. The triple helix is held together by an extensive network of **hydrogen bonds** (Section B2), in particular between the primary amino group of Gly in one helix and the primary carboxyl group of Pro in position X of one of the other helices. In addition, the hydroxyl groups of Hyp residues participate in stabilizing the structure. The relatively inflexible Pro and Hyp also confer rigidity on the collagen structure.

The importance of Gly at every third residue is seen when a **mutation** in the DNA encoding Type I collagen leads to the incorporation of a different amino acid at just one position in

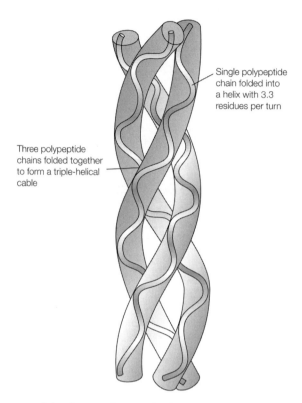

Single polypeptide chain folded into a helix with 3.3 residues per turn

Three polypeptide chains folded together to form a triple-helical cable

Figure 3. Arrangement of the three polypeptide chains in collagen.

the 1000 residue polypeptide chain. For example, if a mutation leads to the incorporation of Cys instead of Gly, the triple helix is disrupted as the -CH$_2$-SH side-chain of Cys is too large to fit in the interior of the triple helix. This leads to a partly unfolded structure that is susceptible to excessive hydroxylation and glycosylation and is not efficiently secreted by the fibroblast cells. This, in turn, results in a defective collagen structure that can give rise to **brittle bones** and **skeletal deformities**. A whole spectrum of such mutations are known, which cause the production of defective collagen and result in **osteogenesis imperfecta** (brittle bones).

Secretion and aggregation

When the collagen polypeptides are synthesized they have additional amino acid residues (100–300) on both their N and C termini that are absent in the mature collagen fiber (Figure 4). These **extension peptides** often contain Cys residues, which are usually absent from the remainder of the polypeptide chain. The extension peptides help to align correctly the three polypeptides as they come together in the triple helix, a process that may be aided by the formation of disulfide bonds between extension peptides on neighboring polypeptide chains. The extension peptides also prevent the premature aggregation of the procollagen triple helices within the cell. On **secretion** out of the fibroblast, the extension peptides are removed by the action of extracellular **peptidases** (Figure 4). The resulting tropocollagen molecules then **aggregate** together in a staggered head-to-tail arrangement in the collagen fiber (Figure 4).

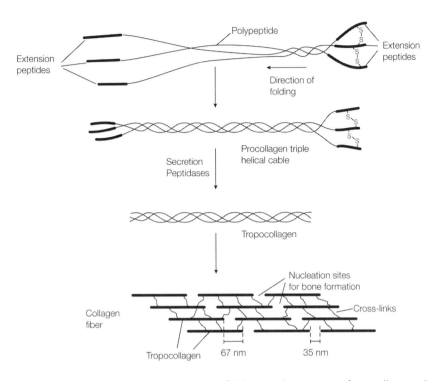

Figure 4. Role of the extension peptides in the folding and secretion of procollagen. Once secreted out of the cell, the extension peptides are removed and the resulting tropocollagen molecules aggregate and are cross-linked to form a fiber.

Cross-links

The strength and rigidity of a collagen fiber is imparted by **covalent cross-links** both between and within the tropocollagen molecules. As there are few, if any, Cys residues in the final mature collagen, these covalent cross-links are not disulfide bonds as commonly found in proteins, but rather are unique cross-links formed between **Lys** and its aldehyde derivative **allysine**. Allysine residues are formed from Lys by the action of the monooxygenase **lysyl oxidase** (Figure 5). This **copper**-containing enzyme requires the coenzyme **pyridoxal phosphate**, derived from vitamin B_6 (Section M2), for activity. The aldehyde group on allysine then reacts spontaneously either with the side-chain amino group of Lys or with other allysine residues on other polypeptide chains to form covalent interchain bonds.

Figure 5. Conversion of lysine to allysine by lysyl oxidase.

The importance of cross-linking to the normal functioning of collagen is demonstrated by the disease **lathyrism**. This occurs in humans and other animals through the ingestion of sweet pea (*Lathyrus odoratus*) seeds, which contain the chemical β-**aminopropionitrile**. This compound irreversibly inhibits lysyl oxidase, thereby preventing the cross-linking of the tropocollagen molecules, resulting in serious abnormalities of the bones, joints and large blood vessels due to the fragile collagen. One collagen deficiency disease, the **Ehlers–Danlos syndrome type V**, is due to a deficiency in lysyl oxidase and results in hypermobile joints and hyperextensibility of the skin.

Bone formation

The regular staggered array of spaces between the ends of the tropocollagen molecules in a collagen fiber (Figure 4) are the **nucleation sites** for the deposition of a form of **calcium phosphate**, **hydroxyapatite**, in bone formation. Further hydroxyapatite is added until the nucleation sites grow and join with one another to form the mature bone structure.

B5 Molecular motors

Key Notes

Molecular motors

Molecular motors or motor proteins bind to cytoskeletal filaments and use energy derived from the hydrolysis of ATP to move along them. The head region or motor domain that hydrolyzes ATP binds to the filament, while the tail region binds the cargo. The major types of motor proteins are the myosins, the kinesins and the dyneins.

Muscle structure

Each cell within vertebrate striated muscle contains within its sarcoplasm many parallel myofibrils, which in turn are made up of repeating sarcomere units. Within the sarcomere are the alternating dark A band and light I band, in the middle of which are the H zone and Z line, respectively. A myofibril contains two types of filaments: the thick filaments consisting of myosin, and the thin filaments consisting of actin, tropomyosin and troponin. When muscle contracts, the thick and thin filaments slide over one another, shortening the length of the sarcomere.

Myosin

The protein myosin consists of two heavy polypeptide chains and two pairs of light chains arranged as a double-headed globular region attached to a two-stranded α-helical coiled-coil. Myosin molecules spontaneously assemble into filaments, hydrolyze ATP and bind actin.

Actin

Actin, the major constituent of the thin filaments, can exist as monomeric globular G-actin or as polymerized fibrous F-actin. The actin filaments are connected to the thick filaments by cross-bridges formed by the S1 heads of myosin.

The generation of force in muscle

The cyclic formation and dissociation of complexes between the actin filaments and the S1 heads of myosin leads to contraction of the muscle. On binding to actin, myosin releases its bound P_i and ADP. This causes a conformational change to occur in the protein, which moves the actin filament along the thick filament. ATP then binds to myosin, displacing the actin. Hydrolysis of the ATP returns the S1 head to its original conformation.

Troponin and tropomyosin

Tropomyosin lies along the thin filament and prevents the association of myosin with actin in the resting state. Ca^{2+} ions released into the sarcoplasm from the sarcoplasmic reticulum in response to a nerve stimulation bind to the TnC subunit of troponin and cause a conformational change in the protein. This movement is transmitted by an allosteric mechanism through the TnI and TnT subunits of troponin to

	tropomyosin, causing the latter to move out of the way and allowing the actin and myosin to associate.
Kinesins	Kinesins move along microtubules and are involved in the movement of vesicles and organelles around the cell and in the segregation of the chromosomes. A kinesin molecule has two heads, each with ATPase activity, which alternately bind to the microtubule thus walking the kinesin and its bound cargo along the microtubule.
Dynein	Eukaryotic cilia are hair-like protrusions on the surface of the cell that consist mainly of microtubules. The microtubule fibers in a cilium are bundled together in a characteristic 9 + 2 arrangement within the axoneme. The outer nine microtubule doublets look like a figure eight with a smaller circle, subfiber A, and a larger circle, subfiber B. Dynein is a very large protein that forms cross-bridges with the B subfibers and possesses ATPase activity. Two dynein arms protrude from subfiber A, which, upon hydrolysis of ATP, move along adjacent B subfibers. Due to extensible nexin links between the doublets, this sliding motion is converted into a local bending of the cilium.
Related topics	(A2) Eukaryotic cells

Molecular motors

Molecular motors or **motor proteins** bind to a cytoskeletal filament and use the energy derived from repeated cycles of ATP hydrolysis to move along it; thus they **convert chemical energy into motion**. There are many different types of motor proteins in eukaryotic cells that differ in the type of filament to which they bind, the direction in which they move along the filament and the cargo they carry. The motor proteins associate with the filaments through a **head region** or **motor domain** that binds and hydrolyzes ATP, while the **tail region** binds the cargo that is transported. There are three types of motor proteins: the **myosins** that bind to actin filaments and are involved in muscle contraction, the **kinesins** that bind to microtubules and are involved in the movement of vesicles and organelles around the cell and in chromosome segregation, and **dyneins** that also bind to microtubules and power the motion of cilia and flagella in eukaryotic cells.

Muscle structure

The best understood force-generating process in biological systems is the contraction of **vertebrate striated muscle**, so named because it appears striated (striped) under phase-contrast microscopy (Section A4). This muscle is composed of numerous multinucleate cells that are bounded by an electrically excitable plasma membrane. Each cell contains within its **sarcoplasm** (cytosol) many parallel **myofibrils**, each approximately 1 μm in diameter. The sarcoplasm is also rich in ATP, creatine phosphate (Section M3) and glycolytic enzymes (Section J3). The functional unit of the myofibril is the **sarcomere**, which repeats every 2.3 μm along the fibril axis (Figure 1a). A dark **A band** and a light **I band** alternate regularly along the length of the myofibril. The central region of the A band, the **H zone**, is less dense than the rest of the band. Within the middle of the I band is a

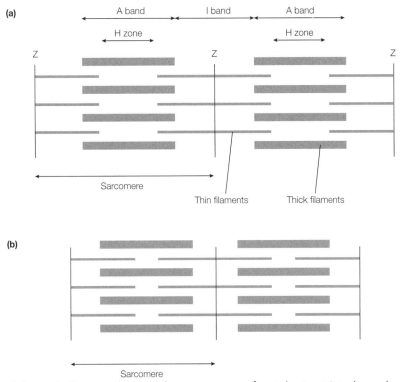

Figure 1. Schematic diagram showing the appearance of vertebrate striated muscle as it appears under phase-contrast microscopy: (a) relaxed; (b) contracted.

very dense narrow **Z line**. A cross-section of a myofibril reveals that there are two types of interacting filaments. The **thick filaments** of diameter approximately 15 nm are found only in the A band (Figure 1a) and consist primarily of the protein **myosin**, while the **thin filaments** of approximately 9 nm diameter contain **actin**, **tropomyosin** and the **troponin complex**.

When muscle contracts, it can shorten by as much as a third of its original length. Information obtained from X-ray crystallographic (Section B2), and light- and electron-microscopic studies (Section A4) led to the proposal of the **sliding filament model** to explain muscle contraction. The thick and thin filaments were seen not to change in length during muscle contraction, but the length of the **sarcomere** was observed to decrease as the thick and thin filaments slide past each other (Figure 1). Thus, as muscle contracts, the sizes of the H zone and the I band are seen to decrease. The force of the contraction is generated by a process that actively moves one type of filament past neighboring filaments of the other type.

Myosin

Myosin is a large protein (520 kDa) consisting of six polypeptide chains: two heavy chains (220 kDa each), and two pairs of light chains (20 kDa each). This large protein has three biological activities:

1. Myosin molecules spontaneously assemble into **filaments** in solutions of physiological ionic strength and pH

2.   Myosin is an **ATPase**, hydrolyzing ATP to ADP and P_i

3.   Myosin binds the **polymerized form of actin**

Myosin consists of a double-headed globular region joined to a long rod. The rod is a two-stranded α-helical coiled-coil formed by the two heavy chains, while the globular heads are also part of each heavy chain with the light chains attached (Figure 2a). Limited **proteolysis** of myosin with the protease **trypsin** results in its dissection into two fragments: **light meromyosin** (**LMM**) and **heavy meromyosin** (**HMM**) (Figure 2b). Functional studies of these two fragments reveal that LMM can still form filaments but lacks ATPase activity, whereas HMM does not form filaments but possesses ATPase activity and can bind to actin. HMM can be further split into two identical **globular subfragments** (**S1**) and one **rod-shaped subfragment** (**S2**) by another protease, **papain** (Figure 2b). The S1 subfragment, whose structure has been determined by X-ray crystallography, contains an ATPase site, an actin-binding site and two light chain-binding sites. The proteolytic cleavage of myosin occurs at **flexible hinge regions** within the protein that separate the globular S1 domains from the rod-like S2 and LMM domains (Figure 2c). These hinges have a crucial role to play in the contraction of muscle.

Actin

Actin, the major constituent of the thin filaments, exists in two forms. In solutions of low ionic strength, it exists as a 42 kDa monomer, termed **G-actin** because of its globular

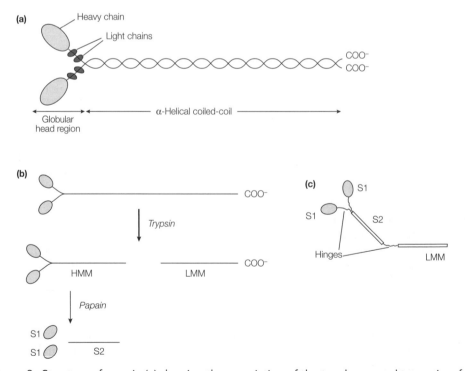

Figure 2. Structure of myosin (a) showing the association of the two heavy and two pairs of light chains; (b) showing the proteolytic fragmentation of myosin; and (c) showing the hinge regions between domains.

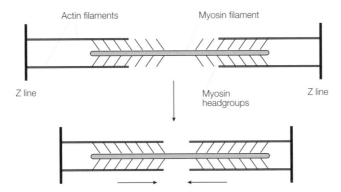

Figure 3. Schematic diagram showing the interaction of the myosin thick filaments and the actin thin filaments during skeletal muscle contraction.

shape. As the ionic strength of the solution rises to that at the physiological level, G-actin polymerizes into a fibrous form, **F-actin**, that resembles the thin filaments found in muscle. Although actin, like myosin, is an **ATPase**, the hydrolysis of ATP is not involved in the contraction–relaxation cycle of muscle but rather in the assembly and disassembly of the actin filament.

On the thick filaments, **cross-bridges** emerge from the filament axis in a regular helical array towards either end, whereas there is a bare region in the middle that is devoid of cross-bridges (Figure 3). In muscle depleted of ATP, the myosin cross-bridges interact with the surrounding actin filaments. The absolute direction of the actin and myosin molecules reverses halfway between the Z lines. Thus, as the two thin filaments that bind the cross-bridges at either end of a thick filament move towards each other, sliding over the thick filament, the distance between the Z lines shortens and the muscle contracts (Figure 3).

The generation of force in muscle

The cyclic formation and dissociation of cross-bridges between actin and the S1 heads of myosin leads to contraction of the muscle because of **conformational changes** that take place in the myosin S1 head. In resting muscle, the S1 heads are unable to interact with the actin in the thin filaments because of steric interference by the regulatory protein **tropomyosin** (Figure 4a). The myosin has bound to it ADP and P_i. When the muscle is stimulated, the tropomyosin moves out of the way, allowing the S1 heads projecting out from the thick filament to attach to the actin in the thin filament (Figure 4b). On binding of myosin–ADP–P_i to actin, first the P_i and then the ADP are released. As the ADP is released, the S1 head undergoes a **conformational change in the hinge region** between the S1 and S2 domains that alters its orientation relative to the actin molecule in the thin filament (Figure 4c). This constitutes the **power stroke** of muscle contraction and results in the thin filament moving a distance of approximately 10 nm relative to the thick filament towards the center of the sarcomere. ATP then binds to the S1 head, which leads to the rapid release of the actin [i.e. dissociation of the thin and thick filaments (Figure 4d)]. The ATP is then hydrolyzed to ADP and P_i by the free S1 head, which is returned to its original conformation ready for another round of attachment (Figure 4e), conformational change and release.

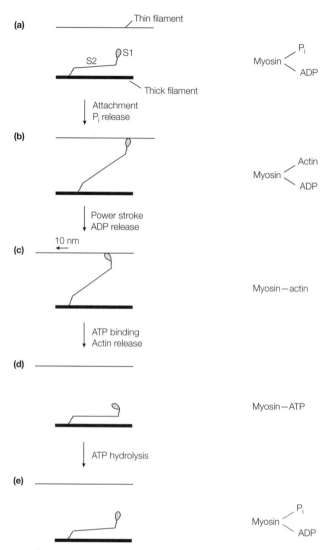

Figure 4. Mechanism for the generation of force in muscle as an S1 head of a myosin thick filament interacts with an actin thin filament.

Troponin and tropomyosin

Troponin and tropomyosin mediate the regulation of muscle contraction in response to **Ca^{2+}**. These two proteins are present in the thin filament, alongside the actin, and constitute about a third of its mass. **Tropomyosin** is an elongated protein of 70 kDa that forms a two-stranded α-helical rod, which lies nearly parallel to the long axis of the thin filament. **Troponin** is a complex of three polypeptide chains: **TnC** (18 kDa), which binds Ca^{2+}; **TnI** (24 kDa), which binds to actin; and **TnT** (37 kDa), which binds to tropomyosin. On muscle stimulation by a nerve impulse, Ca^{2+} ions are released from the **sarcoplasmic reticulum** (a specialized form of the ER found in muscle cells; Section A2) into the cytosol, raising the cytosolic Ca^{2+} concentration from the resting concentration of less than 1 μM to about 10 μM. The Ca^{2+} binds to sites on TnC, causing a **conformational change** in this polypeptide, which is transmitted through the other components of the troponin complex to the tropomyosin. The tropomyosin then moves out of the way, allowing the S1 head of myosin to interact with the actin and initiate a cycle of contraction. Thus, Ca^{2+} controls muscle contraction by an **allosteric mechanism** (Section D5) involving troponin, tropomyosin, actin and myosin.

Kinesins

Kinesins move along **microtubules** formed from the protein **tubulin** (Section A2) and are involved in the movement of vesicles and organelles around the cell and in the segregation of the chromosomes. A kinesin molecule has **two heads**, which operate in tandem: one binds to the microtubule, and then the next one does. So a kinesin molecule essentially walks along the microtubule in a highly processive manner. In contrast to myosin, where ATP binding promotes its dissociation from actin, the addition of ATP strongly increases the affinity of kinesin for microtubules. The cycle begins with both heads in a kinesin molecule in the ADP-bound form. One head binds to the microtubule and releases its ADP, which is replaced by ATP. The binding of ATP results in a **conformational change** that locks the head to the microtubule and pulls the linker between the two heads forward thus throwing the second head along the microtubule. ATP hydrolysis occurs in the first head while the second head interacts with the microtubule. The exchange of ADP for ATP in the second head pulls the first head of the kinesin molecule forward, moving it along the microtubule in a stepwise fashion. The cycle then repeats to move the kinesin molecule and its bound cargo along the microtubule.

Dynein

The hair-like protrusions or **cilia** on the surfaces of certain eukaryotic cells, such as those lining the respiratory passages, consist mainly of **microtubules** (Section A2). Cilia are involved in moving a stream of liquid over the surface of the cell. Free cells such as protozoa and sperm from various species can be propelled by either cilia or a flagellum (Section A1). In eukaryotic cells, flagella differ from cilia only in being much longer. All eukaryotic cilia and flagella have the same basic design; a bundle of fibers called an **axoneme** surrounded by a membrane that is continuous with the plasma membrane (Figure 5). The microtubule fibers in an axoneme are in a characteristic **9 + 2 array**, with a peripheral group of nine pairs of microtubules surrounding two singlet microtubules (Figure 5). Each of the nine outer doublets appears like a figure eight, the smaller circle is termed **subfiber A**, the larger circle, **subfiber B**. Subfiber A is joined to a central sheath by radial spokes, while neighboring microtubule doublets are held together by **nexin** links. Two **dynein** arms emerge from each subfiber A, with all the arms in a cilium pointing in the same direction (Figure 5).

Dynein is a very large protein (1000–2000 kDa) consisting of one, two or three heads depending on the source. The heads of dynein form **cross-bridges** with the B subfibers, and possess **ATPase activity**. The binding of ATP to dynein causes it to dissociate from the B subfiber. On hydrolysis of the ATP to ADP and P_i, the dynein binds again with the B subfiber with the subsequent release of the P_i and ADP. This ATPase cycle leads to the movement of the cilium as the outer doublets of the axoneme slide past each other. The force between adjacent doublets is generated by the dynein cross-bridges. Thus, the dynein arms on subfiber A of one doublet walk along subfiber B of the adjacent doublet. Unlike in muscle, where the myosin and actin filaments slide past each other, in a cilium the radial spokes resist the sliding motion, which instead is converted into a local bending. The highly extensible protein, **nexin**, keeps adjacent doublets together during this process.

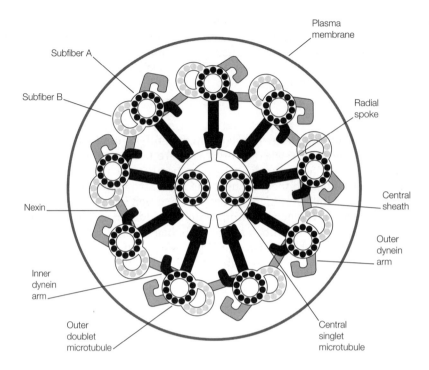

Figure 5. Cross-sectional diagram of a cilium.

A defect or absence in any one of the proteins within the axoneme (e.g. dynein, nexin, etc.) results in cilia that are immotile, so called **immotile-cilia syndrome**. Patients suffering from this disease have chronic pulmonary disorders due to the cilia in the respiratory tract being unable to sweep out bacteria and other foreign particles. In addition, males with this genetic defect are infertile because their sperm are unable to move due to flagella inactivity.

B6 Antibodies

Key Notes

Immune system: an overview	The immune system has two main functions; to recognize invading pathogens and then to trigger pathways that will destroy them. The humoral immune system relies on B lymphocytes to produce soluble antibodies that will bind the foreign antigens. The cellular immune system uses killer T lymphocytes that recognize and destroy invading cells directly. The primary immune response occurs on initial contact with a foreign antigen and results in production of immunoglobulin M (IgM) and then immunoglobulin G (IgG). If the same antigen is encountered again, immunological memory leads to a secondary immune response that produces a much more rapid and larger increase in specific IgG production.
Antibody structure	Each IgG molecule consists of four polypeptide chains (two identical light chains and two identical heavy chains joined by disulfide bonds) and has two antigen-binding sites (i.e. is bivalent). Each light chain and each heavy chain consists of a variable region and a constant region. Each light chain folds into two domains, one for the variable region and one for the constant region. Each IgG heavy chain folds into four domains, one for the variable region and three in the constant region. Papain digests IgG into two Fab fragments (each of which has an antigen-binding site, i.e. is univalent) and one Fc fragment. Pepsin digests IgG to release an $F(ab')_2$ fragment that has two antigen-binding sites.
Five classes of immunoglobulins	Human immunoglobulins exist as IgA, IgD, IgE, IgG and IgM classes, which contain α, δ, ε, γ and μ heavy chains, respectively. IgM is a pentamer that binds to invading microorganisms and activates complement killing of the cells and phagocytosis. IgG is the main antibody found in the blood after antigen stimulation and also has the ability to cross the placenta. IgA mainly functions in body secretions. IgE provides immunity against some parasites but is also responsible for the clinical symptoms of allergic reactions. The exact role of IgD is unknown.
Polyclonal antibodies	A preparation of antibody molecules that arises from several different clones of cells is called a polyclonal antibody. It is a mixture of antibody molecules that bind to different parts (epitopes) of the antigen and with different binding affinities.
Monoclonal antibodies	Antibody produced by a single clone of cells is a monoclonal antibody; all the antibody molecules are identical and bind to the same antigenic site with identical binding affinities.

Antibody synthesis	No complete antibody gene exists in germ-line cells. The genes for light chains and heavy chains assemble by somatic recombination during B-lymphocyte maturation. A B lymphocyte can change the class of antibody being expressed by keeping the same gene sequence that codes for the variable region on the heavy chain but joining this to a new C gene segment. The new heavy chain has a different constant region but the same antigen-binding specificity of the previous heavy chain.	
Related topics	(A4) Cell imaging	(C1) Protein purification
	(B2) Protein structure and function	(C4) Immunodetection

Immune system: an overview

There are two vital functions of the immune system; recognition of an invading pathogen (disease-producing bacteria, fungi, protozoa and viruses) as being distinct from normal body components (and hence treated as foreign) and then the triggering of pathways that lead to destruction of the invader, such as activation of complement (see below) and phagocytic cells that engulf and digest the invading organism (Section E4). The key cells responsible for immunity in vertebrates are white blood cells called lymphocytes, which arise from precursor (stem) cells in the bone marrow. There are two main parts of the immune system, which interact to provide overall protection for the animal:

- The **humoral immune response** (*humor* is an ancient term meaning fluid) relies on the production of soluble proteins called antibodies (or immunoglobulins) by **B lymphocytes** (**B cells**), so called because the cells mature in the bone marrow.

- The **cellular immune response** is mediated by **T lymphocytes** (**T cells**), so called because their maturation from stem cells occurs in the thymus. The intact T lymphocytes are responsible for the recognition and killing of foreign invaders.

In both cellular and humoral immunity, recognition of the foreign invader depends upon the recognition of foreign macromolecules (proteins, carbohydrates, nucleic acids); these foreign components are called **antigens**.

Primary and secondary immune responses

The presence of a foreign antigen stimulates the production of a specific antibody in the bloodstream, which will recognize and bind tightly to it. The antibody molecules first produced after antigen injection are in the **immunoglobulin M** class (IgM molecules). However, about 10 days after antigen injection, the amount of IgM in the bloodstream (the **titer** of antibody) declines and there is a concurrent increase in another class of antibody called **immunoglobulin G** (IgG) (Figure 1). This is called the **primary immune response**.

One of the most important features of the immune system is that, once an animal has encountered a particular pathogen, the system confers protection against future infection. This **immunological memory** means that if the same pathogen or antigen is encountered a second time, perhaps even decades after the previous occurrence, then the system reacts much faster and more dramatically to produce a large titer of specific

IgG to counter the antigen. This is the **secondary immune response** (Figure 1) and is mediated by long-lived memory T cells and memory B cells.

Antibody structure

Light and heavy chains

Each molecule of IgG is Y-shaped and consists of four polypeptide chains joined together by disulfide bonds; two identical copies of **light (L) chains** (25 kDa) and two identical copies of **heavy (H) chains** (50 kDa) (Figure 2A). The N-terminal ends of one heavy chain and its neighboring light chain cooperate to form an **antigen-binding site**, so that the IgG molecule has two binding sites for antigen, that is, it is **bivalent**. Because of this, a single antibody molecule can bind two antigen molecules and so cross-link and precipitate antigens out of solution.

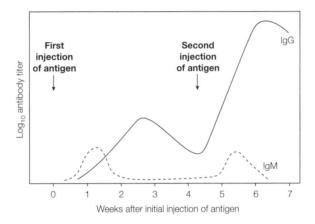

Figure 1. The primary and secondary immune responses to injections of antigen.

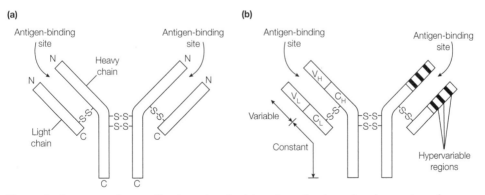

Figure 2. Structure of an antibody molecule: (a) each antibody molecule consists of two identical light chains and two identical heavy chains – the molecule has two antigen-binding sites, each formed by a light chain and a heavy chain; (b) the N-terminal regions of the light and heavy chains are variable in amino acid sequence from antibody to antibody (variable regions; V regions) whilst the C-terminal regions are relatively constant in sequence (constant regions; C regions). The generic terms for these regions in the light chain are V_L and C_L and for the heavy chains are V_H and C_H.

Variable and constant regions

Comparison of the amino acid sequences of many immunoglobulin polypeptides has shown that each light chain has a **variable region** at its N-terminal end and an invariant or **constant region** at its C terminus (Figure 2b). Similarly, each heavy chain has an N-terminal variable region and a C-terminal constant region. Since it is the N-terminal parts of the light and heavy chains that form the antigen-binding site, the variability in amino acid sequence of these regions explains how different sites with different specificities for antigen binding can be formed. In fact, the variability in the variable regions of both light and heavy chains is mainly localized to three **hypervariable regions** in each chain (Figure 2b). In the three-dimensional structure of the immunoglobulin molecule, the hypervariable parts of the light and heavy chains are looped together to form the antigen-binding site. The remaining parts of each variable region stay reasonably constant in sequence, usually do not contact the antigen directly and are called **framework regions**.

Antibody domains

Each light chain consists of two repeating segments of about 110 amino acids that fold into two compact three-dimensional **domains**, one representing the variable region of the light chain and the other domain representing the constant region. Each heavy chain is also made up of repeating units about 110 amino acids long. Since each IgG heavy chain is about 440 amino acids long, it forms four domains – one domain for the variable region and three domains in the constant region. The similarity of amino acid sequence between the various domains suggests that they arose in evolution by gene duplication.

Fab and Fc fragments

Papain, a protease, cuts the IgG molecule to release the two arms of the Y-shaped molecule, each of which has one antigen-binding site and is called an **Fab fragment** (Fragment **a**ntigen **b**inding) (Figure 3). Because Fab fragments have only one antigen-binding site (i.e. are **univalent**), they cannot cross-link antigens. The released stem of the Y-shaped molecule (consisting of the identical C-terminal parts of the two H chains) is named the **Fc fragment** (so called because it readily crystallizes). The Fc fragment carries the effector sites that trigger the destruction of the antigen, for example triggering of the complement system and inducing phagocytosis of pathogens by other white blood cells. In contrast to papain, pepsin (another protease) cuts the IgG molecule to release the two arms of IgG still linked together and hence this fragment has two antigen-binding sites (i.e. is bivalent) and can still cross-link antigens. This is called the **F(ab′)$_2$ fragment** (Figure 3).

Five classes of immunoglobulins

Humans have five different classes of antibody molecule, which differ both in structure and in function. These are called immunoglobulin A (IgA), IgD, IgE, IgG and IgM and each has its own type of heavy chain: α, δ, ε, γ and μ, respectively. Thus IgA molecules have two identical α heavy chains, IgD molecules have two identical δ heavy chains, etc. The human IgG class of antibodies is further divided into four IgG subclasses: IgG$_1$, IgG$_2$, IgG$_3$ and IgG$_4$, having γ_1, γ_2, γ_3 and γ_4 heavy chains, respectively.

The different heavy chains confer different properties and functions on each of the immunoglobulin classes.

- **IgM** has μ heavy chains and exists as a pentamer in combination with another polypeptide called the J chain, which is responsible for initiating the polymerization to

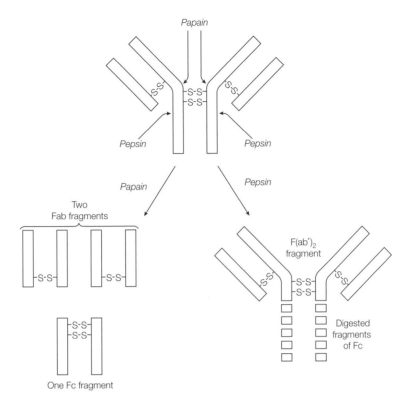

Figure 3. Papain digestion of an antibody molecule yields two univalent Fab fragments and an Fc fragment whereas pepsin digestion yields a bivalent F(ab')$_2$ fragment.

form the pentameric structure. With its large number of antigen-binding sites, each IgM molecule binds very tightly to any pathogen that has multiple copies of the same antigen on its surface. The binding induces the Fc region to activate the complement pathway, which eventually causes the death of the pathogen. IgM also activates macrophages to phagocytose pathogens. Not surprisingly given these functions, IgM is the first antibody produced when an animal responds to a new antigen (Figure 1).

- **IgG** is the main immunoglobulin in the bloodstream late in the primary immune response and particularly during the secondary immune response (Figure 1). Like IgM, it can activate complement and trigger macrophages, but is the only antibody that can pass through the placenta and so provide immunological protection for the fetus. It is also secreted into the mother's milk and is taken up from the gut of the newborn animal into the bloodstream, thus providing continuing protection after birth.

- **IgA** is the main class of antibody in secretions such as tears, saliva, and in secretions of the lungs and the intestine. It is the first line of immunological defense against infection at these sites.

- **IgE** occurs in tissues where, having bound the antigen, it stimulates mast cells to release a range of factors. Some of these in turn activate white blood cells (called eosinophils) to kill various types of parasite. However, the mast cells can also release biologically active amines, including histamine, which cause dilation and increased permeability of blood vessels and lead to the symptoms seen in allergic reactions such as hay fever and asthma.

- **IgD** is found on the surface of mature B lymphocytes and in traces in various body fluids, but its exact function remains unclear.

Two different forms of light chains also exist. Antibody molecules in any of the antibody classes or subclasses can have either two κ light chains or two λ light chains. Unlike the different heavy chains described above, no difference in biological function between κ and λ light chains is known.

Polyclonal antibodies

If an antigen is injected into an animal, a number of antibody-producing cells will bind that antigen, albeit with varying degrees of affinity, and so the antibody that appears in the bloodstream will have arisen from several clones of cells, i.e. it will be a **polyclonal antibody**. Different antibody molecules in a preparation of polyclonal antibody will bind to different parts of the macromolecular antigen and will do so with different binding affinities. The binding region recognized by any one antibody molecule is called an **epitope**. Most antibodies recognize particular surface structures in a protein rather than specific amino acid sequences (i.e. the epitopes are defined by the conformation of the protein antigen). A preparation of polyclonal antibodies will bind to many epitopes on the protein antigen.

Monoclonal antibodies

If a single clone of antibody-producing cells could be isolated, then all of the antibody produced from that clone would be identical; all antibody molecules in such a **monoclonal antibody** preparation would bind to the same antigen epitope.

The problem is that if an individual antibody-producing cell is isolated and grown in culture, its descendants have a limited life-span that severely limits their use for the routine preparation of monoclonal antibodies. In 1975, Milstein and Köhler discovered how monoclonal antibodies of almost any desired antigen specificity can be produced indefinitely and in large quantities. Their method was to fuse a B lymphocyte producing antibody of the desired specificity with a cell derived from a cancerous lymphocyte tumor, called a myeloma cell, which is immortal. The cell fusion is called a **hybridoma**, which is both immortal and secretes the same specific antibody originally encoded by the B lymphocyte.

Monoclonal antibodies produced using this technology are now common tools in research because of their very high specificity. For example, they can be used to locate particular molecules within cells (Section A4) or particular amino acid sequences within proteins (Section C4). If they are first bound to an insoluble matrix, they are also extremely useful for binding to and hence purifying the particular molecule from crude cell extracts or fractions (Section C1). They are also increasingly of use in medicine, both for diagnosis and as therapeutic tools, for example to inactivate bacterial toxins and to treat certain forms of cancer.

Antibody synthesis

The human genome is thought to contain fewer than 10^5 genes, yet a human can make at least 10^{15} different types of antibody in terms of antigen-binding specificity. To account for most of this antibody diversity, the genes exist in separate coding sections and are assembled during B-lymphocyte maturation by a process called **somatic recombination**. This process of assembly takes place in every B lymphocyte. By assembling different

Figure 4. The light chain variable region is encoded by two separate gene segments, V and J. The heavy chain variable region is encoded by three gene segments, V, D and J.

fragments of DNA, completely new immunoglobulin genes can be created and hence this gives an enormous potential reservoir of antibody diversity.

No complete antibody gene exists in germ-line cells. The genes for light chains and heavy chains assemble by somatic recombination during B-lymphocyte maturation. In the germ-line, each light chain gene exists as multiple V and J gene segments upstream of a single C gene segment. During B-lymphocyte differentiation, one V gene segment joins with one J gene segment (VJ joining) to assemble the complete light chain gene, usually by deletion of intervening DNA. The V and J gene sequences encode the variable region while the C gene segment encodes the constant region of the light chain polypeptide (Figure 4).

Heavy chains are encoded by multiple V, J and D gene segments, which lie upstream of a single copy of C gene segments for each of the constant regions of μ, δ, γ, ε and α chains. During B-lymphocyte differentiation, a D gene segment joins a J segment (DJ joining) and then the recombined DJ joins a V gene segment (VDJ joining); the VDJ gene sequence encodes the variable region of the heavy chain polypeptide (Figure 4).

A B lymphocyte can change the class of antibody being expressed by moving a new C gene segment into position after the recombined VDJ segment, deleting the intervening DNA. The new heavy chain has a different constant region but retains the same variable region and so when class switching occurs so that the lymphocyte now makes IgA instead of IgM, the specificity of the antibody for antigens stays the same.

C1 Protein purification

Key Notes

Principles of protein purification

The aim of protein purification is to isolate one particular protein from all the others in the starting material. A combination of fractionation techniques is used that exploits the solubility, size, charge, hydrophobicity or/and specific binding affinity of the protein of interest. A source that is relatively rich in the protein of interest and which is readily available is used as the starting material. The use of recombinant DNA techniques means that large amounts of normally scarce proteins can be obtained by expression in bacterial or eukaryotic cells.

Homogenization and solubilization

The protein has to be obtained in solution prior to its purification. Tissues and cells are disrupted by homogenization or osmotic lysis and then subjected to differential centrifugation to isolate the subcellular fraction in which the protein is located. For membrane-bound proteins, the membrane structure has to be solubilized with a detergent to liberate the protein. Precautions have to be taken in order to prevent proteins being denatured or inactivated during purification by physical or biological factors. These include buffering the pH of the solutions, undertaking the procedures at a low temperature and including protease inhibitors to prevent unwanted proteolysis.

Assay of proteins

In order to monitor the progress of the purification of a protein, it is necessary to have an assay for it. Depending on the protein, the assay may involve measuring its enzyme activity or ligand-binding properties, or may quantify the protein present using antibodies directed against it.

Ammonium sulfate precipitation

The solubility of proteins decreases as the concentration of ammonium sulfate in the solution is increased. The concentration of ammonium sulfate at which a particular protein comes out of solution and precipitates may be sufficiently different from other proteins in the mixture to effect a separation.

Gel filtration chromatography

Gel filtration chromatography separates proteins on the basis of their size and shape using porous beads packed in a column. Large or elongated proteins cannot enter the pores in the beads and elute from the bottom of the column first, whereas smaller proteins can enter the beads, have a larger volume of liquid accessible to them and move through the column more slowly, eluting later.

Ion exchange chromatography	In ion exchange chromatography, proteins are separated on the basis of their net charge. In anion exchange chromatography a column containing positively charged beads is used, to which proteins with a net negative charge will bind, whereas in cation exchange chromatography, negatively charged beads are used, to which proteins with a net positive charge will bind. The bound proteins are then eluted by adding a solution of sodium chloride or by altering the pH of the buffer.
Affinity chromatography	Affinity chromatography exploits the specific binding of a protein for another molecule, its ligand (e.g. antigen for its antibody, so-called immunoaffinity chromatography). The ligand is immobilized on an insoluble support, which is then packed into a column. On adding a mixture of proteins, only the protein of interest binds to the ligand. All other proteins pass straight through the column. The bound protein is then eluted from the immobilized ligand in a highly purified form.
Related topics	(A2) Eukaryotic cells (B1) Amino acid structure (A5) Cell fractionation (C2) Gel electrophoresis

Principles of protein purification

The basic aim in protein purification is to isolate one particular protein of interest from other contaminating proteins so that its structure and/or other properties can be studied. Once a suitable cellular **source** of the protein has been identified, the protein is liberated into solution and then separated from contaminating material by sequential use of a series of different **fractionation techniques** or **separations**. These separations exploit one or more of the following basic properties of the protein: its **solubility**, its **size**, its **charge**, its **hydrophobicity** or its **specific binding affinity**. These separations may be **chromatographic techniques** such as **ion exchange**, **gel filtration** or **affinity chromatography**, **hydrophobic interaction chromatography**, in which the protein binds to a hydrophobic material, or **electrophoretic techniques** such as **isoelectric focusing** (Section C2). Other electrophoretic procedures, mainly **sodium dodecyl sulfate (SDS) polyacrylamide gel electrophoresis (PAGE)** (Section C2), are used to monitor the extent of purification and to determine the molecular mass and subunit composition of the purified protein.

Selection of a protein source

Before attempting to purify a protein, the first thing to consider is the source of **starting material**. Proteins differ in their cellular and tissue distribution, and thus if a protein is known to be abundant in one particular tissue (e.g. kidney) it makes sense to start the purification from this source. Also, some sources are more readily available than others and this should be taken into account too. Nowadays, with the use of **recombinant DNA techniques** (Section I1), even scarce proteins can be expressed (synthesized) in bacteria or eukaryotic cells grown in culture (Section A3) and relatively large amounts of the protein subsequently obtained.

Homogenization and solubilization

Once a suitable source has been identified, the next step is to obtain the protein in solution. For proteins in biological fluids, such as blood serum or cell culture medium, this is already the case, but for many proteins the tissues and cells need to be disrupted and broken open (lysed). **Homogenization** and subsequent **differential centrifugation** of biological samples is detailed in Section A5. In addition to the procedures described there, another simple way of breaking open cells that do not have a rigid cell wall to release the cytosolic contents is **osmotic lysis**. When animal cells are placed in a hypotonic solution (such as water or a buffered solution without added sucrose), the water in the surrounding solution diffuses into the more concentrated cytosol, causing the cell to swell and burst. Differential centrifugation is then employed to remove contaminating subcellular organelles (Section A5). Those proteins that are bound to membranes require a further solubilization step. After isolation by differential centrifugation, the appropriate membrane is treated with a **detergent** such as Triton X-100 to disrupt the lipid bilayer and to release the integral membrane proteins into solution (see Section E2 for more details).

Stabilization of proteins

Throughout the purification procedure, steps have to be taken to ensure that the protein of interest is not **inactivated** or **denatured** either by physical or biological factors. The pH of the solutions used needs to be carefully **buffered** (Section B1) at a pH in which the protein is stable, usually around pH 7. The temperature often needs to be maintained below 25°C (usually around 4°C) to avoid **thermal denaturation** and to minimize the activity of **proteases**. Upon homogenization, **proteases** within the source material that are normally in a different subcellular compartment will be liberated into solution and come into contact with the protein of interest and may degrade it. For example, the acid hydrolases in lysosomes (Section A2) could be liberated into solution and rapidly degrade the protein of interest. Thus, as well as carrying out the procedures at low temperature, **protease inhibitors** (Section D4) are often included in the buffers used in the early stages of the isolation procedure in order to minimize unwanted proteolysis.

Assay of proteins

A suitable means of detecting (**assaying**) the protein must be available to monitor the success of each stage in the purification procedure. The most straightforward **assays** are those for enzymes that catalyze reactions with readily detectable products (for more details on enzyme assays see Section D1). Proteins that are not enzymes may be assayed through the observation of their biological effects. For example, a receptor can be assayed by measuring its ability to bind its specific ligand. Immunological techniques are often used to assay for the protein of interest using antibodies that specifically recognize it [e.g. radioimmunoassay, enzyme-linked immunosorbent assay (ELISA), or Western blot analysis (Section C4)].

Ammonium sulfate precipitation

A commonly employed first separation step is **ammonium sulfate precipitation**. This technique exploits the fact that the **solubility** of most proteins is lowered at high salt concentrations. As the salt concentration is increased, a point is reached where the protein comes out of solution and precipitates. The concentration of salt required for this **salting-out effect** varies from protein to protein, and thus this procedure can be used to fractionate a mixture of proteins. For example, 0.8 M ammonium sulfate precipitates out the clotting protein fibrinogen from blood serum, whereas 2.4 M ammonium sulfate is required to precipitate albumin. However, many other proteins will also precipitate out at these concentrations of ammonium sulfate. Therefore, this is a relatively crude separation technique, although it often provides a convenient concentration step. Salting out is also sometimes used at later stages in a purification procedure to **concentrate** a dilute solution of the protein since the protein precipitates and can then be redissolved in a smaller volume of buffer.

Dialysis

Proteins can be separated from small molecules by dialysis through a **semi-permeable membrane** such as cellophane (cellulose acetate). **Pores** in the membrane allow molecules up to approximately 10 kDa to pass through, whereas larger molecules are retained inside the dialysis bag (Figure 1). As most proteins have molecular masses greater than 10 kDa, this technique is not suitable for fractionating proteins, but is often used to remove small molecules such as salts and ammonium sulfate from a protein solution. It should be noted that at equilibrium, the concentration of small molecules inside a dialysis bag will be equal to that outside (Figure 1b), and so several changes of the surrounding solution are often required to lower the concentration of the small molecule in the protein solution sufficiently.

Gel filtration chromatography

In gel filtration chromatography (**size exclusion chromatography** or **molecular sieve chromatography**), molecules are separated on the basis of their **size and shape**. The protein sample in a small volume is applied to the top of a column of **porous beads** (diameter 0.1 mm) that are made of an insoluble but highly hydrated polymer such as polyacrylamide (Bio-Gel) or the carbohydrates dextran (Sephadex) or agarose (Sepharose) (Figure 2a). Small molecules can enter the **pores** in the beads, whereas larger or more elongated molecules cannot. The smaller molecules therefore have a larger volume of liquid accessible to them; both the liquid surrounding the porous beads and that inside the beads. In contrast, the larger molecules have only the liquid surrounding the beads accessible to them, and thus move through the column faster, emerging out of the bottom (**eluting**) first (Figures 2a and 2b). The smaller molecules move more slowly through the column and elute later. Beads of differing pore sizes are available, allowing proteins of different sizes to be effectively separated. Gel filtration chromatography is often used to desalt a protein sample (for example to remove the ammonium sulfate after ammonium

(a)   (b)

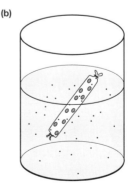

Figure 1. Separation of molecules on the basis of size by dialysis: (a) starting point; (b) at equilibrium.

sulfate precipitation), since the salt enters the porous beads and is eluted late, whereas the protein does not enter the beads and is eluted early. Gel filtration chromatography can also be used to estimate the **molecular mass** of a protein. There is a linear relationship between the relative elution volume of a protein (V_e/V_o where V_e is the elution volume of a given protein and V_o is the void volume of the column, that is the volume of the solvent space surrounding the beads; Figure 2b) and the logarithm of its molecular mass. Thus a 'standard' curve of V_e/V_o against $\log_{10}$ molecular mass can be determined for the column using proteins of known mass. The elution volume of any sample protein then allows its molecular mass to be estimated by reference to its position on the standard curve (Figure 2c).

Ion exchange chromatography

In ion exchange chromatography, proteins are separated on the basis of their **overall (net) charge**. If a protein has a net negative charge at pH 7, it will bind to a column containing positively charged beads, whereas a protein with no charge or a net positive charge will not bind (Figure 3a). The negatively charged proteins bound to such a column can then be eluted by washing the column with an increasing gradient (increasing concentration) of a solution of **sodium chloride** (Na^+ Cl^- ions) at the appropriate pH. The Cl^- ions compete with the protein for the positively charged groups on the column. Proteins having

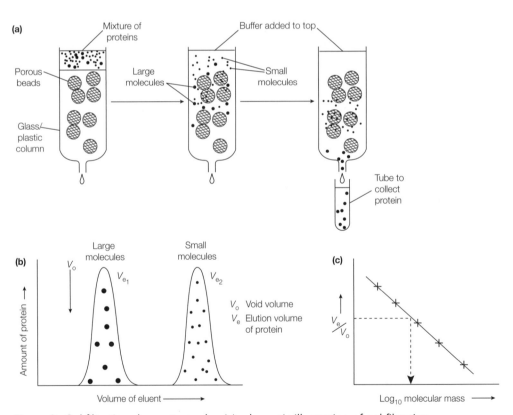

Figure 2. Gel filtration chromatography: (a) schematic illustration of gel filtration chromatography; (b) elution diagram indicating the separation; (c) a plot of relative elution volume versus the logarithm of molecular mass for known proteins, indicating how the molecular mass of an unknown can be read off when its relative elution volume is known.

a low density of negative charge elute first, followed by those with a higher density of negative charge (Figure 3b). Columns containing positively charged diethylaminoethyl (DEAE) groups (such as DEAE-cellulose or DEAE-Sephadex) are used for separation of negatively charged proteins (anionic proteins). This is called **anion exchange chromatography**. Columns containing negatively charged carboxymethyl (CM) groups (such as CM-cellulose or CM-Sephadex) are used for the separation of positively charged proteins (cationic proteins). This is called **cation exchange chromatography**. As an alternative to elution with a gradient of NaCl, proteins can be eluted from anion exchange columns by decreasing the pH of the buffer, and from cation exchange columns by increasing the pH of the buffer, thus altering the ionization state of the amino acid side-chains (Section B1) and hence the net charge on the protein.

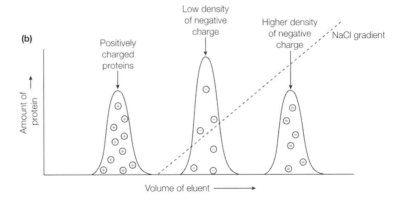

Figure 3. Ion exchange chromatography: (a) schematic illustration of ion exchange chromatography; (b) elution diagram indicating the separation of a protein of net positive charge that does not bind to the positively charged beads and passes straight through the column, and of two proteins with different net negative charges that bind to the positively charged beads and are eluted on increasing the concentration of NaCl applied to the column. The protein with the lower density of negative charge elutes earlier than the protein with the higher density of negative charge.

Affinity chromatography

Affinity chromatography exploits the specific, high affinity, noncovalent binding of a protein to another molecule, the **ligand**. First, the ligand is covalently attached to an inert and porous matrix (such as Sepharose). The protein mixture is then passed down a column containing the **immobilized ligand**. The protein of interest will bind to the ligand, whereas all other proteins pass straight through the column (Figure 4). After extensive washing of the column with buffer to remove nonspecifically bound proteins, the bound protein of interest is released from the immobilized ligand either by adding soluble ligand, which competes with the immobilized ligand for the protein, or by altering the properties of the buffer (changing the pH or salt concentration). If soluble ligand is used to elute the protein from the column, extensive **dialysis** often then has to be used to remove the small ligand from the larger protein. Because this technique exploits the specific, often unique, binding properties of the protein, it is often possible to separate the protein from a mixture of hundreds of other proteins in a single chromatographic step.

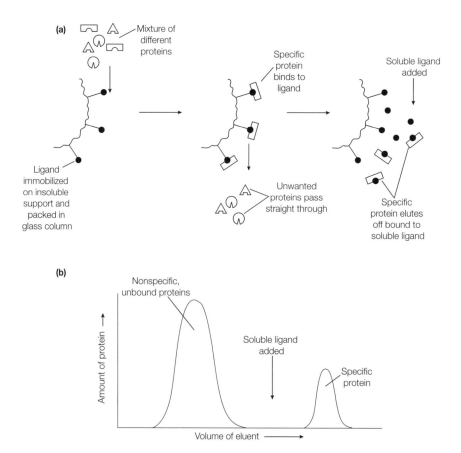

Figure 4. Affinity chromatography: (a) schematic diagram of affinity chromatography; (b) elution diagram indicating that nonspecific proteins that do not bind to the immobilized ligand pass straight through the column, while the specific protein binds to the immobilized ligand and is eluted from the column only on addition of soluble ligand.

Immunoaffinity chromatography in which an antibody is used to purify its antigen is one of the most common forms of affinity chromatography. In immunoaffinity chromatography, a specific **antibody** to a protein **antigen** is coupled to the inert matrix beads, which are then packed in a glass or plastic column. On addition of the impure protein solution to the top of the column, the protein antigen will bind to the antibody on the matrix but other components will flow through the column. The protein antigen can then often be eluted in pure, or almost pure form, by changing the pH of the buffer thus disrupting the antibody-antigen interaction; well over 1000-fold purification is routinely achieved in this single step. If the protein antigen is normally exposed on the plasma membrane of a desired cell type, these cells can be purified from a cell mixture by passing the mixture through the column. Only cells bearing the antigen on their surface will bind and can be eluted subsequently.

Other commonly employed combinations of immobilized ligand and protein to be purified used in affinity chromatographic systems include an **inhibitor** to purify an **enzyme** (Section D4), a **hormone** (e.g. insulin) to purify its **receptor** (Section E5), and a **lectin** (e.g. concanavalin A) to purify a **glycoprotein** (Sections E2 and H5). Advances in recombinant DNA technology (Section I1) mean that proteins can be engineered with specific sequences of amino acids at the C-terminal end, a so-called **tag**. The recombinant tagged protein can then be expressed in a suitable cell system and the affinity of the tag for an immobilized antibody or other molecule exploited to purify the protein.

C2 Gel electrophoresis

Key Notes

Electrophoresis

In polyacrylamide gel electrophoresis (PAGE), proteins are applied to a porous polyacrylamide gel and separated in an electric field on the basis of their net negative charge and their size. Small/more negatively charged proteins migrate further through the gel than larger/less negatively charged proteins.

SDS-PAGE

In SDS-PAGE, the protein sample is treated with a reducing agent to break disulfide bonds and then with the anionic detergent sodium dodecyl sulfate (SDS), which denatures the proteins and covers them with an overall negative charge. The sample is then fractionated by electrophoresis through a polyacrylamide gel. As all the proteins now have an identical charge to mass ratio, they are separated on the basis of their mass. The smallest proteins move farthest. SDS-PAGE can be used to determine the degree of purity of a protein sample, estimate the molecular mass of a protein and deduce the number of polypeptide subunits in a protein.

Isoelectric focusing

In isoelectric focusing, proteins are separated by electrophoresis in a gel containing polyampholytes, which produce a pH gradient. They separate on the basis of their relative content of positively and negatively charged residues. Each protein migrates through the gel until it reaches the point where it has no net charge, its isoelectric point (pI).

Two-dimensional gel electrophoresis

In two-dimensional gel electrophoresis, proteins are subjected first to isoelectric focusing and then in the second direction to SDS-PAGE to produce a two-dimensional pattern of spots separated on the basis of charge and then mass. This technique can be used to compare the proteome of cells under different conditions.

Visualization of proteins in gels

Proteins can be visualized directly in gels by staining them with the dye Coomassie brilliant blue or with a silver stain. Radioactively labeled proteins can be detected by overlaying the gel with X-ray film and observing the darkened areas on the developed autoradiograph that correspond to the radiolabeled proteins. A specific protein of interest can be detected by immunoblot (Western blot) following its transfer from the gel to nitrocellulose using an antibody that specifically recognizes it. This primary antibody is then detected with either a radiolabeled or enzyme-linked secondary antibody.

Related topics	(B1) Amino acid structure	(C3) Protein sequencing and
	(B2) Protein structure and function	peptide synthesis
		(C4) Immunodetection
	(C1) Protein purification	

Electrophoresis

When placed in an **electric field**, molecules with a net charge, such as proteins, will move towards one electrode or the other, a phenomenon known as **electrophoresis**. The greater the net charge the faster the molecule will move. In **polyacrylamide gel electrophoresis (PAGE)**, the electrophoretic separation is carried out in a gel, which serves as a molecular sieve. Small molecules move readily through the pores in the gel, whereas larger molecules are retarded. The gels are commonly made of **polyacrylamide**, which is chemically inert and which is readily formed by the polymerization of acrylamide. The pore sizes in the gel can be controlled by choosing appropriate concentrations of acrylamide and the cross-linking reagent, methylene bisacrylamide. The higher the concentration of acrylamide used, the smaller the pore size in the final gel. The gel is usually cast between two glass plates separated by a distance of 0.5–1.0 mm (Figure 1). The protein sample is added to wells in the top of the gel, which are formed by placing a plastic comb in the gel solution before it sets (Figure 1). A blue dye (bromophenol blue) is mixed with the protein sample to aid its loading on to the gel. Because bromophenol blue is a small molecule, it also migrates quickly through the gel during electrophoresis and so indicates the progress of electrophoresis.

Figure 1. PAGE. The protein samples are loaded into the sample wells formed in the top of the gel. An electric field is applied across the gel from top to bottom and the proteins migrate down through the gel. The smaller the protein the further it will migrate.

SDS-PAGE

In **sodium dodecyl sulfate (SDS)-PAGE**, the proteins are **denatured** and coated with an **overall negative charge** (due to bound SDS molecules) and thus the basis for their separation is only their **mass**. The protein mixture is first treated with a **reducing agent** such as 2-mercaptoethanol or dithiothreitol to break all the disulfide bonds (Figure 2) (Section B2). The strong **anionic detergent SDS** is then added, which disrupts nearly all the noncovalent interactions in the protein, unfolding the polypeptide chain. Approximately one molecule of SDS binds via its hydrophobic alkyl chain to the polypeptide backbone for every two amino acid residues, which gives the denatured protein a large net negative charge that is proportional to its mass. The SDS/protein mixture is then applied to sample

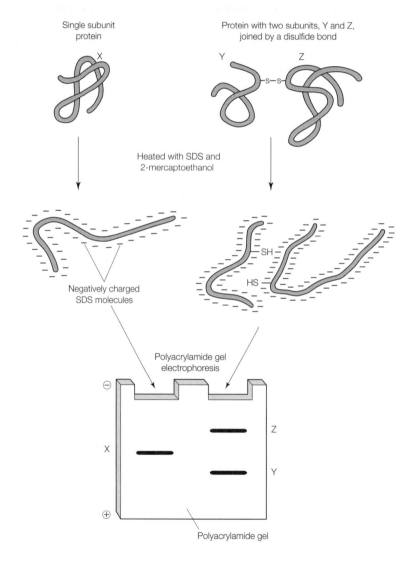

Figure 2. SDS-PAGE. The protein mixture is heated in the presence of 2-mercaptoethanol, which breaks any disulfide bonds, and SDS. The unfolded polypeptide chains are coated with the negatively charged molecules of SDS and will migrate towards the anode on PAGE. Smaller polypeptides migrate further through the gel than larger ones.

Figure 3. SDS-PAGE. (a) Appearance of proteins after electrophoresis on a SDS-polyacrylamide gel. Lane 1, proteins (markers) of known molecular mass; lane 2, unpurified mixture of proteins; lane 3, partially purified protein; lane 4, protein purified to apparent homogeneity/purity. (b) Determination of the molecular mass of an unknown protein by comparison of its electrophoretic mobility (distance migrated) with those of proteins (markers) of known molecular mass.

wells in the top of a **polyacrylamide gel** (Figure 1). The buffer, which is the same in both the upper and lower reservoirs and in the gel, has a pH of approximately 9, such that the proteins have **net negative charge** and will migrate towards the anode in the lower reservoir. An electric current (approximately 300 V) is applied across the gel from top to bottom for 30–90 min in order to move the proteins through the gel (Figure 1). After carrying out electrophoresis, the gel is removed from the apparatus and the proteins visualized (Figure 3a). Small proteins move furthest through the gel, whereas large ones move more slowly as they are held back by the cross-linking in the gel. Under these conditions, the mobility of most polypeptide chains is linearly proportional to the logarithm of their mass. Thus, if proteins of known molecular mass are electrophoresed alongside the samples, the mass of the unknown proteins can be determined as there is a linear relationship between $\log_{10}$ of molecular mass and distance migrated through the gel (Figure 3b). Proteins that differ in mass by about 2% (e.g. 40 and 41 kDa; a difference of approximately 10 amino acid residues) can be distinguished under appropriate conditions. SDS-PAGE is a rapid, sensitive and widely used technique, which can be used to determine the degree of purity of a protein sample, to estimate the molecular mass of an unknown protein and to deduce the number of polypeptide subunits within a protein.

Isoelectric focusing

Isoelectric focusing electrophoretically separates proteins on the basis of their relative content of positively and negatively charged groups. When a protein is at its **isoelectric point (pI)** (Section B1), its **net charge is zero** and hence it will not move in an electric field. In isoelectric focusing, a polyacrylamide gel is used, which has large pores (so as not to impede protein migration) and contains a mixture of **polyampholytes** (small multicharged polymers that have many pI values). If an electric field is applied to the gel, the polyampholytes migrate and produce a **pH gradient**. To separate proteins by isoelectric focusing, they are electrophoresed through such a gel. Each protein will migrate through the gel until it reaches a position at which the pH is equal to its pI (Figure 4). If a protein diffuses away from this position, its net charge will change as it moves into a region of different pH and the resulting electrophoretic forces will move it back to its isoelectric position. In this way, each protein is focused into a narrow band (as thin as 0.01 pH unit) about its pI.

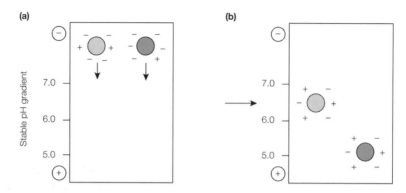

Figure 4. Isoelectric focusing: (a) before applying an electric current; (b) after applying an electric current the proteins migrate to a position at which their net charge is zero (pI).

Two-dimensional gel electrophoresis

Isoelectric focusing can be combined with SDS-PAGE to obtain very high-resolution separations in a procedure known as **two-dimensional gel electrophoresis**. The protein sample is first subjected to isoelectric focusing in a narrow strip of gel containing poly-ampholytes (Figure 4). This isoelectric focusing gel strip is then placed on top of an SDS-polyacrylamide gel and electrophoresed to produce a two-dimensional pattern of spots in which the proteins have been **separated in the horizontal direction on the basis of their pI**, and **in the vertical direction on the basis of their mass** (Figure 5). The over-all result is that proteins are separated both on the basis of their size and their charge. Thus two proteins that have very similar or identical pIs, and produce a single band by isoelectric focusing, if they have different molecular masses will produce two spots by two-dimensional gel electrophoresis (Figure 5). Similarly, proteins with similar or identi-cal molecular masses, which would produce a single band by SDS-PAGE, will also pro-duce two spots if they have different pIs because of the initial separation by isoelectric focusing. The **high-resolution separation** of proteins in a complex mixture that can be achieved by two-dimensional gel electrophoresis makes this technique extremely useful for comparing the **proteome** (the entire complement of proteins in a cell or organism) of cells or tissues under different conditions, e.g. differentiated versus undifferentiated cells, and is often used prior to the analysis of individual protein spots by **mass spec-trometry** (Section C3).

Visualization of proteins in gels

As most proteins are not directly visible on gels to the naked eye, a method has to be employed in order to visualize them following electrophoresis. The most commonly used protein stain is the dye **Coomassie brilliant blue**. After electrophoresis, the gel contain-ing the separated proteins is immersed in an acidic alcoholic solution of the dye. This **denatures** the proteins, **fixes** them in the gel so that they do not wash out, and allows the dye to bind to them. After washing away excess dye, the proteins are visible as discrete blue bands (Figure 3a). As little as 0.1–1.0 µg of a protein in a gel can be visualized using Coomassie brilliant blue. A more sensitive general protein stain involves soaking the gel in a **silver** salt solution. However, this technique is rather more difficult to apply. If the protein sample is **radioactive**, the proteins can be visualized indirectly by overlaying the gel with a sheet of **X-ray film**. With time (hours to weeks depending on the radioactivity

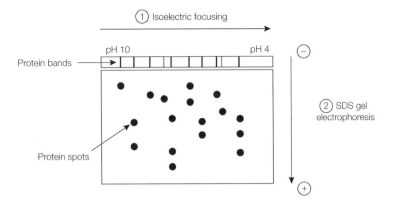

Figure 5. Two-dimensional gel electrophoresis. The protein sample is first subjected to isoelectric focusing in one dimension and then to SDS-PAGE in the second dimension.

of the sample proteins), the radiation emitted will cause a darkening of the film. Upon development of the film, the resulting **autoradiograph** will have darkened areas corresponding to the positions of the radiolabeled proteins.

Another way of visualizing the protein of interest is to use an **antibody** against the protein in an **immunoblot** (**Western blot**) (Section C4 for more detail). For this technique, the proteins have to be transferred out of the gel on to a sheet of **nitrocellulose** or **nylon membrane**. This is accomplished by overlaying the gel with the nitrocellulose and **blotting** the protein on to it by applying an electric current. The nitrocellulose then has an exact image of the pattern of proteins that was in the gel. The excess binding sites on the nitrocellulose are then blocked with a nonspecific protein solution such as milk powder, before placing the nitrocellulose in a solution containing the antibody that recognizes the protein of interest (the **primary antibody**). After removing excess unbound antibody, the primary antibody that is now specifically bound to the protein of interest is detected with either a **radiolabeled**, **fluorescent** or **enzyme-coupled secondary antibody**. Finally, the secondary antibody is detected either by placing the nitrocellulose against a sheet of X-ray film (if a radiolabeled secondary antibody has been used), by using a fluorescence detector or by adding to the nitrocellulose a solution of a substrate that is converted into a colored insoluble product by the enzyme that is coupled to the secondary antibody.

C3 Protein sequencing and peptide synthesis

Key Notes

Amino acid composition analysis	The number of each type of amino acid in a protein can be determined by acid hydrolysis and separation of the individual amino acids by ion exchange chromatography. The amino acids are detected by colorimetric reaction with, for example, ninhydrin or fluorescamine.
Edman degradation	The N-terminal amino acid of a protein can be determined by reacting the protein with dansyl chloride or fluorodinitrobenzene prior to acid hydrolysis. The amino acid sequence of a protein can be determined by Edman degradation, which sequentially removes one residue at a time from the N-terminus.
Sequencing strategy	In order to sequence an entire protein, the polypeptide chain has to be broken down into smaller fragments using either chemicals (e.g. cyanogen bromide) or enzymes (e.g. trypsin). The resulting smaller fragments are then sequenced by Edman degradation. The complete sequence is assembled by analyzing overlapping fragments generated by cleaving the polypeptide with different reagents.
Mass spectrometry	Matrix-assisted laser desorption ionization-time-of-flight (MALDI-TOF) spectrometry is used to determine the precise mass of peptides. The peptides are immobilized in an organic matrix and then blasted with a laser, causing them to be ejected in the form of an ionized gas. The ionized peptides in the gas are then accelerated in an electric field and separated. Tandem mass spectrometry (MS-MS) uses two mass spectrometers in tandem to fragment the peptides further. The resulting peptide map is diagnostic of the protein and is referred to as the protein's fingerprint.
Proteomics	Proteomics is the study of the entire complement of proteins, the proteome, in a cell or organism.
Recombinant DNA technology	The sequence of a protein can be determined using recombinant DNA technology to identify and sequence the piece of DNA encoding the protein. The amino acid sequence of the protein can then be deduced from its DNA sequence using the genetic code.

Information derived from protein sequences	The amino acid sequence of a protein not only reveals the primary structure of the protein but also information on possible protein families or groups and evolutionary relationships, potential gene duplication(s) and possible post-translational modifications. In addition, a knowledge of the amino acid sequence can be used to generate specific antibodies.	
Peptide synthesis	In solid phase peptide synthesis, polypeptides are chemically synthesized by addition of free amino acids to a tethered peptide. To prevent unwanted reactions, the α-amino group and reactive side-chain groups of the free amino acids are chemically protected (blocked), and then deprotected (deblocked) once the amino acid is attached to the growing polypeptide chain.	
Related topics	(B1) Amino acid structure (B2) Protein structure and function	(C1) Protein purification (C2) Gel electrophoresis

Amino acid composition analysis

The number of each type of **amino acid** in a protein sample can be determined by amino acid composition analysis. The purified protein sample is hydrolyzed into its constituent amino acids by heating it in 6 M HCl at 110°C for 24 h in an evacuated and sealed tube. The resulting mixture (**hydrolysate**) of amino acids is subjected to **ion exchange chromatography** (Section C1) on a column of sulfonated polystyrene to separate out the **20 standard amino acids** (Section B1). The separated amino acids are then detected and quantified by reacting them with **ninhydrin**. The α-amino acids produce a blue color, whereas the imino acid Pro produces a yellow color. The amount of each amino acid in an unknown sample can be determined by comparison of the optical absorbance with a known amount of each of the individual amino acids in a standard sample. With ninhydrin, as little as 10 nmol of an amino acid can be detected. A more sensitive detection system (detecting down to 10 pmol of an amino acid) uses **fluorescamine** to react with the α-amino group to form a fluorescent product. Amino acid composition analysis indicates the number of each amino acid residue in a peptide, but it does not provide information on the sequence of the amino acids. For example, the amino acid composition of the oligopeptide:

Val-Phe-Asp-Lys-Gly-Phe-Val-Glu-Arg

would be:

(Arg, Asp, Glu, Gly, Lys, Phe$_2$, Val$_2$)

where the parentheses and the commas between each amino acid denote that this is the amino acid composition, not the sequence.

Edman degradation

The **amino-terminal** (N-terminal) residue of a protein can be identified by reacting the protein with a compound that forms a stable covalent link with the free α-amino group,

prior to hydrolysis with 6 M HCl. The labeled N-terminal amino acid can then be identified by comparison of its chromatographic properties with standard amino acid derivatives. Commonly used reagents for N-terminal analysis are **fluorodinitrobenzene** and
dansyl chloride. If this technique was applied to the oligopeptide above, the N-terminal residue would be identified as Val, but the remainder of the sequence would still be
unknown. Further reaction with dansyl chloride would not reveal the next residue in the
sequence since the peptide is totally degraded in the acid hydrolysis step.

This problem was overcome by Pehr Edman who devised a method for labeling the
N-terminal residue and then cleaving it from the rest of the peptide without breaking
the peptide bonds between the other amino acids. In so-called **Edman degradation**, one
residue at a time is sequentially removed from the N-terminal end of a peptide or protein
and identified. The uncharged N-terminal amino group of the protein is reacted with
phenyl isothiocyanate to form a phenylthiocarbamoyl derivative, which is then released
from the rest of the protein as a cyclic **phenylthiohydantoin (PTH) amino acid** under
mildly acidic conditions (Figure 1). This milder cleavage reaction leaves the remainder
of the peptide intact, available for another round of labeling and release. The released
PTH amino acid is identified by **high performance liquid chromatography** (HPLC). This
sequencing technique has been **automated** and refined so that upwards of 50 residues
from the N-terminus of a protein can be sequenced from picomole quantities of material.

Sequencing strategy

An 'average' sized protein of 50 kDa would contain approximately 500 amino acids. Thus,
even with large amounts of highly purified material, only about the N-terminal one-tenth

Figure 1. Edman degradation. The N-terminal amino acid is labeled with phenyl
isothiocyanate. Upon mild acid hydrolysis, this residue is released as a PTH derivative and the
peptide is shortened by one residue, ready for another round of labeling and release.

of the protein can be sequenced by Edman degradation. In order to sequence a larger protein, the first step is to **cleave** it into **smaller fragments** of 20–100 residues, which are then separated and sequenced. Specific cleavage can be achieved by **chemical** or **enzymatic** methods. For example, the chemical **cyanogen bromide** (CNBr) cleaves polypeptide chains on the C-terminal side of Met residues, whereas the enzymes **trypsin** and **chymotrypsin** cleave on the C-terminal side of basic (Arg, Lys) and aromatic (Phe, Trp, Tyr) residues, respectively. On digestion with trypsin, a protein with six Lys and five Arg would yield 12 tryptic peptides, each of which would end with Arg or Lys, apart from the C-terminal peptide. The peptide fragments obtained by specific chemical or enzymatic cleavage are then separated by **chromatography** (e.g. ion exchange chromatography; Section C1) and the sequence of each in turn determined by Edman degradation.

Although the sequence of each peptide fragment would now be known, the order of these fragments in the polypeptide chain would not. The next stage is to generate **overlapping fragments** by cleaving another sample of the original polypeptide chain with a different chemical or enzyme (e.g. chymotrypsin), separating the fragments and then sequencing them. These **chymotryptic peptides** will overlap one or more of the **tryptic peptides**, enabling the order of the fragments to be established (Figure 2). In this way, the entire length of the polypeptide chain can be sequenced.

To sequence the polypeptides in a **multisubunit protein**, the individual polypeptide chains must first be dissociated by disrupting the **noncovalent interactions** with **denaturing agents** such as **urea** or **guanidine hydrochloride**. The **disulfide bonds** in the protein also have to be broken by reduction with **2-mercaptoethanol** or **dithiothreitol**. To prevent the cysteine residues recombining, **iodoacetate** is added to form stable S-carboxymethyl derivatives. The individual polypeptide chains then have to be separated by, for example, ion exchange chromatography (Section C1) before sequencing each. Nowadays, as little as picomole amounts of proteins can be sequenced following their separation by SDS-PAGE either using the polyacrylamide gel containing the protein directly, or following their transfer to nitrocellulose (Section C2).

To summarize the key steps in the chemical sequencing of a protein:

1. Purify the protein.

2. Reduce any disulfide bonds and block their reoxidation with iodoacetate.

Figure 2. The use of overlapping fragments to determine the sequence of a peptide. The protein is first digested with trypsin and the resulting peptides separated and sequenced. The protein is separately digested with chymotrypsin and the resulting peptides again separated and sequenced. The order of the peptide fragments in the protein can be determined by comparing the sequences obtained.

3. Cleave the protein with cyanogen bromide or a protease (e.g. trypsin).

4. Separate the fragments and sequence them.

5. Repeat steps 3 and 4 using a different chemical reagent or protease in order to generate overlapping fragments and look for overlaps between the two sets of sequences in order to construct the full sequence.

Mass spectrometry

The precise mass of intact proteins and peptides derived from them can be determined by **mass spectrometry**. This is a very sensitive technique that requires only very small amounts of material. The most commonly used mass spectrometric method is called **matrix-assisted laser desorption ionization-time-of-flight (MALDI-TOF) spectrometry**. In this method, peptides are first mixed with an organic acid and then dried onto a ceramic or metal slide. The sample is then blasted with a **laser**, which causes the peptides to be ejected from the slide in the form of an ionized gas in which each molecule carries one or more positive charges (Figure 3a). The ionized peptides are then accelerated in an electric field and fly towards a detector. The time it takes for them to reach the detector is determined by their mass and their charge; large peptides move more slowly, and more highly charged peptides move more quickly. The precise mass is then determined by analysis of those peptides with a single charge. The resulting peptide masses or **fingerprint** can be used to search **protein sequence databases** for matches with theoretical masses calculated for all theoretically cleaved peptides for all proteins in a sequenced genome (Figure 3b). Thus, the identity of the original protein and its sequence can readily be determined by a combination of mass spectrometry and protein sequence database searching.

A variation of this method can be used directly to determine the sequences of the individual peptides. Following trypsin digestion of the purified protein and determination of their masses by mass spectrometry as above, each peptide is further fragmented at the peptide bonds and the masses of these fragments measured in a coupled second mass spectrometer. This is so-called **tandem mass (MS-MS) spectrometry**. The mass differences between the fragments can be used to construct a partial amino acid sequence, which, in turn, can be used to search protein sequence databases or provide the means for cloning the gene.

Sequencing of proteins by mass spectrometry has several advantages over traditional chemical Edman sequencing.

● Much smaller amounts of material are required.

● The sequence of the peptide can be obtained in only a few minutes compared with the hour required for just one cycle of Edman degradation.

● Mass spectrometry can be used to sequence several polypeptides in a mixture, alleviating the need to completely purify the sample prior to analysis.

● Mass spectrometry can be used to determine the sequence of peptides which have **blocked N-termini**, such as pyroglutamate, a derivative of Glu in which the side-chain carboxyl group forms an amide bond with its primary amino group (a common eukaryotic post-translational modification that prevents Edman degradation) and to characterize other **post-translational modifications** such as glycosylation and phosphorylation.

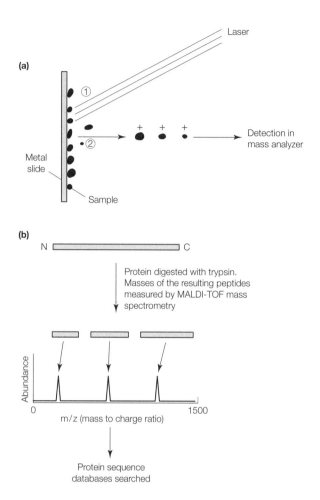

Figure 3. Mass spectrometry to determine protein sequence. (a) In a MALDI-TOF mass spectrometer, pulses of light from a laser ionize a peptide mixture that is absorbed on a metal slide ①. An electric field accelerates the molecules in the sample towards the detector ②. The time to the detector is inversely proportional to the mass of the protein. (b) Use of mass spectrometry to identify proteins. The protein of interest is digested with trypsin and the resulting peptide fragments are loaded into the mass spectrometer where their masses are measured. Sequence databases are then searched to identify the protein whose calculated tryptic digest profile matches the experimentally determined data.

Proteomics

The **proteome** is the entire complement of proteins in a cell or organism. **Proteomics** is the large-scale effort to identify and characterize all of the proteins encoded in an organism's genome, including their post-translational modifications. Increasingly, in the field of proteomics, proteins resolved by two-dimensional gel electrophoresis (Section C2) are subjected to trypsin digestion and the extremely accurate molecular masses of the peptides produced from mass spectrometry are used as a '**fingerprint**' to identify the protein from **databases** of real or predicted tryptic peptide sizes (Figure 3).

Recombinant DNA technology

Although numerous proteins have been sequenced by Edman degradation and mass spectrometry, the determination of the complete sequences of large proteins by these methods is a demanding and time-consuming process. **Recombinant DNA technology** (Section I1) has enabled the sequences of even very large proteins or of proteins that are difficult to purify to be determined by first sequencing the stretch of **DNA** encoding the protein and then using the **genetic code** to decipher the protein sequence (Section H1). Even so, some direct protein sequence data are often required to confirm that the protein sequence obtained is the correct one and to identify any post-translational modifications on the protein. Thus, protein sequencing and DNA sequencing techniques are often used together to determine the complete sequence of a protein.

Information derived from protein sequences

The amino acid sequence can provide information over and above the **primary structure** of the protein.

1. The sequence of interest can be compared with other known sequences to see whether there are similarities. For example, the sequences of hemoglobin and myoglobin indicate that they belong to the globin group or **family of proteins** (Section B3).

2. The comparison of the sequences of the same protein in different species can provide information about **evolutionary relationships**.

3. The presence of repeating stretches of sequence would indicate that the protein may have arisen by **gene duplication** (e.g. in antibody molecules; Section B6).

4. Within the amino acid sequence there may be specific sequences which act as signals for the **post-translational processing** of the protein (e.g. glycosylation or proteolytic processing; Section H5).

5. The amino acid sequence data can be used to prepare **antibodies** specific for the protein of interest, which can be used to study its structure and function (Section C4).

6. The amino acid sequence can be used for designing **DNA probes** that are specific for the gene encoding the protein (Sections I3 and I4).

Peptide synthesis

Polypeptides can be **chemically synthesized** by covalently linking amino acids to the end of a growing polypeptide chain. In **solid phase peptide synthesis**, the growing polypeptide chain is covalently anchored at its C-terminus to an insoluble support such as polystyrene beads. The next amino acid in the sequence has to react with the free α-amino group on the tethered peptide, but it has a free α-amino group itself, which will also react. To overcome this problem, the free amino acid has its α-amino group **chemically protected** (blocked) so that it does not react with other molecules. Once the new amino acid is coupled, its now N-terminal α-amino group is **deprotected** (deblocked) so that the next peptide bond can be formed. Every cycle of amino acid addition therefore requires a **coupling step** and a **deblocking step**. In addition, reactive side-chain groups must also be blocked to prevent unwanted reactions occurring.

C4 Immunodetection

Key Notes

Immunodetection methods	Because of the high specificity of an antibody for its epitope, an antibody raised against a particular protein antigen can be used to determine the location of that antigen in a cell (immunocytochemistry), to detect and quantify the antigen in a complex mixture (ELISA) or to detect proteins following SDS-PAGE (Western blotting).
Immunocyto-chemistry	The antibody against the antigen of interest is fluorescently labeled to allow the location of the antigen in a cell to be visualized by immunofluorescence or immuno-electron microscopy.
ELISA	A sandwich ELISA can be used to detect and quantify the amount of a specific protein antigen in a sample. The antibody is bound to an inert polymer support, then exposed to the sample. Unbound protein is washed away and a second antibody that reacts with the antigen at a different epitope is added. The second antibody used is one that is fluorescently labeled or has an enzyme attached to it that converts a colorless substrate into a colored product. The amount of second antibody bound, and hence the amount of protein antigen present in the original sample, is determined by quantification of the intensity of color or fluorescence produced. An indirect ELISA, in which the antigen is bound to an inert polymer, can be used to detect the presence in a sample of an antibody against the antigen.
Immuno (Western) blotting	Protein samples are separated by SDS-PAGE and the separated proteins are then transferred (blotted) to a nitrocellulose or nylon sheet. This is incubated with specific antibody to the protein and then unbound antibody is washed away. Those proteins in the gel that bind the antibody are detected either by autoradiography (if the specific antibody was radiolabeled) or by using a second labeled antibody that binds to the primary antibody.
Related topics	(A4) Cell imaging (C2) Gel electrophoresis (C1) Protein purification (B6) Antibodies

---

Immunodetection methods

The availability of an **antibody** (immunoglobulin) against a specific **antigen** (Section B6) offers the opportunity to use that antibody in a range of immunological detection methods. The site recognized by an antibody on the antigen is called the **antigenic determinant** or **epitope**. Both **monoclonal** and **polyclonal** antibodies (Section B6) can be used for immunodetection of the desired antigen. The high specificity of an antibody for its epitope allows it to be used as a reagent for determining the location of the antigen in a cell (**immunocytochemistry**), to detect and quantify an antigen in a complex mixture (**ELISA**), and to detect proteins following their separation by SDS-PAGE (**Western blotting**).

Immunocytochemistry

In immunocytochemistry a **fluorescent label** (e.g. fluorescein, which emits green light) is coupled to the antibody, which is then added to the fixed cell. The binding of the antibody to its antigen can then be visualized by detecting its fluorescence by **immunofluorescence light microscopy** (for details see Section A4). Even higher resolution can be achieved using antibody to which electron-dense particles, such as colloidal gold, have been coupled and then viewing the binding sites using electron microscopy (Section A4). Indeed, **immuno-electron microscopy** can map the position of protein antigens within cells and within macromolecular complexes such as ribosomes.

ELISA

Specific antibodies can be used to detect and quantify the amount of the corresponding antigen in a biological sample. Several types of immunological assays exist. A popular version is **enzyme-linked immunosorbent assay (ELISA)**, which can readily detect and quantify less than a nanogram of a specific antigenic protein. The **sandwich ELISA** allows both the detection and the quantitation of the antigen. This method uses a plastic tray that has molded wells in it (a **microtiter tray**) where the antibody has been coupled to the plastic forming the wells (Figure 1). Samples containing the antigen to be assayed are added to the wells. If antigen is present that is recognized by the antibody, it becomes bound (Figure 1). The wells are then washed to remove unbound protein and incubated with a second antibody that recognizes the protein but at a different epitope than the first antibody (Figure 1). The second antibody is attached either to a fluorescent group or to an enzyme that can catalyze the conversion of a colorless substrate into a colored product. The intensity of the color or fluorescence produced for each sample is then measured to determine the amount of antigen present in each sample. Several machines are commercially available that scan the wells of microtiter plates following ELISA, measure the absorbance or fluorescence and quantify the amount of antigen bound in each well.

An alternative format, the **indirect ELISA**, can be used to detect the presence of antibody against a particular antigen in a biological sample. The antigen itself is bound in the wells of the microtiter tray, to which the sample containing the antibody that recognizes the antigen is added. The presence of the bound antibody can then be detected by adding a secondary fluorescent antibody that in this case recognizes the first antibody (see below). This technique is the basis of the test for **human immunodeficiency virus** (HIV) infection where a viral protein is immobilized and the presence in an individual's blood sample of antibodies against it can be measured.

Immuno (Western) blotting

This technique can be used for detection of one or more proteins in a mixture. The sample is electrophoresed on an SDS–polyacrylamide gel (SDS-PAGE; Section C2) that separates the proteins on the basis of size, resulting in a series of protein bands down the gel (Figure 2). Because the gel matrix does not let large proteins such as antibodies enter readily, the sample proteins must be first transferred to a more accessible medium.

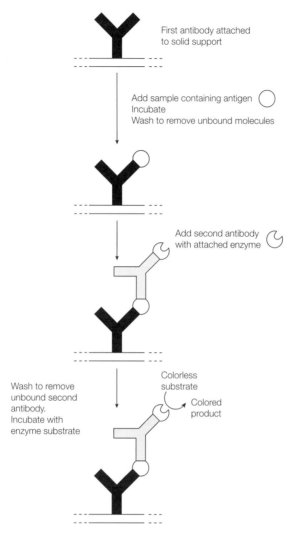

First antibody attached
to solid support

Add sample containing antigen
Incubate
Wash to remove unbound molecules

Add second antibody
with attached enzyme

Wash to remove
unbound second
antibody.
Incubate with
enzyme substrate

Colorless
substrate

Colored
product

Figure 1. ELISA using a second antibody with an attached (conjugated) enzyme that converts a colorless substrate to a colored product.

This process is called **blotting**. The gel is placed next to a nitrocellulose or nylon sheet and an electric field is applied so that proteins migrate from the gel to the sheet where they become bound. The nitrocellulose sheet is incubated with a protein such as bovine serum albumin to bind to nonspecific protein-binding sites and hence prevent spurious binding of antibody molecules in subsequent steps. This step is said to 'block' nonspecific binding sites. The nitrocellulose sheet is then reacted with labeled antibody, unreacted antibody is washed away and those protein bands that have bound the antibody become visible and hence are identified (Figure 2). The method of visualization depends on how the antibody was labeled. If the labeling is by the incorporation of a radiolabel (e.g. ^{125}I), then autoradiography is carried out to detect the radioactive protein bands (Figure 2). Alternatively, the antibody may be detected by incubating the sheet with a second antibody that recognizes the first antibody (e.g. if the first specific antibody was raised in rabbits, the second antibody could be a goat anti-rabbit antibody). The second antibody could be radiolabeled and its binding detected by autoradiography or it could be conjugated to an enzyme that generates a colored product as in ELISA (Figure 1). A similar technique is used for the blotting of DNA (Southern blotting; Section I3) or RNA (Northern blotting; Section I3).

Figure 2. Immuno (Western) blotting using a radiolabeled antibody: (a) sodium dodecyl sulfate-polyacrylamide gel electrophoresis, yielding polypeptides separated into discrete bands; (b) nitrocellulose or nylon membrane onto which the protein bands have been blotted (i.e. a Western blot); (c) autoradiograph after incubating the Western blot with radiolabeled antibody, washing away unbound antibody and placing the membrane against X-ray film.

D1 Introduction to enzymes

Key Notes

Enzymes as catalysts	Enzymes are catalysts that change the rate of a reaction without being changed themselves. They are highly specific and their activity can be regulated. Virtually all enzymes are proteins, although some catalytically active RNAs have been identified.
Active site	The active site is the region of the enzyme that binds the substrate, to form an enzyme–substrate complex, and transforms it into product. The active site is a three-dimensional entity, often a cleft or crevice on the surface of the protein, in which the substrate is bound by multiple weak interactions. Two models have been proposed to explain how an enzyme binds its substrate: the lock-and-key model and the induced-fit model.
Substrate specificity	The substrate specificity of an enzyme is determined by the properties and spatial arrangement of the amino acid residues forming the active site.
Enzyme classification	Enzymes are classified into six major groups on the basis of the type of reaction that they catalyze. Each enzyme has a unique classification number.
Enzyme assays	An enzyme assay measures the conversion of substrate to product, under conditions of cofactors, pH and temperature at which the enzyme is optimally active. High substrate concentrations are used so that the initial reaction rate is proportional to the enzyme concentration. Either the rate of appearance of product or the rate of disappearance of substrate is measured, often by following the change in absorbance using a spectrophotometer.
Linked enzyme assays	If neither the substrates nor products of an enzyme-catalyzed reaction absorb light at an appropriate wavelength, the enzyme can be assayed by linking it to another enzyme-catalyzed reaction that does involve a change in absorbance. The second enzyme must be in excess, so that the rate-limiting step in the linked assay is the action of the first enzyme.
Coenzymes and prosthetic groups	Some enzymes require the presence of cofactors, small nonprotein units, to function. Cofactors may be inorganic ions or complex organic molecules called coenzymes. A cofactor that is covalently attached to the enzyme is called a prosthetic group. A holoenzyme is the catalytically active form of the enzyme with its cofactor, whereas an apoenzyme is the protein part on its own. Many coenzymes are derived

	from dietary vitamin precursors, and deficiencies in them lead to certain diseases.
Isozymes	Isozymes are different forms of an enzyme, which catalyze the same reaction, but which exhibit different physical or kinetic properties. The isozymes of lactate dehydrogenase (LDH) can be separated electrophoretically and can be used clinically to diagnose a myocardial infarction.
Related topics	(A4) Cell imaging (D3) Enzyme kinetics (B2) Protein structure and (D4) Enzyme inhibition function (D5) Regulation of (D2) Thermodynamics enzyme activity

Enzymes as catalysts

Enzymes are **catalysts** that increase the rate of a chemical reaction without being changed themselves in the process. In the absence of an enzyme, the reaction may hardly proceed at all, whereas in its presence the rate can be increased up to 10^{17}-fold. Enzyme-catalyzed reactions usually take place under relatively mild conditions (temperatures well below 100°C, atmospheric pressure and neutral pH) as compared with the corresponding chemical reactions. Enzymes are also **highly specific** with respect to the substrates on which they act and the products that they form. In addition, enzyme activity can be **regulated**, varying in response to the concentration of substrates or other molecules (Section D5). Nearly all enzymes are **proteins**, although a few catalytically active **RNA molecules** (**ribozymes**) have been identified.

Active site

The **active site** of an enzyme is the region that **binds the substrate** and converts it into product. It is usually a relatively small part of the whole enzyme molecule and is a **three-dimensional entity** formed by amino acid residues that can lie far apart in the linear polypeptide chain (Section B2). The active site is often a cleft or crevice on the surface of the enzyme that forms a predominantly nonpolar environment, which enhances the binding of the substrate. The substrate(s) is bound in the active site by **multiple weak forces** (electrostatic interactions, hydrogen bonds, van der Waals forces, hydrophobic interactions; see Section B2) and in some cases by reversible covalent bonds. Having bound the substrate molecule, and formed an **enzyme–substrate complex**, catalytically active residues within the active site of the enzyme act on the substrate molecule to transform it first into the transition state complex (Section D2) and then into product, which is released into solution. The enzyme is now free to bind another molecule of substrate and begin its catalytic cycle again.

Originally, two models were proposed to explain how an enzyme binds its substrate. In the **lock-and-key** model proposed by Emil Fischer in 1894, the shape of the substrate and the active site of the enzyme are thought to fit together like a key into its lock (Figure 1a). The two shapes are considered as rigid and fixed, and perfectly complement each other when brought together in the right alignment (also referred to as **geometric complementarity**). In the **induced-fit model** proposed in 1958 by Daniel E. Koshland, Jr., the binding of substrate **induces a conformational change** in the active site of the enzyme (Figure 1b). In addition, the enzyme may distort the substrate, forcing it into a conformation similar

Figure 1. Binding of a substrate to an enzyme: (a) lock-and-key model; (b) induced-fit model.

to that of the transition state (Section D2). For example, the binding of **glucose** to **hexokinase** induces a conformational change in the structure of the enzyme such that the active site assumes a shape that is complementary to the substrate (glucose) only after it has bound to the enzyme. The reality is that different enzymes show features of both models, with some complementarity and some conformational change.

Substrate specificity

The properties and spatial arrangement of the amino acid residues forming the active site of an enzyme will determine which molecules can bind and be substrates for that enzyme. **Substrate specificity** is often determined by changes in relatively few amino acids in the active site. This is clearly seen in the three digestive enzymes **trypsin, chymotrypsin** and **elastase** (Section D5). These three enzymes belong to a family of enzymes called the **serine proteases** – 'serine' because they have a serine residue in the active site that is critically involved in catalysis and 'proteases' because they catalyze the hydrolysis of peptide bonds in proteins. The three enzymes cleave peptide bonds in protein substrates on the carboxyl side of certain amino acid residues.

Trypsin cleaves on the carboxyl side of positively charged Lys or Arg residues, chymotrypsin cleaves on the carboxyl side of bulky aromatic and hydrophobic amino acid residues, and elastase cleaves on the carboxyl side of residues with small uncharged side-chains. Their differing specificities are determined by the nature of the amino acid groups in their substrate binding sites, which are complementary to the substrates upon which they act. Thus trypsin has a negatively charged Asp residue in its substrate binding site, which interacts with the positive charge on the Lys and Arg side-chains of the substrate (Figure 2a). Chymotrypsin has amino acid residues with small side-chains, such as Gly and Ser, in its substrate binding site that allow access of the bulky side-chain of the substrate (Figure 2b). In contrast, elastase has the relatively large uncharged amino acid side-chains of two Val residues protruding into its substrate binding site, preventing access of all but the small side-chains on Gly and Ala (Figure 2c).

Enzyme classification

Many enzymes are named by adding the suffix '-ase' to the name of their substrate. Thus **urease** is the enzyme that catalyzes the hydrolysis of urea, and **fructose-1,6-bisphosphatase** hydrolyzes fructose-1,6-bisphosphate. However, other enzymes, such as trypsin and chymotrypsin, have names that do not denote their substrate. Some enzymes have several alternative names. To rationalize enzyme names, a system of **enzyme nomenclature** has been internationally agreed. This system places all enzymes into one of **six major classes** based on the type of reaction catalyzed (Table 1). Each enzyme is then uniquely identified with a four-part classification number. Thus trypsin has the Enzyme

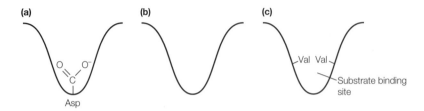

Figure 2. Schematic representation of the substrate binding sites in the serine proteases: (a) trypsin, (b) chymotrypsin and (c) elastase.

Commission (EC) number 3.4.21.4, where the first number (3) denotes that it is a hydrolase, the second number (4) that it is a protease that hydrolyzes peptide bonds, the third number (21) that it is a serine protease with a critical serine residue at the active site, and the fourth number (4) indicates that it was the fourth enzyme to be assigned to this class. For comparison, chymotrypsin has the EC number 3.4.21.1, and elastase 3.4.21.36.

Enzyme assays

The amount of enzyme protein present can be determined (**assayed**) in terms of the catalytic effect it produces, that is the conversion of substrate to product. In order to **assay** (monitor the activity of) an enzyme, the overall equation of the reaction being catalyzed must be known, and an analytical procedure must be available for determining either the disappearance of substrate or the appearance of product. In addition, one must take into account whether the enzyme requires any **cofactors**, and the **pH** and **temperature** at which the enzyme is optimally active (Section D3). For mammalian enzymes, this is usually in the range 25–37°C. Finally, it is essential that the rate of the reaction being assayed is a measure of the enzyme activity present and is not limited by an insufficient supply of substrate. Therefore, very high substrate concentrations are generally required so that the **initial reaction rate**, which is determined experimentally, is proportional to the enzyme concentration (Section D3).

Table 1. International classification of enzymes

Class	Name	Type of reaction catalyzed		Example
1	Oxidoreductases	Transfer of electrons	$A^- + B \rightarrow A + B^-$	Alcohol dehydrogenase
2	Transferases	Transfer of functional groups	$A\text{–}B + C \rightarrow A + B\text{–}C$	Hexokinase
3	Hydrolases	Hydrolysis reactions	$A\text{–}B + H_2O \rightarrow A\text{–}H + B\text{–}OH$	Trypsin
4	Lyases	Addition or removal of a group to form a double bond	$\begin{array}{c} A\text{–}B \rightarrow A\text{=}B + X\text{–}Y \\ \mid \ \mid \\ X \ Y \end{array}$	Pyruvate decarboxylase
5	Isomerases	Transfer of groups within a molecule	$\begin{array}{c} A\text{–}B \rightarrow A\text{–}B \\ \mid \ \mid \quad \mid \ \mid \\ X \ Y \quad Y \ X \end{array}$	Maleate isomerase
6	Ligases (or synthases)	Bond formation coupled to ATP hydrolysis	$A + B \rightarrow A\text{–}B$	Pyruvate carboxylase

ATP, adenosine triphosphate.

An enzyme is most conveniently assayed by measuring the **rate of appearance of product** or the **rate of disappearance of substrate**. If the substrate (or product) **absorbs light** at a specific wavelength, then changes in the concentration of these molecules can be measured by following the **change of absorbance** at this wavelength using a **spectrophotometer**. If the substrate (or product) **fluoresces** (Section A4), then changes in the concentration can be measured by following the **change in fluorescence** using a **fluorimeter**. Since absorbance (or fluorescence) is proportional to concentration, the rate of change in absorbance (or fluorescence) is proportional to the rate of enzyme activity in moles of substrate used (or product formed) per unit time.

Two common molecules used for absorbance measurement in enzyme assays are the coenzymes **reduced nicotinamide adenine dinucleotide (NADH)** and **reduced nicotinamide adenine dinucleotide phosphate (NADPH)** (below), which each absorb in the ultraviolet (UV) region at 340 nm. Thus, if NADH or NADPH is produced during the course of the reaction there will be a corresponding increase in absorbance at 340 nm, whilst if the reaction involves the oxidation of NADH or NADPH to NAD^+ or $NADP^+$, respectively, there will be a corresponding decrease in absorbance, since these oxidized forms do not absorb at 340 nm. One example is **lactate dehydrogenase** whose activity with lactate as substrate can be assayed by following the increase in absorbance at 340 nm, according to the following equation:

$$CH_3CH(OH)COO^- + NAD^+ \rightleftharpoons CH_3COCOO^- + NADH + H^+$$

lactate   pyruvate

Linked enzyme assays

Numerous reactions do not involve substrates or products that absorb light at a suitable wavelength. In this case it is often possible to assay the enzyme that catalyzes this reaction by **linking** (or **coupling**) it to a second enzyme reaction that does involve a characteristic absorbance change. For example, the action of the enzyme **glucose oxidase**, which is often used to measure the concentration of glucose in the blood of diabetic patients, does not result in a change in absorbance upon conversion of substrates to products (Figure 3). However, the hydrogen peroxide produced in this reaction can be acted on by a second enzyme, **peroxidase**, which simultaneously converts a colorless compound into a colored one (**chromogen**) whose absorbance can be easily measured (Figure 3).

If the activity of the first enzyme (glucose oxidase) is to be measured accurately, the second enzyme (peroxidase) and its cosubstrates or coenzymes (below) must be **in excess** so as not to be the **rate-limiting step** of the linked assay. This will ensure that the rate of production of the colored chromogen is proportional to the rate of production of H_2O_2, whose production in turn is proportional to the activity of glucose oxidase.

Figure 3. A linked enzyme assay with glucose oxidase and peroxidase can be used to measure the amount of glucose in a blood sample.

Coenzymes and prosthetic groups

Many enzymes require the presence of small, nonprotein units or **cofactors** to carry out their particular reaction. Cofactors may be either one or more **inorganic ions**, such as Zn^{2+} or Fe^{2+}, or a complex organic molecule called a **coenzyme**. A metal or coenzyme that is covalently attached to the enzyme is called a **prosthetic group** (cf. heme in hemoglobin; see Section B3). A complete catalytically active enzyme together with its coenzyme or metal ion is called a **holoenzyme**. The protein part of the enzyme on its own without its cofactor is termed an **apoenzyme**. Some coenzymes, such as NAD$^+$, are bound and released by the enzyme during its catalytic cycle and in effect function as cosubstrates. Many coenzymes are derived from **vitamin precursors** (Table 2), which are often essential components of the organism's diet, thus giving rise to **deficiency diseases** when in inadequate supply.

Nicotinamide adenine dinucleotide (**NAD$^+$**) and **nicotinamide adenine dinucleotide phosphate** (**NADP$^+$**) coenzymes are based on a common structure consisting of the base adenine, two ribose sugars linked by phosphate groups and a nicotinamide ring (Figure 4). NADP$^+$ differs from NAD$^+$ in having an additional phosphate group attached to one of the ribose sugars (Figure 4). These two coenzymes share a common function as they both act as carriers of electrons and are involved in **oxidation–reduction reactions**. NAD$^+$ is more commonly used in **catabolic** (breakdown) reactions, whilst NADP$^+$ is used in **anabolic** (biosynthetic) reactions. The reactive part of both molecules is the **nicotinamide ring**, which exists in either a reduced or an oxidized form, and so acts to accept or donate electrons in an enzymic reaction. The reaction also involves the transfer of protons, according to the equation:

$$NAD^+ + H^+ + 2e^- \rightleftharpoons NADH$$

Table 2. Some common coenzymes, their vitamin precursors and deficiency diseases

Coenzyme	Precursor	Deficiency disease
Coenzyme A	Pantothenic acid	Hypertension
FAD, FMN	Riboflavin (vitamin B$_2$)	Dermatitis
NAD$^+$, NADP$^+$	Niacin (nicotinic acid)	Pellagra (dermatitis, depression, diarrhea)
Thiamine pyrophosphate	Thiamine (vitamin B$_1$)	Beriberi (weight loss, heart problems, neurological dysfunction)
Tetrahydrofolate	Folic acid	Anemia, neural tube defects in development
Deoxyadenosyl cobalamin	Cobalamin (vitamin B$_{12}$)	Pernicious anemia
–	Vitamin C (ascorbic acid)	Scurvy (swollen and bleeding gums, subdermal hemorrhages)
Pyridoxal phosphate	Pyridoxine (vitamin B$_6$)	Depression, confusion, convulsions

FAD, flavin adenine dinucleotide; FMN, flavin mononucleotide; NAD$^+$, nicotinamide adenine dinucleotide; NADP$^+$, nicotinamide adenine dinucleotide phosphate.

Figure 4. The structures of the coenzymes NAD^+ and $NADP^+$.

Flavin adenine dinucleotide (**FAD**) and **flavin mononucleotide** (**FMN**) are also carriers of electrons and have related chemical structures (Figure 5). Both of these coenzymes consist of a **flavin mononucleotide unit**, which contains the reactive site. FAD has an additional sugar group and an adenine base, which complete its structure. FAD and FMN react with two protons, as well as two electrons, in alternating between the reduced and oxidized state:

$$FAD + 2H^+ + 2e^- \rightleftharpoons FADH_2$$

Figure 5. The structures of the coenzymes FAD and FMN.

Isozymes

Isozymes (isoenzymes) are different forms of an enzyme, which **catalyze the same reaction**, but which exhibit **different physical or kinetic properties**, such as isoelectric point, pH optimum, substrate affinity or effect of inhibitors. Different isozyme forms of a given enzyme are usually derived from different genes and often occur in different tissues of the body.

An example of an enzyme that has different isozyme forms is **lactate dehydrogenase** (LDH), which catalyzes the reversible conversion of pyruvate into lactate in the presence of the coenzyme NADH (above). LDH is a tetramer of two different types of **subunits**, called H and M, which have small differences in amino acid sequence. The two subunits can combine randomly with each other, forming five isozymes that have the compositions H_4, H_3M, H_2M_2, HM_3 and M_4. The five isozymes can be resolved electrophoretically (Section C2). M subunits predominate in skeletal muscle and liver, whereas H subunits predominate in the heart. H_4 and H_3M isozymes are found predominantly in the heart and red blood cells; H_2M_2 is found predominantly in the brain; while HM_3 and M_4 are found predominantly in the liver and skeletal muscle. Thus, the isozyme pattern is characteristic of a particular tissue, a factor that is of immense diagnostic importance in medicine. **Myocardial infarction**, **infectious hepatitis** and **muscle diseases** involve cell death of the affected tissue, with release of the cell contents into the blood. As LDH is a soluble, cytosolic protein it is readily released in these conditions. Under normal circumstances there is little LDH in the blood. Therefore, the pattern of LDH isozymes in the blood is indicative of the tissue that released the isozymes and so can be used to diagnose a condition and to monitor the progress of treatment. For the **clinical diagnosis** of a myocardial infarction, in addition to the pattern of LDH isozymes, other enzymes, including **creatine kinase** and **aspartate aminotransferase** are routinely measured, along with an electrocardiogram (ECG).

D2 Thermodynamics

Key Notes

Thermodynamics   A knowledge of thermodynamics, which is the description of the relationships among the various forms of energy and how energy affects matter, enables one to determine whether a physical process is possible. The first and second laws of thermodynamics are combined in the thermodynamic function, free energy (G). The unit of energy is the Joule (J) or the calorie (cal).

Activation energy and transition state   For a biochemical reaction to proceed, the energy barrier needed to transform the substrate molecules into the transition state has to be overcome. The transition state has the highest free energy in the reaction pathway. The difference in free energy between the substrate and the transition state is termed the Gibbs free energy of activation ($\Delta G^{\ddagger}$). An enzyme stabilizes the transition state and lowers $\Delta G^{\ddagger}$, thus increasing the rate at which the reaction occurs.

Free energy change   The difference in energy level between the substrates and products is termed the change in Gibbs free energy (ΔG). A negative ΔG indicates that the reaction is thermodynamically favorable in the direction indicated, whereas a positive ΔG indicates that the reaction is not thermodynamically favorable and requires an input of energy to proceed in the direction indicated. An energetically unfavorable reaction is often driven by linking it to an energetically favorable reaction, such as the hydrolysis of adenosine triphosphate (ATP).

Chemical equilibria   A chemical reaction often exists in a state of dynamic equilibrium. The equilibrium constant (K) defines the ratio of the concentrations of substrates and products at equilibrium. Enzymes do not alter the equilibrium position, but do accelerate the attainment of the equilibrium position by accelerating the forward and reverse reactions.

Related topics   (D1) Introduction to enzymes   (D4) Enzyme inhibition
(D3) Enzyme kinetics   (D5) Regulation of enzyme activity

Thermodynamics

A knowledge of thermodynamics enables one to determine whether a physical process is possible, and is required for understanding why proteins fold to their native conformation, why some enzyme-catalyzed reactions require an input of energy, how muscles generate mechanical force, etc. **Thermodynamics** (Greek: *therme*, heat; *dynamis*, power) is the description of the relationships among the various forms of energy and how energy affects matter on the macroscopic level. As it applies to biochemistry, thermodynamics is most often concerned with describing the conditions under which processes occur spontaneously (by themselves).

In thermodynamics, a **system** is the matter within a defined region. The matter in the rest of the universe is called the **surroundings**. **The first law of thermodynamics**, a mathematical statement of the law of conservation of energy, states that the total energy of a system and its surroundings is a constant:

$$\Delta E = E_B - E_A = Q - W$$

in which E_A is the energy of the system at the start of a process and E_B at the end of the process. Q is the heat absorbed by the system and W is the work done by the system. The change in energy of a system depends only on the initial and final states and not on how it reached that state. Processes in which the system releases heat (i.e. have a negative Q) are known as **exothermic** processes; those in which the system gains heat (i.e. have a positive Q) are known as **endothermic**. The SI **unit of energy** is the Joule (J), although the calorie (cal) is still often used (1 kcal=4.184 kJ).

The first law of thermodynamics cannot be used to predict whether a reaction can occur spontaneously, as some spontaneous reactions have a positive ΔE. Therefore a function different from ΔE is required. One such function is **entropy** (S), which is a measure of the degree of randomness or disorder of a system. The entropy of a system increases (ΔS is positive) when the system becomes more disordered. **The second law of thermodynamics** states that a process can occur spontaneously only if the sum of the entropies of the system and its surroundings increases (or that the universe tends towards maximum disorder), that is:

$$(\Delta S_{system} + \Delta S_{surroundings}) > 0 \text{ for a spontaneous process.}$$

However, using entropy as a criterion of whether a biochemical process can occur spontaneously is difficult, as the entropy changes of chemical reactions are not readily measured, and the entropy change of both the system and its surroundings must be known. These difficulties are overcome by using a different thermodynamic function, **free energy** (G), proposed by Josiah Willard Gibbs, which combines the first and second laws of thermodynamics:

$$\Delta G = \Delta H - T\Delta S$$

in which ΔG is the free energy of a system undergoing a transformation at constant pressure (P) and temperature (T), ΔH is the change in **enthalpy** (heat content) of this system, and ΔS is the change in the entropy of this system. The enthalpy change is given by:

$$\Delta H = \Delta E + P\Delta V.$$

The volume change (ΔV) is small for nearly all biochemical reactions, and so ΔH is nearly equal to ΔE. Therefore

$$\Delta G = \Delta E - T\Delta S.$$

Thus, the ΔG of a reaction depends both on the change in internal energy and on the change in entropy of the system. The change in free energy ΔG of a reaction is a valuable criterion of whether that reaction can occur spontaneously:

- A reaction can occur spontaneously only if ΔG is negative

- A system is at equilibrium if ΔG is zero

- A reaction cannot occur spontaneously if ΔG is positive. An input of energy is required to drive such a reaction

- The ΔG of a reaction is independent of the path of the transformation

- ΔG provides no information about the rate of a reaction

Activation energy and transition state

The energy changes that take place during the course of a particular biochemical reaction are shown in Figure 1. In all reactions, there is an **energy barrier** that has to be overcome in order for the reaction to proceed. This is the energy needed to transform the substrate molecules into the **transition state** – an unstable chemical form partway between the substrates and the products. The transition state has the **highest free energy** of any component in the reaction pathway. The **Gibbs free energy of activation** ($\Delta G^{\ddagger}$) is equal to the difference in free energy between the transition state and the substrate (Figure 1). An enzyme works by **stabilizing the transition state** of a chemical reaction and decreasing $\Delta G^{\ddagger}$ (Figure 1). The enzyme does not alter the energy levels of the substrates or the products. Thus an enzyme increases the rate at which the reaction occurs, but has no effect on the overall change in energy of the reaction.

Free energy change

The change in **Gibbs free energy** (ΔG) dictates whether a reaction will be energetically favorable or not. Figure 1 shows an example where the overall energy change of the reaction makes it energetically favorable (i.e. the products are at a lower energy level than the substrates and ΔG is negative). It should be noted that ΔG is unrelated to $\Delta G^{\ddagger}$. The ΔG of a reaction is independent of the path of the reaction, and it provides no information about the rate of a reaction since the rate of the reaction is governed by $\Delta G^{\ddagger}$. A negative ΔG indicates that the reaction is thermodynamically favorable in the direction indicated (i.e. that it is likely to occur without an input of energy), whereas a positive ΔG indicates

Figure 1. The energy changes taking place during the course of a biochemical reaction.

Figure 2. Structure of ATP, ADP, AMP and adenosine.

that the reaction is not thermodynamically favorable and requires an **input of energy** to proceed in the direction indicated. In biochemical systems, this input of energy is often achieved by **coupling** the energetically unfavorable reaction with a more energetically favorable one (**coupled reactions**).

It is often convenient to refer to ΔG under a standard set of conditions, defined as when the substrates and products of a reaction are all present at concentrations of 1.0 M and the reaction is taking place at a constant pH of 7.0. Under these conditions a slightly different value for ΔG is found, and this is called $\Delta G^{0\prime}$. An example of an energetically favorable reaction, which has a large negative $\Delta G^{0\prime}$ and is commonly used to drive less energetically favorable reactions, is the hydrolysis of **adenosine triphosphate** (ATP; Figure 2) to form adenosine diphosphate (ADP) and free inorganic phosphate (P_i):

$$\text{ATP} + H_2O \rightarrow \text{ADP} + P_i \qquad \Delta G^{0\prime} = -30.5 \text{ kJ mol}^{-1}$$
$$-7.3 \text{ kcal mol}^{-1}$$

Chemical equilibria

A chemical reaction usually exists in a state of **dynamic equilibrium**, where although new molecules of substrate and product are continually being transformed and formed, the ratio of substrate to product remains at a constant value.

Consider the reaction:

$$A \underset{10^{-6} \text{ s}^{-1}}{\overset{10^{-4} \text{ s}^{-1}}{\rightleftharpoons}} B$$

where the rate of the forward reaction is 10^{-4} per second (s^{-1}) and the rate of the reverse reaction is 10^{-6} s^{-1}. At equilibrium the ratio of the concentrations of the substrate and product gives a constant value, known as the **equilibrium constant** (K). The equilibrium constant for a given reaction is defined as:

$$K = \frac{[\text{products}]_{eq}}{[\text{reactants}]_{eq}} = \frac{[B]_{eq}}{[A]_{eq}}$$

where square brackets indicate concentration. The equilibrium constant is also given by the ratio of the forward reaction rate (k_f) and the reverse reaction rate (k_b):

$$K = \frac{k_f}{k_b} = \frac{10^{-4}}{10^{-6}} = 100$$

Thus, for the above reaction at equilibrium, there is 100 times more of product B than there is of substrate A, regardless of whether there is enzyme present or not. This is because enzymes do not alter the equilibrium position of a reaction, but accelerate the forward and reverse reactions to the same extent. In other words, **enzymes accelerate the attainment of the equilibrium position** but do not shift its position. For the hypothetical reaction shown above, in the absence of added enzyme the reaction may take over an hour to reach the equilibrium position, whereas in the presence of enzyme the equilibrium position may be reached in less than 1 s.

D3 Enzyme kinetics

Key Notes

Enzyme velocity

Enzyme activity is commonly expressed by the initial rate (V_0) of the reaction being catalyzed. The units of V_0 are µmol min^{-1}, which can also be represented by the enzyme unit (U), where 1 µmol min^{-1}=1 U. The term activity (or total activity) refers to the total units of enzyme in a sample, whereas specific activity is the number of units per milligram of protein (units mg^{-1}).

Substrate and enzyme concentration

At low substrate concentrations ([S]) a doubling of [S] leads to a doubling of V_0, whereas at higher [S] the enzyme becomes saturated and there is no further increase in V_0. A graph of V_0 against [S] will give a hyperbolic curve. When [S] is saturating, a doubling of the enzyme concentration leads to a doubling of V_0.

Temperature

Temperature affects the rate of an enzyme-catalyzed reaction by increasing the thermal energy of the substrate molecules. This increases the proportion of molecules with sufficient energy to overcome the activation barrier and hence increases the rate of the reaction. In addition, the thermal energy of the component molecules of the enzyme is increased, which leads to an increased rate of denaturation of the enzyme protein due to the disruption of the noncovalent interactions holding the structure together.

pH

Each enzyme has an optimum pH at which the rate of the reaction that it catalyzes is at its maximum. Slight deviations in the pH from the optimum lead to a decrease in the reaction rate. Larger deviations in pH lead to denaturation of the enzyme due to changes in the ionization of amino acid residues and the disruption of noncovalent interactions.

Michaelis–Menten model

The Michaelis–Menten model uses the following concept of enzyme catalysis:

$$E + S \underset{k_2}{\overset{k_1}{\rightleftharpoons}} ES \xrightarrow{k_3} E + P$$

where the rate constants k_1, k_2 and k_3 describe the rates associated with each step of the catalytic process. At low [S], V_0 is directly proportional to [S], while at high[S] the velocity tends towards a maximum velocity (V_{max}). The Michaelis–Menten equation:

$$V_0 = \frac{V_{max} \cdot [S]}{K_m + [S]}$$

describes these observations and predicts a hyperbolic curve of V_0 against [S]. The Michaelis constant, K_m, is equal to the sum of the rates of breakdown of the enzyme–substrate complex over its rate of formation, and is a measure of the affinity of an enzyme for its substrate.

Lineweaver–Burk plot	V_{max} and K_m can be determined experimentally by measuring V_0 at different substrate concentrations, and then plotting $1/V_0$ against $1/[S]$ in a double reciprocal or Lineweaver–Burk plot. The intercept on the y-axis is equal to $1/V_{max}$, the intercept on the x-axis is equal to $-1/K_m$ and the slope of the line is equal to K_m/V_{max}.
Related topics	(B2) Protein structure and function (D4) Enzyme inhibition (D1) Introduction to enzymes (D5) Regulation of enzyme (D2) Thermodynamics activity

Enzyme velocity

The rate of an enzyme-catalyzed reaction is often called its velocity. Enzyme velocities are normally reported as values at time zero (**initial velocity**, symbol V_0; μmol min^{-1}), since the rate is fastest at the point where no product is yet present. This is because the substrate concentration is greatest before any substrate has been transformed to product, because enzymes may be subject to **feedback inhibition** by their own products and/or because with a reversible reaction the products will fuel the reverse reaction. Experimentally V_0 is measured before more than approximately 10% of the substrate has been converted to product in order to minimize such complicating factors. A typical **plot of product formed against time** for an enzyme-catalyzed reaction shows an initial period of rapid product formation, which gives the linear portion of the plot (Figure 1). This is followed by a slowing down of the enzyme rate as substrate is used up and/or as the enzyme loses activity. V_0 is obtained by drawing a straight line through the linear part of the curve, starting at the zero time-point (Figure 1). The slope of this straight line is equal to V_0.

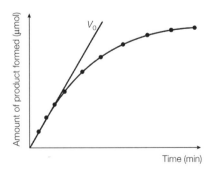

Figure 1. The relationship between product formation and time for an enzyme-catalyzed reaction.

Enzyme units

Enzyme activity may be expressed in a number of ways. The commonest is by the initial rate (V_0) of the reaction being catalyzed (e.g. μmol of substrate transformed per minute; μmol min^{-1}). There is also a standard unit of enzyme activity, the **enzyme unit** (**U**). An enzyme unit is that amount of enzyme that will catalyze the transformation of 1 μmol of substrate per minute at 25°C under optimal conditions for that enzyme (i.e. 1 μmol min^{-1}=1 U). The term **activity** (or **total activity**) refers to the total units of enzyme in the sample, whereas the **specific activity** is the number of enzyme units per milligram of protein (units mg^{-1}). The specific activity is a measure of the purity of an enzyme; during the purification of the enzyme, its specific activity increases and becomes maximal and constant when the enzyme is pure.

Substrate and enzyme concentration

The normal pattern of dependence of enzyme rate on **substrate concentration** ([S]) is that at low substrate concentrations a doubling of [S] will lead to a doubling of the initial velocity (V_0). However, at higher substrate concentrations the enzyme becomes **saturated**, and further increases in [S] lead to very small changes in V_0. This occurs because at saturating substrate concentrations effectively all of the enzyme molecules have bound substrate. The overall enzyme rate is now dependent on the rate at which the product can dissociate from the enzyme, and adding further substrate will not affect this. The shape of the resulting graph when V_0 is plotted against [S] is called a **hyperbolic curve** (Figure 2).

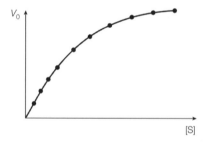

Figure 2. The relationship between [S] and V_0.

In situations where the substrate concentration is saturating (i.e. all the enzyme molecules are bound to substrate), a doubling of the **enzyme concentration** will lead to a doubling of V_0. This gives a straight line graph when V_0 is plotted against enzyme concentration.

Temperature

Temperature affects the rate of enzyme-catalyzed reactions in two ways. First, a rise in temperature increases the **thermal energy** of the substrate molecules. This raises the proportion of substrate molecules with sufficient energy to overcome the Gibbs free energy of activation ($\Delta G^{\ddagger}$) (Section D2), and hence increases the rate of the reaction. However, a second effect comes into play at higher temperatures. Increasing the thermal energy of the molecules that make up the protein structure of the enzyme itself will increase the chances of breaking the multiple weak, noncovalent interactions (hydrogen bonds, van der Waals forces, etc.), which hold the three-dimensional structure of the

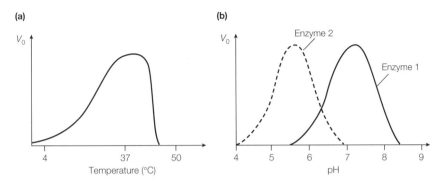

Figure 3. The effect of (a) temperature and (b) pH on enzyme activity.

enzyme together (Section B2). Ultimately, this will lead to the **denaturation** (unfolding) of the enzyme, but even small changes in the three-dimensional shape of the enzyme can alter the structure of the active site and lead to a decrease in catalytic activity. The overall effect of a rise in temperature on the reaction rate of the enzyme is a balance between these two opposing effects. A graph of temperature plotted against V_0 will therefore show a curve, with a well-defined temperature optimum (Figure 3a). For many mammalian enzymes, this is around 37°C, but there are also organisms that have enzymes adapted to working at considerably higher or lower temperatures. For example, ***Taq* polymerase** that is used in the polymerase chain reaction (Section I6) is found in a bacterium that lives at high temperatures in hot springs, and thus is adapted to work optimally at high temperatures.

pH

Each enzyme has an **optimum pH** at which the rate of the reaction that it catalyzes is at its maximum. Small deviations in pH from the optimum value lead to decreased activity due to changes in the ionization of groups at the active site of the enzyme. Larger deviations in pH lead to the **denaturation** of the enzyme protein itself, due to interference with the many weak noncovalent bonds maintaining its three-dimensional structure. A graph of V_0 plotted against pH will usually give a bell-shaped curve (Figure 3b). Many enzymes have a pH optimum of around 6.8, but there is great diversity in the pH optima of enzymes, due to the different environments in which they are adapted to work. For example, the digestive enzyme **pepsin** is adapted to work at the acidic pH of the stomach (around pH 2.0).

Michaelis–Menten model

The **Michaelis–Menten model** uses the following concept of enzyme catalysis:

$$E + S \underset{k_2}{\overset{k_1}{\rightleftharpoons}} ES \xrightarrow{k_3} E + P$$

The enzyme (E) combines with its substrate (S) to form an **enzyme–substrate complex** (ES). The ES complex can dissociate again to form E+S, or can proceed chemically to form E and the product P. The **rate constants** k_1, k_2 and k_3 describe the rates associated with each step of the catalytic process. It is assumed that there is no significant rate for the backward reaction of enzyme and product (E+P) being converted to ES complex. [ES] remains approximately constant until nearly all the substrate is used, hence the rate of synthesis of ES equals its rate of consumption over most of the course of the reaction; that is, [ES] maintains a **steady state**. From the observation of the properties of many enzymes it was known that the initial velocity (V_0) at low substrate concentrations is directly proportional to [S], while at high substrate concentrations the velocity tends towards a maximum value, that is the rate becomes independent of [S] (Figure 4a). This maximum velocity is called V_{max} (units of μmol min^{-1}). The **initial velocity** (V_0) is the velocity measured experimentally before more than approximately 10% of the substrate has been converted to product in order to minimize such complicating factors as the effects of reversible reactions, inhibition of the enzyme by product, and progressive inactivation of the enzyme (see above).

Michaelis and Menten derived an equation to describe these observations, the **Michaelis–Menten equation**:

$$V_0 = \frac{V_{max} \cdot [S]}{K_m + [S]}$$

The equation describes a **hyperbolic curve** of the type shown for the experimental data in Figure 1. In deriving the equation, Michaelis and Menten defined a new constant, K_m, the **Michaelis constant** [units: Molar (i.e. per mole), M]:

$$K_m = \frac{k_2 + k_3}{k_1}$$

K_m is a measure of the **stability of the ES complex**, being equal to the sum of the rates of breakdown of ES over its rate of formation. For many enzymes, k_2 is much greater than k_3. Under these circumstances, K_m becomes a measure of the **affinity** of an enzyme for its substrate since its value depends on the relative values of k_1 and k_2 for ES formation and dissociation, respectively. A high K_m indicates weak substrate binding (k_2 predominant over k_1), a low K_m indicates strong substrate binding (k_1 predominant over k_2). For most enzymes, K_m lies between 10^{-1} and 10^{-7} M. K_m can be determined experimentally by the fact that its value is equivalent to the substrate concentration at which the velocity is equal to half of V_{max}.

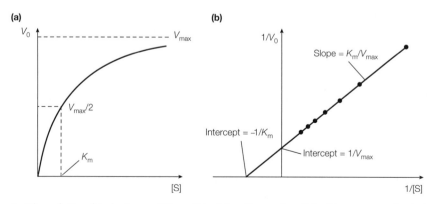

Figure 4. The relationship between [S] and V_0: (a) a direct plot; (b) a Lineweaver–Burk double reciprocal plot.

Lineweaver–Burk plot

Because V_{max} is achieved at infinite substrate concentration, it is impossible to estimate V_{max} (and hence K_m) from a hyperbolic plot as shown in Figure 4a. However, V_{max} and K_m can be determined experimentally by measuring V_0 at different substrate concentrations (Figure 1). Then a **double reciprocal** or **Lineweaver–Burk** plot of $1/V_0$ against $1/[S]$ is made (Figure 4b). Alternatively, a suitable curve-fitting program on a computer can be used on the data in Figure 4a. The Lineweaver–Burk plot is a derivation of the Michaelis–Menten equation:

$$\frac{1}{V_0} = \frac{1}{V_{max}} + \frac{K_m}{V_{max}} \cdot \frac{1}{[S]}$$

which gives a straight line, with the **intercept on the y-axis equal to $1/V_{max}$**, and the **intercept on the x-axis equal to $-1/K_m$**. The slope of the line is equal to K_m/V_{max} (Figure 4b). The Lineweaver–Burk plot is also a useful way of determining how an inhibitor binds to an enzyme (Section D4). The K_m and V_{max} can also be determined from an **Eadie–Hofstee plot** of $V_0/[S]$ against V_0, where the intercept on the x-axis equals V_{max} and the slope of the line is equal to $-1/K_m$.

Although the Michaelis–Menten model provides a very good model of the experimental data for many enzymes, a few enzymes do not conform to Michaelis–Menten kinetics. These enzymes, such as aspartate transcarbamoylase, are called **allosteric enzymes** (Section D5).

D4 Enzyme inhibition

Key Notes

Enzyme inhibition   The catalytic rate of an enzyme can be lowered by inhibitor molecules. Many inhibitors exist, including normal body metabolites, foreign drugs and toxins. Enzyme inhibition can be of two main types: irreversible or reversible. Reversible inhibition can be subdivided into competitive and noncompetitive.

Irreversible inhibition   An irreversible inhibitor binds tightly, often covalently, to amino acid residues at the active site of the enzyme, permanently inactivating the enzyme. Examples of irreversible inhibitors are diisopropylphosphofluoridate (DIPF), iodoacetamide, aspirin and penicillin.

Reversible competitive inhibition   A competitive inhibitor competes with the substrate molecules for binding to the active site of the enzyme. At high substrate concentration, the effect of a competitive inhibitor can be overcome. On a Lineweaver–Burk plot, a competitive inhibitor can be seen to increase the K_m but leave V_{max} unchanged.

Reversible noncompetitive inhibition   A noncompetitive inhibitor binds at a site other than the active site of the enzyme and decreases its catalytic rate by causing a conformational change in the three-dimensional shape of the enzyme. The effect of a noncompetitive inhibitor cannot be overcome at high substrate concentrations. On a Lineweaver–Burk plot a noncompetitive inhibitor can be seen to decrease V_{max} but leave K_m unchanged.

Related topics   (D1) Introduction to enzymes   (D5) Regulation of enzyme activity
(D3) Enzyme kinetics

Enzyme inhibition

Many types of molecule exist which are capable of interfering with the activity of an individual enzyme. Any molecule that acts directly on an enzyme to lower its catalytic rate is called an **inhibitor**. Some enzyme inhibitors are normal body metabolites that inhibit a particular enzyme as part of the normal metabolic control of a pathway. Other inhibitors may be foreign substances, such as drugs or toxins, where the effect of enzyme inhibition could be either therapeutic or, at the other extreme, lethal. Enzyme inhibition may be of two main types: **irreversible** or **reversible**, with reversible inhibition itself being subdivided into **competitive** and **noncompetitive** inhibition. Reversible inhibition can be overcome by removing the inhibitor from the enzyme, for example by dialysis (Section C1), but this is not possible for irreversible inhibition, by definition.

Irreversible inhibition

Inhibitors that bind **irreversibly** to an enzyme often form a **covalent bond** to a reactive amino acid residue at or near the active site, and permanently inactivate the enzyme. Susceptible amino acid residues include Ser and Cys residues, which have reactive -OH and -SH groups, respectively. The compound **diisopropylphosphofluoridate** (DIPF) and the structurally related **sarin**, components of nerve gases, react with a Ser residue in the active site of the enzyme **acetylcholinesterase**, irreversibly inhibiting the enzyme and preventing the transmission of nerve impulses (Figure 1a) (Section E6). **Iodoacetamide** modifies Cys residues and hence may be used as a diagnostic tool in determining whether one or more Cys residues are required for enzyme activity (Figure 1b). **Aspirin** acts by covalently modifying the enzyme **prostaglandin H2 synthase**, reducing the synthesis of inflammatory signals (Section K1). The **antibiotic penicillin** irreversibly inhibits the **glycopeptide transpeptidase** enzyme that forms the cross-links in the bacterial cell wall by covalently attaching to a Ser residue in the active site of the enzyme (Section A1). The penicillin accesses the active site of the enzyme because it mimics the D-Ala-D-Ala moiety of the normal substrate. The transpeptidase then covalently links the penicillin to its active site Ser residue. Thus penicillin acts as a **suicide inhibitor**.

Reversible competitive inhibition

A **competitive inhibitor** typically has close structural similarities to the normal substrate for the enzyme. Thus it competes with substrate molecules to bind to the active site (Figure 2a). The enzyme may bind either a substrate molecule or an inhibitor molecule, but not both at the same time (Figure 2b). The competitive inhibitor binds **reversibly to the active site**. At **high substrate concentrations**, the action of a competitive inhibitor is overcome because a sufficiently high substrate concentration will successfully compete out the inhibitor molecule in binding to the active site. Thus there is no change in the V_{max} of the enzyme but the apparent affinity of the enzyme for its substrate decreases in the presence of the competitive inhibitor, and hence K_m increases.

A good example of competitive inhibition is provided by **succinate dehydrogenase**. This enzyme uses **succinate** as its substrate and is competitively inhibited by **malonate**, which differs from succinate in having one rather than two methylene groups (Figure 3).

Figure 1. Structure and mechanism of action of: (a) DIPF; (b) iodoacetamide.

Figure 2. The characteristics of competitive inhibition: (a) a competitive inhibitor competes with the substrate for binding at the active site of the enzyme; (b) the enzyme can bind either substrate or the competitive inhibitor but not both; (c) Lineweaver–Burk plot showing the effect of a competitive inhibitor on K_m and V_{max}.

Many drugs work by mimicking the structure of the substrate of a target enzyme, and hence act as competitive inhibitors of the enzyme. For example, **methotrexate**, a structural analog of the coenzyme tetrahydrofolate (Section D1), which is required by the enzyme **dihydrofolate reductase** to synthesize purines and pyrimidines, is used to treat cancer. Inhibitors of **angiotensin-converting enzyme**, such as captopril, are used in the treatment of hypertension (high blood pressure). Many of the drugs used to treat **human immunodeficiency virus** (**HIV**), which causes acquired immunodeficiency syndrome (AIDS), are competitive inhibitors of either the **reverse transcriptase** (e.g. AZT or zidovudine) or of the **HIV protease** (e.g. saquinavir, ritonavir).

Competitive inhibition is the principle behind the use of ethanol to treat **methanol poisoning**. Although methanol itself is only mildly toxic, it is converted by the liver enzyme **alcohol dehydrogenase** to the highly toxic formaldehyde, which even in small amounts causes blindness and death. Ethanol competes with methanol for binding to the active site of alcohol dehydrogenase, thereby slowing the conversion of methanol to formaldehyde (the ethanol is converted to the readily metabolized acetaldehyde). Thus, on administration of ethanol, a large proportion of the methanol is harmlessly excreted from the body in the urine. The same principle underlies the use of ethanol to treat **antifreeze** (ethylene glycol) **poisoning**.

Competitive inhibition can be recognized by using a Lineweaver–Burk plot. V_0 is measured at different substrate concentrations in the presence of a fixed concentration of inhibitor. A competitive inhibitor increases the slope of the line on the Lineweaver–Burk plot, and alters the intercept on the x-axis (since K_m **is increased**), but leaves the intercept on the y-axis unchanged (since V_{max} **remains constant**; Figure 2c).

COO⁻		COO⁻	
CH₂ + FAD	*Succinate*	CH + FADH₂	
CH₂	*dehydrogenase*	CH	
COO⁻		COO⁻	
Succinate		Fumarate	

$$\begin{array}{c} COO^- \\ | \\ CH_2 + FAD \\ | \\ CH_2 \\ | \\ COO^- \end{array} \xrightarrow[dehydrogenase]{Succinate} \begin{array}{c} COO^- \\ | \\ CH \\ || \\ CH \\ | \\ COO^- \end{array}$$

Succinate   Fumarate

$$\begin{array}{c} COO^- \\ | \\ CH_2 \\ | \\ COO^- \end{array} \xrightarrow[dehydrogenase]{Succinate} \text{No reaction}$$

Malonate

Figure 3. Inhibition of succinate dehydrogenase by malonate.

Reversible noncompetitive inhibition

A **noncompetitive inhibitor** binds **reversibly at a site other than the active site** (Figure 4a) and causes a change in the overall three-dimensional shape of the enzyme that leads to a decrease in catalytic activity. Since the inhibitor binds at a different site from the substrate, the enzyme may bind the inhibitor, the substrate or both the inhibitor and substrate together (Figure 4b). The effects of a noncompetitive inhibitor cannot be overcome by increasing the substrate concentration, so there is a decrease in V_{max}. In noncompetitive inhibition, the affinity of the enzyme for the substrate is unchanged and so K_m remains the same. An example of noncompetitive inhibition is the action of **pepstatin** on the enzyme **renin**.

Noncompetitive inhibition can be recognized on a Lineweaver–Burk plot, since it increases the slope of the experimental line, and alters the intercept on the y-axis (since V_{max} **is decreased**), but leaves the intercept on the x-axis unchanged (since K_m **remains constant**; Figure 4c).

Figure 4. The characteristics of noncompetitive inhibition: (a) a noncompetitive inhibitor binds at a site distinct from the active site; (b) the enzyme can bind either substrate or the noncompetitive inhibitor or both; (c) Lineweaver–Burk plot showing the effect of a noncompetitive inhibitor on K_m and V_{max}.

D5 Regulation of enzyme activity

Key Notes

Feedback regulation

The rates of enzyme-catalyzed reactions in biological systems are altered by activators and inhibitors, collectively known as effector molecules. In metabolic pathways, the end-product often feedback-inhibits the committed step earlier on in the same pathway to prevent the build-up of intermediates and the unnecessary use of metabolites and energy. For branched metabolic pathways a process of sequential feedback inhibition often operates.

Allosteric enzymes

A plot of V_0 against [S] for an allosteric enzyme gives a sigmoidal-shaped curve. Allosteric enzymes are usually multisubunit proteins with an active site on each subunit, which cooperatively bind substrate molecules, such that the binding of substrate at one active site induces a conformational change in the enzyme, which alters the affinity of the other active sites for substrate. Two models have been proposed to explain the allosteric behavior of enzymes, the concerted or symmetry model and the sequential model. Allosteric enzymes may be controlled by effector molecules (activators or inhibitors) that bind to a site other than the active site and alter the rate of enzyme activity. Aspartate transcarbamoylase is an allosteric enzyme, which catalyzes the committed step in pyrimidine biosynthesis. This enzyme consists of six catalytic subunits each with an active site and six regulatory subunits to which the allosteric effectors cytosine triphosphate (CTP) and ATP bind. Aspartate transcarbamoylase is feedback-inhibited by the end-product of the pathway, CTP, which acts as an allosteric inhibitor. In contrast, ATP, an intermediate earlier in the pathway, acts as an allosteric activator.

Reversible covalent modification

The activity of many enzymes is altered by the reversible making and breaking of a covalent bond between the enzyme and a small nonprotein group. The most common such modification is the addition and removal of a phosphate group; phosphorylation and dephosphorylation, respectively. Phosphorylation is catalyzed by protein kinases, often using ATP as the phosphate donor, whereas dephosphorylation is catalyzed by protein phosphatases. Approximately 30% of human proteins are subject to phosphorylation.

Proteolytic activation	Some enzymes are synthesized as larger inactive precursors called proenzymes or zymogens. These are activated by the irreversible hydrolysis of one or more peptide bonds. The pancreatic proteases trypsin, chymotrypsin and elastase are all derived from zymogen precursors (trypsinogen, chymotrypsinogen and proelastase, respectively) by proteolytic activation. Premature activation of these zymogens leads to the condition of acute pancreatitis. The blood clotting cascade and apoptosis (programmed cell death) also involve a series of zymogen activations that brings about a large amplification of the original signal.
Regulation of enzyme synthesis and breakdown	The amount of enzyme present is a balance between the rates of its synthesis and degradation. The level of induction or repression of the gene encoding the enzyme, and the rate of degradation of its mRNA, will alter the rate of synthesis of the enzyme protein. Once the enzyme protein has been synthesized, the rate of its breakdown (half-life) can also be altered as a means of regulating enzyme activity.
Related topics	(B3) Myoglobin and hemoglobin (D1) Introduction to enzymes (D3) Enzyme kinetics — (D4) Enzyme inhibition (J7) Control of glycogen metabolism

Feedback regulation

In biological systems, the rates of many enzymes are altered by the presence of other molecules such as activators and inhibitors (collectively known as **effectors**). A common theme in the control of metabolic pathways is when an enzyme early on in the pathway is inhibited by an end-product of the metabolic pathway in which it is involved. This is called **feedback inhibition** and often takes place at the **committed step** in the pathway (conversion of A to B in Figure 1a). The committed step is the first step to produce an intermediate which is unique to the pathway in question, and therefore normally commits the metabolite to further metabolism along that pathway. Control of the enzyme that carries out the committed step of a metabolic pathway conserves the metabolic energy supply of the organism, and prevents the build-up of large quantities of unwanted metabolic intermediates further along the pathway. Such feedback inhibition is seen, for example, in the pathways involved in the biosynthesis of amino acids (Section M2).

As many metabolic pathways are **branched**, feedback inhibition must allow the synthesis of one product of a branched pathway to proceed even when another is present in excess. Here a process of **sequential feedback inhibition** may operate where the end-product of one branch of a pathway will inhibit the first enzyme after the branchpoint (the conversion of C to D or C to E in Figure 1b). When this branchpoint intermediate builds up, it in turn inhibits the first committed step of the whole pathway (conversion of A to B

Figure 1. Feedback inhibition (a) and sequential feedback inhibition (b) in metabolic pathways.

in Figure 1b). Since the end-product of a metabolic pathway involving multiple enzyme reactions is unlikely to resemble the starting compound structurally, the end-product will bind to the enzyme at the control point at a site other than the active site. Such enzymes are usually **allosteric enzymes**.

Allosteric enzymes

A plot of V_0 against [S] for an **allosteric enzyme** gives a **sigmoidal curve** rather than the hyperbolic plots predicted by the Michaelis–Menten equation (as seen for the binding of O_2 to hemoglobin; Section B3, Figure 3). The curve has a steep section in the middle of the substrate concentration range, reflecting the rapid increase in enzyme velocity that occurs over a narrow range of substrate concentrations. This allows allosteric enzymes to be particularly sensitive to small changes in substrate concentration within the physiological range. In allosteric enzymes, the binding of a substrate molecule to one active site affects the binding of substrate molecules to other active sites in the enzyme; the different active sites are said to behave **cooperatively** in binding and acting on substrate molecules (cf. the binding of O_2 to the four subunits of hemoglobin; Section B3). Thus allosteric enzymes are often multisubunit proteins (i.e. oligomeric), with one or more active sites on each subunit. The binding of substrate at one active site **induces a conformational change** in the protein that is conveyed to the other active sites, altering their affinity for substrate molecules.

Two models have been put forward to account for the allosteric effects observed in proteins. In the **symmetry** or **concerted model**, first proposed by Jacques Monod, Jeffries Wyman and Jean-Pierre Changeaux [sometimes referred to as the **Monod-Wyman-Changeaux (MWC) model**], the subunits of an allosteric enzyme can exist in one of only two states, T and R (Figure 2a). T-state subunits are in a tense state that is compact and relatively inactive, while R-state subunits are in a relaxed, expanded, active state with higher affinity for the substrate; no intermediate states are allowed. In the absence of bound substrate, the equilibrium favors the T-state. As substrate binds to each active site in the T-state, the equilibrium shifts towards the R-state. All of the subunits change conformation in a concerted manner, which implies that the conformation of each subunit is constrained by its association with the other subunits; in other words, there are no oligomeric forms of the enzyme that simultaneously contain R- and T-state subunits and the molecular symmetry of the protein is conserved during the conformational change (Figure 2a).

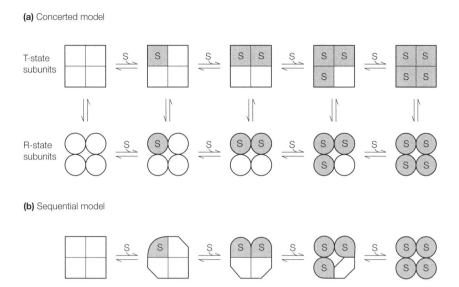

Figure 2. Models of allosterism: (a) the concerted or symmetry model; the squares and circles represent the T- and R-states, respectively; (b) the sequential model; substrate binding progressively induces conformational changes in the subunits.

In the alternative **sequential model**, first proposed by Daniel Koshland, sequential changes in structure take place within an oligomeric enzyme as the individual active sites are occupied (Figure 2b). The binding of substrate to one site influences the substrate affinity of neighboring active sites without necessarily inducing a transition encompassing the whole enzyme, such that the molecular symmetry of the whole protein is not necessarily conserved (Figure 2b). The sequential model builds upon the induced-fit hypothesis of enzyme–substrate interaction, whereas the concerted model implicitly assumes the lock-and-key model of substrate binding to the enzyme's active site (Section D1). In the sequential model, substrate binding induces a conformational change in a subunit and cooperative interactions arise through the influence that these conformational changes have on neighboring subunits. The strengths of these interactions depend on the degree of mechanical coupling between subunits. In the sequential model, the enzyme–substrate binding affinity varies with the number of bound substrate molecules, whereas in the concerted model this affinity depends only on the quaternary state of the enzyme. The results of studies of a number of allosteric proteins suggest that most behave according to a combination of the concerted and sequential models.

Allosteric enzymes may be controlled by **effector** molecules (activators and inhibitors) that bind to the enzyme at a site other than the active site (either on the same subunit or on a different subunit), thereby causing a change in the conformation of the active site, which alters the rate of enzyme activity (cf. the binding of CO_2, H^+ and 2,3-bisphosphoglycerate to hemoglobin; see Section B3). An **allosteric activator** increases the rate of enzyme activity, while an **allosteric inhibitor** decreases the activity of the enzyme.

Aspartate transcarbamoylase

Aspartate transcarbamoylase (aspartate carbamoyltransferase; **ATCase**), a key enzyme in **pyrimidine biosynthesis** (Section F1), provides a good example of allosteric regulation.

ATCase catalyzes the formation of **N-carbamoylaspartate** from aspartate and carbamoyl phosphate, and is the committed step in pyrimidine biosynthesis (Figure 3). The binding of the two substrates aspartate and carbamoyl phosphate is cooperative, as shown by the **sigmoidal curve** of V_0 against substrate concentration (Figure 4).

Figure 3. Formation of N-carbamoylaspartate by ATCase is the committed step in pyrimidine biosynthesis and a key control point.

Figure 4. Plot of initial velocity (V_0) against substrate concentration for the allosteric enzyme aspartate transcarbamoylase.

ATCase consists of six **catalytic subunits** and six **regulatory subunits**. The enzyme is **feedback-inhibited** by the end-product of the pathway, **cytosine triphosphate** (**CTP**; Section F1), which acts as an **allosteric inhibitor** (Figure 3). This molecule binds to the regulatory subunits and causes a decrease in the catalytic activity of ATCase by decreasing the affinity of the catalytic subunits for substrate molecules. In contrast, **ATP**, one of the intermediates earlier on in the pathway, acts as an **allosteric activator**, enhancing the affinity of ATCase for its substrates and leading to an increase in activity (Figure 3). ATP competes with the same binding site on the regulatory subunit as CTP. High levels of ATP signal to the cell that energy is available for DNA replication, and so ATCase is activated, resulting in the synthesis of the required pyrimidine nucleotides. When pyrimidines are abundant, the high levels of CTP inhibit ATCase, preventing needless synthesis of N-carbamoylaspartate and subsequent intermediates in the pathway.

Reversible covalent modification

Reversible covalent modification is the making and breaking of a covalent bond between a nonprotein group and an enzyme molecule. Although a range of nonprotein groups may be reversibly attached to enzymes that affect their activity, the most common modification is the addition and removal of a **phosphate group** (**phosphorylation** and **dephosphorylation**, respectively). Phosphorylation is catalyzed by **protein kinases**, often using ATP as the phosphate donor, and dephosphorylation is catalyzed by **protein phosphatases** (Figure 5). The addition and removal of the phosphate group causes changes in the **tertiary structure** of the enzyme that alter its catalytic activity. One class of protein kinases transfers the phosphate specifically on to the hydroxyl group of Ser or Thr residues on the target enzyme [serine/threonine protein kinases, typified by 3′,5′-cyclic adenosine monophosphate (cAMP)-dependent protein kinase], while a second class transfers the phosphate on to the hydroxyl group of Tyr residues (tyrosine kinases). Protein phosphatases catalyze the hydrolysis of phosphate groups from proteins to regenerate the unmodified hydroxyl group of the amino acid and release P_i (Figure 5).

A phosphorylated enzyme may be either more or less active than its dephosphorylated form. Thus phosphorylation/dephosphorylation may be used as a rapid, reversible switch to turn a metabolic pathway on or off according to the needs of the cell. For example, **glycogen phosphorylase**, an enzyme involved in glycogen breakdown, is active in its phosphorylated form, and **glycogen synthase**, involved in glycogen synthesis, is most active in its dephosphorylated form (Section J7).

Other types of reversible covalent modification that are used to regulate the activity of certain enzymes include **adenylylation** (the transfer of adenylate from ATP as seen in glutamine synthetase) and **ADP-ribosylation** (the transfer of an ADP-ribosyl moiety from NAD^+ as seen in RNA polymerase).

Figure 5. The reversible phosphorylation and dephosphorylation of an enzyme.

Proteolytic activation

Several enzymes are synthesized as larger inactive precursor forms called proenzymes or **zymogens**. Activation of zymogens involves **irreversible hydrolysis** of one or more **peptide bonds**.

Pancreatic proteases

The digestive enzymes **trypsin**, **chymotrypsin** and **elastase** (Section D1) are produced as zymogens in the **pancreas**. They are then transported to the small intestine as their zymogen forms and activated there by cleavage of specific peptide bonds. Trypsin is synthesized initially as the zymogen **trypsinogen**. It is cleaved (and hence activated) in the intestine by the enzyme **enteropeptidase**, which is only produced in the intestine. Once activated, trypsin can cleave and activate further trypsinogen molecules as well as other zymogens, such as **chymotrypsinogen** and **proelastase** (Figure 6).

As well as synthesizing and secreting the zymogens, the pancreas also synthesizes a small (6 kDa) **trypsin-inhibitor protein** (**pancreatic trypsin inhibitor**). This inhibitor protein binds very tightly to the active site of trypsin, preventing the pancreas from being destroyed by prematurely activated trypsin molecules. If this safety mechanism fails, for example because of a blocked pancreatic duct, the zymogens can become activated and literally digest the pancreas, a condition known as **acute pancreatitis**.

Blood clotting cascade

Another example of the occurrence of inactive zymogens is found in the enzymes involved in the **blood clotting cascade**. Here the whole process of blood clotting is brought about by a **series of zymogen activations**.

Apoptosis

Apoptosis, or **programmed cell death**, is executed by a group of proteases called **caspases**. In response to specific pro-apoptotic signals, the inactive procaspases are proteolytically activated to their active form. These activated caspases then act on other procaspases, as well as on other cellular proteins, to bring about cell death.

Zymogen activation may produce a **large amplification** of the initial signal, as a single activated enzyme may act on many thousands of substrate molecules to bring about further activation. Since proteolytic cleavage does not require ATP, zymogen cleavage is

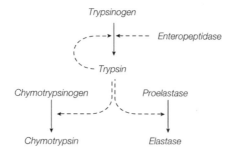

Figure 6. The central role of trypsin in activating the pancreatic zymogens.

a particularly appropriate mechanism for activation of proteins outside cells. However, unlike the covalent modification of an enzyme (see above), zymogen activation is not reversible. Once activated, the enzyme stays active.

Regulation of enzyme synthesis and breakdown

The amount of a particular enzyme present in a cell or tissue changes according to the rates of its **synthesis** and **degradation**.

Factors affecting the rate of synthesis include the level of **induction or repression of the gene** encoding the enzyme (Sections G3 and G4) and also the **rate of degradation of the mRNA** produced from that gene. Many key enzymes at control points in metabolic pathways have particularly short-lived mRNAs and the rate of enzyme synthesis is thus readily controlled by factors that affect the rate of gene transcription.

The rate of degradation of an enzyme is reflected in its **half-life** – the time taken for 50% of the protein to be degraded. Most enzymes that are important in metabolic regulation have short half-lives, and are termed **labile** enzymes.

E1 Membrane lipids

Key Notes

Membranes

Membranes form boundaries around the cell and around distinct subcellular compartments. They act as selectively permeable barriers and are involved in signaling processes. All membranes contain varying amounts of lipid and protein and some contain small amounts of carbohydrate.

Membrane lipids

In membranes, the three major classes of lipids are the glycerophospholipids, the sphingolipids and the sterols. The glycerophospholipids have a glycerol backbone that is attached to two fatty acid hydrocarbon chains and a phosphorylated headgroup, and include phosphatidylcholine, phosphatidylethanolamine and phosphatidylserine. The sphingolipids are based on sphingosine to which a single fatty acid chain is attached and either a phosphorylated headgroup (sphingomyelin) or one or more sugar residues (cerebrosides and gangliosides, the glycosphingolipids). The major sterol in animal plasma membranes is cholesterol, while the structurally related stigmasterol and β-sitosterol are found in plants.

Fatty acid chains

The fatty acid chains of glycerophospholipids and sphingolipids consist of long chains of carbon atoms, which are usually unbranched and have an even number of carbon atoms (e.g. palmitate C16, stearate C18). The chains are either fully saturated with hydrogen atoms or have one or more unsaturated double bonds that are in the *cis* configuration (e.g. oleate C18:1 with one double bond).

Lipid bilayer

Membrane lipids are amphipathic since they contain both hydrophilic and hydrophobic regions. In the glycerophospholipids and the sphingolipids the fatty acid hydrocarbon chains are hydrophobic whereas the polar headgroups are hydrophilic. In aqueous solution, the amphipathic lipids arrange themselves into extensive bimolecular sheets (bilayers). The structure of the bilayer is maintained by multiple noncovalent interactions between neighboring fatty acid chains and between the polar headgroups of the lipids. In biological membranes, there is an asymmetrical distribution of lipids between the inner and outer leaflets of the bilayer.

Membrane fluidity

Lipids are relatively free to move within the plane of the bilayer by either rotational or lateral motion, but do not readily flip from one side of the bilayer to the other (transverse motion). Increasing the length of the fatty acid chains or decreasing the number of unsaturated double bonds in the fatty acid chains leads to a decrease in the fluidity of the membrane. In animal membranes, increasing the amount of cholesterol also decreases the fluidity of the membrane.

Related topics	(A1) Prokaryotic cells	(E5) Signal transduction
	(A2) Eukaryotic cells	(E6) Nerve function
	(E2) Membrane structure	(K1) Fatty acid structure and
	(E3) Membrane transport:	function
	small molecules	(K5) Cholesterol

Membranes

Membranes form boundaries both around the cell (the plasma membrane) and around distinct subcellular compartments (e.g. nucleus, mitochondria, lysosomes, etc.) (Sections A1 and A2). They act as **selectively permeable barriers** allowing the inside environment of the cell or organelle to differ from that outside (Sections E3). Membranes are involved in **signaling processes**; they contain specific receptors for external stimuli and are involved in both chemical and electrical signal generation (Sections E5 and E6). All membranes contain two basic components: **lipids** and **proteins**. Some membranes also contain **carbohydrate**. The composition of lipid, protein and carbohydrate varies from one membrane to another. For example, the inner mitochondrial membrane has a larger amount of protein than lipid due to the presence of numerous protein complexes involved in oxidative phosphorylation and electron transfer (Section L2), whereas the myelin sheath membrane of nerve cells, which serves to insulate the cell electrically, has a larger proportion of lipid (Section E6).

Membrane lipids

Lipids were originally classified as biological substances that were insoluble in water but soluble in organic solvents such as chloroform and methanol. In addition to being structural components of membranes, lipids have several other biological roles. They serve as fuel molecules (Section K2), as concentrated energy stores (e.g. triacylglycerol, Section K4) and as signaling molecules (Section E5). Within membranes, there are three major types of lipid: the **glycerophospholipids**, the **sphingolipids** and the **sterols**.

Glycerophospholipids

The glycerophospholipids are made up of three components: a **phosphorylated headgroup**, a three-carbon **glycerol backbone** and two **hydrocarbon fatty acid chains** (Figure 1). The phosphorylated headgroup is attached to carbon-3 of the glycerol backbone, while the two fatty acid chains are attached to the other two carbon atoms. The simplest glycerophospholipid is phosphatidate (diacylglycerol 3-phosphate), which has only a phosphoric acid group esterified to carbon-3 of the glycerol. Although phosphatidate itself is present in small amounts in membranes, the major glycerophospholipids are derived from it. In these other lipids the phosphate is further esterified to the hydroxyl group of one of several alcohols (choline, ethanolamine, glycerol, inositol or serine). The major glycerophospholipids found in membranes include phosphatidylcholine, phosphatidylethanolamine, phosphatidylglycerol, phosphatidylinositol and phosphatidylserine (Figure 1). Diphosphatidylglycerol (or cardiolipin) (Figure 1) is found predominantly in the inner mitochondrial membrane.

Figure 1. Structures of membrane glycerophospholipids. R_1 and R_2 represent hydrocarbon chains of fatty acids. The glycerol backbone is shaded.

Sphingolipids

Sphingomyelins, the commonest sphingolipids, have a **sphingosine backbone** (Figure 2a) in place of the glycerol in glycerophospholipids. Like the glycerophospholipids, they also have a phosphorylated headgroup (either choline or ethanolamine) and two hydrocarbon chains (Figure 2a). One of the hydrocarbon chains comes from the sphingosine molecule, the other is a fatty acid as found in the glycerophospholipids except that it is bonded via an amide bond in sphingolipids. The sphingomyelins are present in the plasma membrane of most cells and are particularly abundant in the myelin sheath that surrounds nerve cells. The **glycosphingolipids**, such as the **cerebrosides** and **gangliosides**, are also derived from sphingosine, but in place of the phosphorylated headgroup they have one or more sugar residues. The galactocerebrosides have a single galactose residue (Figure 2a) and are found predominantly in the neuronal cell membranes of the brain. The gangliosides GM_1, GM_2 and GM_3 have several sugar residues including at least

(a)

Figure 2. Structures of (a) the sphingolipids sphingomyelin and galactocerebroside; (b) cholesterol. R_1 represents the hydrocarbon chain of fatty acids.

one sialic acid (*N*-acetylneuraminic acid) residue and are a major constituent of most mammalian plasma membranes, being particularly abundant in brain cells.

Sterols

The sterol **cholesterol** (Figure 2b) is a major constituent of animal plasma membranes but is absent from prokaryotes. The fused ring system of cholesterol means that it is more rigid than other membrane lipids. As well as being an important component of membranes, cholesterol is the metabolic precursor of the steroid hormones (Section K5). Plants contain little cholesterol but have instead a number of other sterols, mainly **stigmasterol** and **β-sitosterol**, which differ from cholesterol only in their aliphatic side-chains.

Fatty acid chains

The two fatty acid chains of glycerophospholipids and the single fatty acid chain in sphingolipids consist of long chains of carbon atoms. Usually these chains have an even number of carbon atoms (e.g. palmitate, C16; stearate, C18) and are unbranched. The chains can either be fully saturated with hydrogen atoms or unsaturated and have one or more double bonds that are usually in the *cis* configuration (e.g. oleate C18:1, which has 18 carbon atoms and one double bond; arachidonic acid C20:4, which has 20 carbon atoms and four double bonds) (for more details see Section K1). The two fatty acid chains on a glycerophospholipid are usually not identical [e.g. 1-stearoyl-2-oleoyl-3-phosphatidylcholine (Figure 3)]. In the sphingolipids, the unbranched hydrocarbon chain of sphingosine also consists of 18 carbon atoms but has a double bond in the *trans* configuration (Figure 2a).

$$H_3C - (CH_2)_{16} - \overset{\overset{\displaystyle O}{\|}}{C} - O - CH_2$$

$$H_3C - (CH_2)_7 - \underset{\underset{\displaystyle H}{|}}{C} = \underset{\underset{\displaystyle H}{|}}{C} - (CH_2)_7 - \overset{\overset{\displaystyle O}{\|}}{C} - O - CH$$

$$H_2C - O - \overset{\overset{\displaystyle O}{\|}}{\underset{\underset{\displaystyle O^-}{|}}{P}} - O - CH_2 - CH_2 - \overset{\overset{\displaystyle CH_3}{|}}{\underset{\underset{\displaystyle CH_3}{|}}{N^+}} - CH_3$$

Figure 3. Structure of 1-stearoyl-2-oleoyl-3-phosphatidylcholine showing the saturated stearoyl (C18:0) and the mono-unsaturated oleoyl (C18:1) hydrocarbon chains. The hydrogen atoms surrounding the C=C bond are in the *cis* configuration.

Lipid bilayer

Amphipathic (or amphiphilic) molecules contain both **hydrophilic** (water-loving) and **hydrophobic** (water-hating) regions. Membrane lipids are amphipathic molecules as they are made up of hydrophobic fatty acid chains and a hydrophilic polar headgroup. In the glycerophospholipids, the two hydrocarbon chains are hydrophobic whereas the glycerol backbone and the phosphorylated headgroup are hydrophilic. In the sphingolipids, the fatty acid chain and the hydrocarbon chain of the sphingosine are hydrophobic whereas the phosphorylated or sugar headgroup is hydrophilic. In the case of cholesterol, the entire molecule apart from the hydroxyl group on carbon-3 is hydrophobic in nature.

In aqueous solution, amphipathic molecules will orientate themselves in such a way as to prevent the hydrophobic region coming into contact with the water molecules. In the case of those fatty acids that contain only one hydrocarbon chain (such as sodium palmitate, a constituent of soap), the molecules form a spherical micellar structure (diameter usually >20 nm) in which the hydrophobic fatty acid chains are hidden inside the **micelle** and the hydrophilic headgroups interact with the surrounding water molecules (Figure 4a). Because the two fatty acid chains of phospholipids are too bulky to fit into the interior of a micelle, the favored structure for most phospholipids in aqueous solution is a two-dimensional bimolecular sheet or **lipid bilayer** (Figure 4b). Such lipid bilayers, in which the phospholipid molecules are orientated with their hydrophobic chains in the interior of the structure and their hydrophilic headgroups on the surfaces, can be relatively large structures of up to about 1 mm^2 in area but are only two molecules thick (6–10 nm). The two layers of lipids in the bilayer are referred to as the **inner and outer leaflets**. In biological membranes, the individual lipid species are **asymmetrically distributed** between the two leaflets. For example, in the plasma membrane of erythrocytes, sphingomyelin and phosphatidylcholine are preferentially located in the outer leaflet, whereas phosphatidylethanolamine and phosphatidylserine are mainly in the inner leaflet.

Figure 4. Structure of (a) a micelle and (b) a lipid bilayer.

Lipid bilayers will spontaneously self-assemble in aqueous solution. The major driving force behind this is the **hydrophobic effect** – the hydrophobic fatty acid chains avoid coming into contact with the water molecules. Once formed, the bilayer structure is maintained by multiple noncovalent interactions including hydrophobic interactions and van der Waals forces between the hydrocarbon chains, charge interactions and hydrogen bonding between the polar headgroups, and hydrogen bonding between the headgroups and the surrounding water molecules (see Section B2 for a fuller description of these noncovalent interactions).

Membrane fluidity

Because there are no covalent bonds between the lipids in the bilayer, the membrane is not a static structure but has **fluidity**. The lipids are generally free to move within the plane of the inner or outer leaflet of the bilayer by either **rotational** or **lateral movement** (Figure 5). However, they cannot readily flip from one leaflet of the bilayer to the other, so-called **transverse movement**, due to the unfavorable energetics involved in moving a hydrophilic headgroup through the hydrophobic interior of the bilayer.

The fluidity of the bilayer can be altered in a number of ways. Upon heating above a characteristic **transition temperature**, the lipid bilayer will change from a gel-like consistency to a more fluid-like consistency. This transition temperature depends on the length of the fatty acid chains and on their degree of unsaturation. If the length of the fatty acid chains is increased, the fluidity of the bilayer will decrease due to the larger propensity for noncovalent interactions between the hydrocarbon chains. In contrast, if the degree of unsaturation in the fatty acid chains is increased, the fluidity of the bilayer will increase. This is because the double bonds that are in the *cis* configuration kink the hydrocarbon chain and disrupt the highly ordered packing of the fatty acid chains, thus reducing the number of interactions between neighboring lipids. An important regulator of membrane fluidity in mammalian systems is **cholesterol**. At physiological temperature (37°C), increasing the amount of cholesterol in the bilayer will lead to a decrease in the fluidity of the membrane since the rigid steroid ring system interferes with the lateral movement of the fatty acid chains.

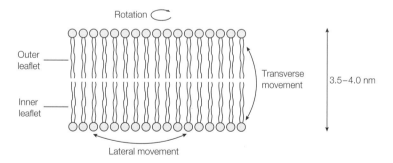

Figure 5. Movement of lipids in membranes.

E2 Membrane structure

Key Notes

Integral membrane proteins	Integral membrane proteins are associated with the hydrophobic core of the lipid bilayer. Most integral membrane proteins have one (e.g. glycophorin) or more (e.g. bacteriorhodopsin) regions of the polypeptide chain that traverse the lipid bilayer. These transmembrane regions consist mainly of amino acids with hydrophobic side-chains that fold into an α-helix and interact noncovalently with the surrounding lipids. A few transmembrane proteins traverse the membrane with a structure formed from β-strands. Some integral membrane proteins do not traverse the membrane but are attached covalently to a membrane lipid. Integral membrane proteins are asymmetrically distributed across the bilayer and are usually free to move in the plane of the bilayer but are not able to flip from one side of the membrane to the other.
Peripheral membrane proteins	No part of a peripheral membrane protein interacts with the hydrophobic interior of the bilayer. Instead, peripheral membrane proteins are bound to the surface of the membrane by noncovalent ionic and hydrogen bonds and can be removed by washing the membrane with a solution of either high ionic strength or high pH. The cytoskeleton that covers the cytosolic surface of the plasma membrane is made up of a number of peripheral proteins and is important in maintaining and altering the shape of the cell.
Fluid mosaic model of membrane structure	The fluid mosaic model describes the structure of biological membranes, in which the membranes are considered as two-dimensional solutions of orientated lipids and globular proteins. Many proteins are free to move laterally in the plane of the bilayer and this movement can be monitored using the technique fluorescence recovery after photobleaching (FRAP).
Lipid domains	Within biological membranes, lipids and proteins cluster together in discrete domains. Lipid rafts are domains of the plasma membrane that are enriched in cholesterol, sphingomyelin and glycosphingolipids, as well as lipid-modified proteins.
Membrane protein purification and reconstitution	The first step in the purification of integral membrane proteins is the disruption of the membrane structure by solubilizing it with a detergent (e.g. Triton X-100). Once solubilized, the hydrophobic region of the protein is coated with a layer of amphipathic detergent molecules, allowing the protein to remain in aqueous solution and be purified,

	as for a soluble globular protein. Once purified, integral membrane proteins can be reconstituted into artificial lipid vesicles (liposomes) in order to study their function.
Membrane carbohydrate	Sugar residues are found only on the extracellular side of biological membranes attached either to lipids to form glycolipids or to proteins to form glycoproteins. The carbohydrate forms a protective coat on the outer surface of the cell and is involved in intercellular recognition.
Related topics	(A2) Eukaryotic cells (E1) Membrane lipids (A4) Cell imaging (E5) Signal transduction (B2) Protein structure and (H5) Protein glycosylation function (C1) Protein purification

Integral membrane proteins

Membrane proteins are classified as either **peripheral** (extrinsic) or **integral** (intrinsic) depending on how tightly they are associated with the membrane. Integral membrane proteins are tightly bound to the membrane through interactions with the hydrophobic core of the bilayer (see Figure 4) and can be extracted from it only by using agents that disrupt the membrane structure, such as **organic solvents** (e.g. chloroform) or **detergents** (see below). Most integral membrane proteins have one or more regions of the polypeptide chain that span the lipid bilayer and interact noncovalently with the hydrophobic fatty acid chains. However, some are anchored in the membrane by a covalently attached fatty acid or hydrocarbon chain (see below). Like lipids, integral membrane proteins are **amphipathic**, having both hydrophobic and hydrophilic regions, and are **asymmetrically distributed** across the bilayer (Section E1).

Glycophorin – a transmembrane protein

Because **erythrocytes** (red blood cells) do not contain any subcellular organelles (they are essentially a membranous sac for carrying hemoglobin; Section B3), their plasma membrane is a convenient model system for studies of membrane structure as it can readily be isolated from other membranes and intracellular components. One of the major proteins in the plasma membrane of erythrocytes is **glycophorin A**; a 131-amino acid protein that was the first integral membrane protein to be sequenced (Section C3). This revealed that the polypeptide chain of glycophorin consists of three domains:

1. An N-terminal region on the extracellular side of the membrane rich in polar and charged residues that contains all the N- and O-linked **glycosylation sites** (Section H5)

2. A central region rich in nonpolar residues that is buried in the hydrophobic core of the bilayer

3. A C-terminal region rich in polar and charged residues that is exposed on the cytosolic side of the membrane (Figure 1a)

As with the majority of transmembrane proteins, the hydrophobic membrane-spanning region consists mainly of amino acid residues with hydrophobic side-chains that are folded in an α-**helical conformation** (Section B2). As each amino acid residue adds

Figure 1. Integral membrane proteins: (a) single membrane-spanning region (e.g. glycophorin); (b) multiple membrane-spanning regions (e.g. bacteriorhodopsin).

0.15 nm to the length of an α-helix, a helix of 25 residues would have a length of 3.75 nm, just enough to span the hydrophobic core of the bilayer. The hydrophobic side-chains of the residues in the helix protrude outwards from the helix axis to interact via hydrophobic bonds with the fatty acid chains. Either side of this hydrophobic α-helix are clusters of amino acids with charged side-chains, which interact noncovalently with opposite charges on the polar headgroups of the membrane lipids.

Multiple membrane-spanning proteins

Some transmembrane proteins have **multiple membrane-spanning α-helices**. **Bacterio-rhodopsin**, a protein found in a photosynthetic bacterium, captures energy from light and uses it to pump protons across the bacterial membrane. Like numerous other integral proteins, such as the **G protein-coupled receptors** (Section E5), the polypeptide chain of bacteriorhodopsin loops backwards and forwards across the lipid bilayer seven times (Figure 1b). Each of the **seven transmembrane α-helices** is linked to the next by a short hydrophilic region of the polypeptide chain that is exposed either on the extracellular or cytosolic side of the membrane. Other multiple membrane-spanning proteins have from two to up to 14 transmembrane helices. For example, the **anion exchange band 3 protein** of the erythrocyte plasma membrane that transports Cl⁻ and HCO_3^- across the membrane loops backwards and forwards across the lipid bilayer up to 14 times.

A few transmembrane proteins, for example the **porin protein** in the outer membrane of *E. coli* bacteria, do not contain α-helices but instead traverse the membrane with a structure formed from β-**strands**. Each β-strand is hydrogen bonded to its neighbor in an antiparallel arrangement, forming a single β-sheet (Section B2). This β-sheet curls up to form a hollow cylinder that forms a channel through the bilayer.

Lipid-anchored proteins

A significant number of integral membrane proteins in eukaryotes do not contain transmembrane regions but are anchored in one or other leaflet of the bilayer through covalent attachment to a hydrocarbon chain. Several proteins, including the **prion protein** (the causative agent of mad cow disease), are stably anchored at the cell surface through covalent linkage of their C-terminal amino acid to the headgroup of a phosphatidylinositol lipid via an ethanolamine–phosphate–trimannose bridge, so-called **glycosyl phosphatidylinositol (GPI)-anchored** proteins (Figure 2a). This complex structure is built up by sequential addition of the individual sugar residues and ethanolamine phosphate to phosphatidylinositol. A C-terminal hydrophobic signal peptide is removed from the protein in the lumen of the RER and the preformed GPI anchor added to the newly exposed C-terminal amino acid (see Section H4 for details).

Other proteins are transiently attached to the cytosolic face of the membrane either by amide linkage of a myristate (C14:0) molecule to an N-terminal Gly residue (**myristoylated** proteins; Figure 2b), or by thioether linkage of a 15-carbon **farnesyl** or a 20-carbon **geranylgeranyl** polyunsaturated hydrocarbon to a C-terminal Cys residue (**prenylated** proteins; Figure 2c). Farnesyl and geranylgeranyl are synthesized from

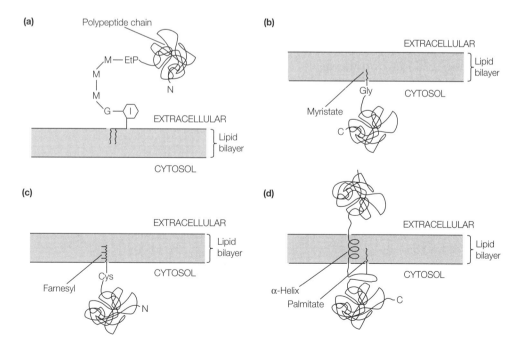

Figure 2. Lipid-modified proteins: (a) a glycosyl phosphatidylinositol-anchored protein (I, inositol; G, glucosamine; M, mannose; EtP, ethanolamine phosphate); (b) a myristoylated protein; (c) a prenylated protein; (d) a palmitoylated protein.

isopentenyl pyrophosphate, the precursor of cholesterol (Section K5). Some proteins are also modified on Cys residues with covalently attached palmitate (C16:0) (**palmitoylated proteins**). These include some with membrane-spanning polypeptides (Figure 2d), some prenylated proteins and some myristoylated proteins. Many of the proteins involved in cell signaling, such as the G proteins and the Ras family of proteins are lipid modified (Section E5).

Peripheral membrane proteins

Peripheral membrane proteins are less tightly bound to the lipid bilayer than integral membrane proteins and can be readily removed by washing the membranes with a solution of **high ionic strength** (e.g. 1 M NaCl) or **high pH**. These procedures leave the lipid bilayer intact but disrupt the ionic and hydrogen bond interactions that hold the peripheral proteins on the surface of the membrane. No part of a peripheral membrane protein interacts with the hydrophobic core of the bilayer. Peripheral membrane proteins can be found either on the outer or the inner surface of the bilayer and can be associated with the membrane through noncovalent interactions with either the lipid headgroups and/or other proteins in the membrane (see Figure 4). Once removed from the membrane, peripheral proteins behave as water-soluble globular proteins and can be purified as such (Section C1).

Cytoskeleton

The cytosolic surface of the erythrocyte plasma membrane is covered by a network of peripheral membrane proteins that make up the **cytoskeleton** (Figure 3) (Section A2). The major component of this cytoskeleton is **spectrin**, which folds into a triple-stranded **α-helical coiled-coil** to form long chains. The spectrin chains are attached to the plasma membrane through interactions with two other peripheral proteins, **ankyrin** and **protein band 4.1**. Ankyrin forms a cross-link between spectrin and the cytosolic domain of the integral anion exchanger band 3 protein, while band 4.1 promotes the binding of **actin filaments** (Section A2) to the spectrin chains linking them to the cytosolic domain of glycophorin. The cytoskeleton gives the erythrocyte plasma membrane great strength and flexibility, and is important in maintaining and altering the shape of the cell. In other mammalian cells, the cytoskeleton has similar functions but consists of numerous other proteins and criss-crosses throughout the cytoplasm (Section A2).

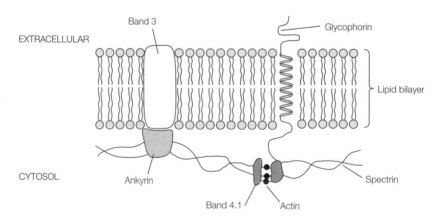

Figure 3. The erythrocyte cytoskeleton.

Fluid mosaic model of membrane structure

In 1972, S. Jonathan Singer and Garth Nicholson proposed the **fluid mosaic model** for the overall structure of biological membranes, in which the membranes can be viewed as two-dimensional solutions of orientated lipids and globular proteins (Figure 4). The integral membrane proteins can be considered as 'icebergs' floating in a two-dimensional lipid 'sea'. They proposed that the bilayer organization of the lipids would act both as a solvent for the amphipathic integral membrane proteins and as a **permeability barrier**. They also proposed that some lipids may interact with certain membrane proteins, that these interactions would be essential for the normal functioning of the protein, and that membrane proteins would be free to diffuse laterally in the plane of the bilayer unless restricted in some way, but would not be able to flip from one side of the bilayer to the other. This model is now supported by a wide variety of experimental observations.

Integral membrane protein movement and distribution

Many proteins are free to **move laterally** in the plane of the bilayer. One experiment used to show this involved fusing cultured mouse cells with human cells under appropriate conditions to form a hybrid cell known as a **heterokaryon** (Figure 5a). The mouse cells were labeled with mouse protein-specific antibodies to which the green-fluorescing dye **fluorescein** had been covalently attached, whilst the human cells were labeled with the red-fluorescing dye **rhodamine** (Section A4). Upon cell fusion, the mouse and human proteins as seen under the **fluorescence microscope** (Section A4) were segregated on the two halves of the heterokaryon (Figure 5a). After 40 min at 37°C, however, the mouse and human proteins had completely intermingled. Lowering the temperature to below 15°C inhibited this process, indicating that the proteins are free to diffuse laterally in the membrane and that this movement is slowed as the temperature is lowered. It should be noted, though, that some integral membrane proteins are not free to move laterally in the membrane because they interact with the cytoskeleton inside the cell (see below).

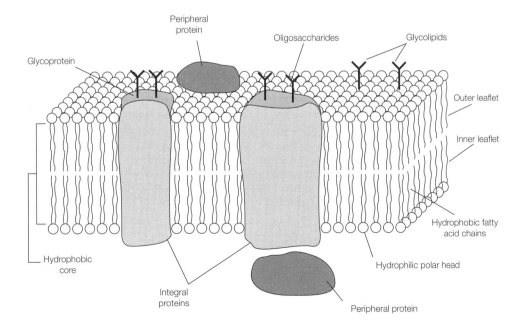

Figure 4. The fluid mosaic model of membrane structure.

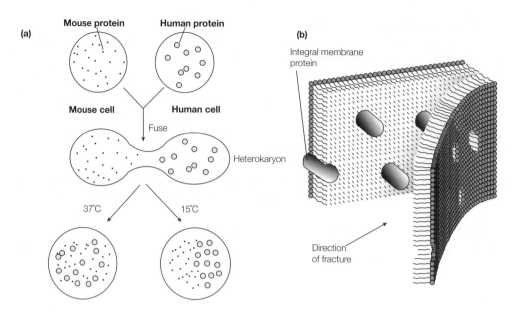

Figure 5. (a) Movement and (b) distribution (as shown by freeze–fracture electron microscopy) of integral membrane proteins.

The lateral movement of proteins in a membrane can also be visualized by fluorescence microscopy (Section A4) through the use of the technique **fluorescence recovery after photobleaching** (**FRAP**). First, the membrane protein is specifically labeled with a fluorescent compound (Section A4) and then, while visualizing the cell surface through a fluorescence microscope, the fluorescent molecules in a small region are destroyed (bleached) by a very intense light pulse from a laser. The fluorescence of this bleached region is subsequently monitored as a function of time, as other fluorescently labeled proteins move into the bleached region of the membrane. The rate of recovery of the fluorescence depends on the lateral mobility of the fluorescently labeled protein.

The distribution of proteins in membranes can be revealed by electron microscopy using the **freeze–fracture technique** (Figure 5b). In this technique, a membrane specimen is rapidly frozen to the temperature of liquid nitrogen (–196°C) and then fractured by a sharp blow. The bilayer often splits into monolayers, revealing the interior. The exposed surface is then coated with a film of carbon and shadowed with platinum in order for the surface to be viewed in the electron microscope (Section A4). The fractured surface of the membrane is revealed to have numerous randomly distributed protuberances that correspond to integral transmembrane proteins.

Lipid domains

Biological membranes are not just homogeneous mixtures of lipids and proteins. Within them are discrete **domains** in which certain lipids and proteins cluster together to form structural and functional units. For example, the clustering together of **cholesterol**, **sphingomyelin** and **glycosphingolipids** in the outer leaflet of the bilayer, with other lipids in the inner leaflet gives rise to **lipid rafts**. Various lipid-modified proteins, such as GPI-anchored proteins in the outer leaflet and myristoylated and palmitoylated proteins

(a)

(b)

Figure 6. Structures of (a) Triton X-100 and (b) octyl glucoside.

in the inner leaflet, associate with lipid rafts and these structures have been implicated in various cellular processes, including **cell signaling** (Section E5) and the transport of proteins and lipids from the Golgi to the plasma membrane.

Membrane protein purification and reconstitution

The first step in the **purification** of an integral membrane protein is to disrupt its inter-actions with other proteins and the lipids in the membrane. This is commonly achieved by adding a **detergent**, which solubilizes the membrane. In order to solubilize the mem-brane but not denature the protein, gentle detergents such as Triton X-100 or octyl gluco-side are used (Figure 6), rather than the harsh detergent SDS. As the detergent molecules are themselves amphipathic, they readily intercalate into the lipid bilayer and disrupt the hydrophobic interactions.

Once solubilized, the hydrophobic region of the integral protein is coated with a layer of detergent molecules, which enables the protein to remain in solution (Figure 7a). The solubilized protein can then be purified as for a water-soluble globular protein (Section C1) as long as detergent is kept in the buffers to prevent aggregation and loss of the pro-tein. Once purified, an integral membrane protein can be reincorporated (reconstituted) into artificial lipid vesicles (**liposomes**) in order to study its function (Figure 7b). If phos-pholipids are added to the protein in detergent solution and the detergent dialyzed away, phospholipid vesicles containing the protein will spontaneously form. The function of the protein can then be studied. For example, if the **Ca^{2+}-ATPase** is reconstituted into lipid vesicles, its function (i.e. transport of Ca^{2+} upon ATP hydrolysis) can be studied by monitoring Ca^{2+} on the inside of the vesicle upon addition of Ca^{2+} and ATP to the outside (Figure 7b).

Membrane carbohydrate

The extracellular surface of the plasma membrane is often covered with a protective coat of **carbohydrate**. The sugar residues of this carbohydrate coat can be attached either to certain lipids such as the **glycosphingolipids** (Section E1), or to the polypeptide chains of peripheral or integral membrane proteins. These **glycolipids** and **glycoproteins** are abundant in the plasma membrane of eukaryotic cells but are virtually absent from most intracellular membranes, particularly the inner mitochondrial membrane and the chloroplast lamellae. In glycoproteins the sugar residues can be attached to the poly-peptide chain either through the hydroxyl group in the side-chain of Ser or Thr residues as **O-linked oligosaccharides**, or through the amide group in the side-chain of Asn as **N-linked oligosaccharides** (Section H5). The carbohydrate on the extracellular face of the membrane not only serves a **protective role** but is also involved in **intercellular rec-ognition** and in maintaining the asymmetry of the membrane.

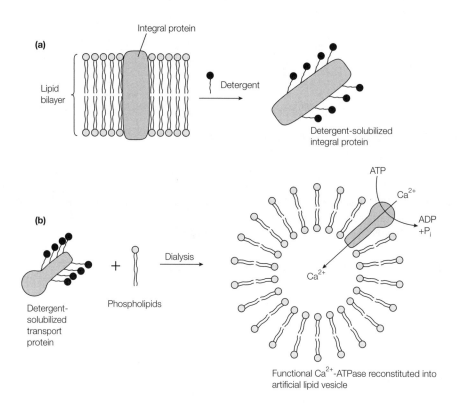

Figure 7. (a) Detergent solubilization and (b) reconstitution into artificial lipid vesicles of an integral membrane protein.

E3 Membrane transport: small molecules

Key Notes

Membrane permeability

The plasma membrane is a selectively permeable barrier. Some small, lipophilic molecules can pass directly through the bilayer unaided (nonmediated transport), whereas hydrophilic/charged molecules require the presence of integral membrane transport proteins (mediated transport).

Passive transport

The movement of molecules across a membrane by passive transport does not require an input of metabolic energy. The molecule moves from a high concentration to a lower concentration.

Simple diffusion

Passive transport by simple diffusion (e.g. of water, gases, urea, steroid hormones) does not require the presence of integral membrane proteins and the rate of movement is directly proportional to the concentration gradient of the molecule across the membrane.

Facilitated diffusion

Passive transport by facilitated diffusion requires the presence of specific integral membrane proteins (channels, transporters) to facilitate the movement of the molecule (e.g. glucose, other sugars, amino acids) across the membrane. The transport protein (e.g. the erythrocyte glucose transporter) is specific for a particular molecule. In some cells, water passes through channel proteins called aquaporins.

Active transport

Active transport of a molecule across a membrane against its concentration gradient requires an input of metabolic energy.

ATP-driven active transport (primary active transport)

In the case of ATP-driven active transport, the energy required for the transport of the molecule (Na^+, K^+, Ca^{2+} or H^+) across the membrane is derived from the coupled hydrolysis of ATP (e.g. Na^+/K^+-ATPase), an example of a P-type ATPase. The cardiac glycosides, digitalin and ouabain, which are widely used in the treatment of congestive heart failure, inhibit the Na^+/K^+-ATPase.

Ion-driven active transport (secondary active transport)

In ion-driven active transport, the movement of the molecule to be transported across the membrane is coupled to the movement of an ion (e.g. Na^+, H^+) down its concentration gradient. If both the molecule to be transported and the ion move in the same direction across the membrane, the process is called symport (e.g. $Na^+/$glucose transporter); if the

	molecule and the ion move in opposite directions it is called antiport (e.g. Na^+–Ca^{2+} exchanger).
Glucose transport into intestinal epithelial cells	The transport of glucose across the polarized epithelial lining cells of the intestine involves its transport across the apical membrane by the Na^+/glucose symporter, with the energy for the movement of the glucose coming from the movement of Na^+ down its concentration gradient. The concentration of Na^+ ions inside the cell is maintained at a low level by the action of a Na^+/K^+-ATPase on the basolateral membrane. The glucose then exits from the cell across the basolateral membrane by facilitated diffusion through the glucose transporter. The movement of Na^+ and glucose across the cell sets up a difference in osmotic pressure causing water to follow by simple diffusion, which forms the basis of oral rehydration therapy.
Related topics	(E1) Membrane lipids (E6) Nerve function (E2) Membrane structure

Membrane permeability

A pure phospholipid bilayer, with its hydrophobic interior (Section E1), is **permeable** to water, gases (O_2, CO_2, N_2), small uncharged polar molecules (e.g. urea, ethanol) and larger lipophilic molecules (e.g. steroid hormones). However, it is **impermeable** to large uncharged polar molecules (e.g. glucose), ions (Na^+, K^+, Cl^-, Ca^{2+}) and charged polar molecules (e.g. amino acids, ATP, glucose 6-phosphate). The first group of molecules can cross a biological membrane unaided and without an input of energy (**nonmediated transport**), whereas the latter group require the presence of **integral membrane transport proteins** and, in some cases, an **input of energy** to travel through the otherwise impermeable membrane barrier (**mediated transport**). Thus the plasma membrane and the membranes of internal organelles are **selectively permeable barriers**, maintaining a distinct internal environment.

Passive transport

The passive transport of molecules across a membrane does not require an input of metabolic energy. The rate of transport (diffusion) is proportional to the concentration gradient of the molecule across the membrane. There are two types of passive transport: **simple diffusion** and **facilitated diffusion**.

Simple diffusion

Only relatively small uncharged or hydrophobic molecules (H_2O, O_2, CO_2, other gases, urea and ethanol) or larger lipophilic molecules (e.g. steroid hormones, which are derivatives of cholesterol; Section K5) cross the lipid bilayer by simple diffusion. No membrane proteins are involved, so there is no specificity. The molecule in aqueous solution on one side of the membrane dissolves into the lipid bilayer, crosses it, and then dissolves into the aqueous solution on the opposite side. The rate of diffusion is directly proportional to the concentration gradient of the molecule across the membrane and the process is not saturable (Figure 1a).

Figure 1. Kinetics of (a) simple and (b) facilitated diffusion.

Facilitated diffusion

Unlike simple diffusion, the facilitated (or carrier-mediated) diffusion of a molecule across a biological membrane is dependent on **specific integral membrane proteins** (Section E2), variously called **channels**, **carriers**, **permeases** and **transporters**. The molecule to be transported binds to the integral membrane protein on one side of the membrane; the protein then undergoes a **conformational change**, transports the molecule across the membrane and then releases it on the other side. Molecules transported across membranes in this way include hydrophilic molecules such as glucose, other sugars and amino acids. The transport proteins are **specific** for one particular molecule or a group of structurally similar molecules. The transport proteins are capable of being saturated, display Michaelis–Menten-type binding kinetics (K_m and V_{max}) (Figure 1b), and are influenced by temperature, pH and inhibitor molecules in a similar manner to enzymes (Sections D1 and D3).

An example of facilitated diffusion is the uptake of **glucose** into erythrocytes by the glucose transporter. The **erythrocyte glucose transporter** or **uniporter** (**GLUT1**) is an integral membrane protein of mass 45 kDa that is asymmetrically orientated in the plasma membrane. This uniporter protein structure traverses the membrane with 12 α-helices (Section E2), which form a central pore through which the glucose molecule is passed upon **conformational changes** in the protein (Figure 2). All the steps in the transport of glucose into the cell are **freely reversible**, the direction of movement of glucose being dictated by the relative concentrations of glucose on either side of the membrane. In order to maintain the concentration gradient across the membrane, the glucose is rapidly phosphorylated inside the cell by hexokinase to glucose 6-phosphate (Section J3), which is no longer a substrate for the glucose transporter. The erythrocyte glucose transporter is highly specific for D-glucose (K_m=1.5 mM), the nonbiological L-isomer being transported at a barely measurable rate. D-Mannose and D-galactose, which differ from D-glucose in the configuration at one carbon atom (Section J1), are transported at intermediate rates. Thus the transporter has a higher affinity for glucose than for other sugars.

Although **water** readily traverses membranes because of its small size, high concentration and lack of a complete charge, many cells, such as erythrocytes and those in the kidney, contain water-channel proteins, **aquaporins**, which accelerate the osmotic flow of water. Each aquaporin protein is a tetramer of identical 28 kDa subunits, with each subunit containing six transmembrane α-helices. The water molecules move through a central pore in each subunit. Aquaporins allow cells to move large quantities of water rapidly across their plasma membrane.

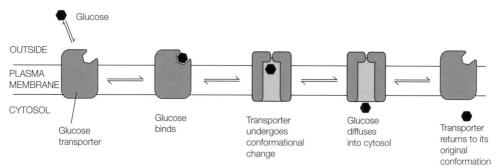

Figure 2. Facilitated diffusion of glucose into erythrocytes.

Active transport

Active transport, in which a molecule is transported from a low to a high concentration, requires an **input of metabolic energy**. This can be derived either from direct coupling to the **hydrolysis of ATP** or by coupling to the **movement of an ion** down its concentration gradient.

ATP-driven active transport (primary active transport)

In this case, the energy required for the transport of the molecule across the membrane is derived from the **coupled hydrolysis of ATP**. Several families of ATP-dependent transporters have been identified, including the **P-type ATPases** and **ABC transporters**. A well-studied example of a P-type ATPase is the **Na⁺/K⁺-ATPase** involved in the movement of Na⁺ and K⁺ ions across the plasma membrane of eukaryotes. All cells maintain a high internal concentration of K⁺ and a low internal concentration of Na⁺. The resulting **Na⁺/K⁺ gradient** across the plasma membrane is important for the active transport of certain molecules, and the maintenance of the **membrane electrical potential** (Section E6). The **cystic fibrosis transmembrane conductance regulator** (**CFTR**) is an example of an ABC transporter. This protein, which is defective in individuals with the inherited disease cystic fibrosis, transports Cl⁻ out of the cell. The movement across the membrane of Na⁺, K⁺, Ca²⁺ and H⁺, as well as a number of other molecules, is directly coupled to the hydrolysis of ATP.

Structure and action of the Na⁺/K⁺-ATPase

The Na⁺/K⁺-ATPase is an integral membrane protein consisting of a 110 kDa α subunit containing the ion binding sites and a 55 kDa β subunit of unknown function, that form a heterotetramer $(\alpha\beta)_2$ (Figure 3). Upon hydrolysis of one molecule of ATP to ADP and P_i (the P_i transiently binds to an aspartyl residue in the protein), the protein undergoes a conformational change and three Na⁺ ions are pumped out of the cell across the plasma membrane and two K⁺ ions are pumped in the opposite direction into the cell. Both ions are being moved up their concentration gradients across the membrane; hence the requirement for an input of energy. No transport occurs unless ATP is hydrolyzed, and no ATP is hydrolyzed if there is no Na⁺ and K⁺ to transport (i.e. it is a **coupled system**). The **cardiac glycosides** or **cardiotonic steroids**, **digitalin** (found in extracts of purple foxglove leaves) and **ouabain**, inhibit the Na⁺/K⁺-ATPase. The resultant diminished Na⁺ gradient inhibits the Na⁺–Ca²⁺ exchanger (see below), resulting in an increase in intracellular Ca²⁺, which enhances the contractility of heart muscle. These drugs are widely used in the treatment of **congestive heart failure**.

Figure 3. The Na^+/K^+-ATPase.

Ion-driven active transport (secondary active transport)

In this case, the movement of the molecule to be transported across the membrane is coupled to the **movement of an ion**, usually either Na^+ or H^+. The energy for the movement of the molecule across the membrane against its concentration gradient comes from the movement of the ion down its concentration (electrochemical) gradient. If both the molecule and the ion move in the same direction, it is termed **symport**, and the protein involved in the process is called a **symporter** (e.g. Na^+/glucose transporter; Figure 4a); if the molecule and the ion move in opposite directions, it is termed **antiport**, and the protein involved in the process is called an **antiporter** (e.g. the Na^+–Ca^{2+} exchanger; Figure 4b).

Glucose transport into intestinal epithelial cells

The cells lining the lumen of the intestine are **polarized**, that is they have two distinct sides or domains that have different lipid and protein compositions. The **apical** or **brush border** membrane facing the lumen is highly folded into **microvilli** to increase the surface area available for the absorption of nutrients. The rest of the plasma membrane, the **basolateral** surface, is in contact with neighboring cells and the blood capillaries (Figure 5). Movement between adjacent epithelial cells is prevented by the formation of **tight junctions** around the cells near the apical domain. Thus any nutrient molecules in the lumen of the intestine have to pass through the cytosol of the epithelial cell in order to enter the blood.

Glucose (or other sugars and amino acids) is transported across the apical membrane from a relatively low concentration in the lumen of the intestine to a relatively high concentration in the cytosol of the epithelial cell by a **Na^+/glucose symporter** (Figure 5). This is a form of **ion-driven (secondary) active transport**; the energy for the movement of glucose against its concentration gradient comes from the movement of Na^+ down its concentration gradient. The blood flowing through the capillaries on the basolateral side

Figure 4. Ion-driven cotransport mechanisms. (a) Symport process involving a symporter (e.g. Na^+/glucose transporter); (b) antiport process involving an antiporter (e.g. Na^+–Ca^{2+} exchanger).

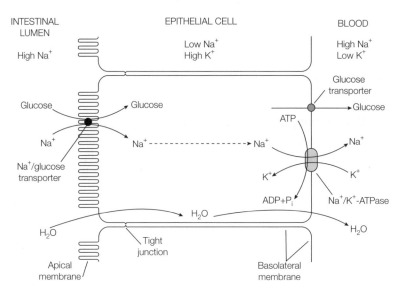

Figure 5. Transport of glucose, Na^+ and water across intestinal epithelial cells.

of the epithelial cell maintains a concentration gradient of glucose across this membrane allowing the glucose to move out of the cell by **facilitated diffusion** through a **glucose transporter (a uniporter)**, which is similar to the erythrocyte glucose transporter (see above). The relatively low concentration of Na^+ inside the epithelial cell is maintained by a **Na^+/K^+-ATPase** (see above) on the basolateral membrane, an example of **ATP-driven (primary) active transport** (Figure 5).

Oral rehydration therapy

In **oral rehydration therapy** (**ORT**) a solution of glucose and salt (NaCl) is administered to the patient as a remedy for dehydration. This is a simple, inexpensive but extremely important treatment, which has saved the lives of many infants in developing countries who would have otherwise died of the effects of dehydration, usually associated with diarrhea. Glucose enhances Na^+ uptake through the $Na^+/$glucose symporter, which in turn enhances water movement by osmosis across the intestinal epithelial cells into the blood. Simply drinking water is ineffective as the large intestine is usually secreting instead of absorbing both water and Na^+ ions, which are then reabsorbed. However, in someone suffering from diarrhea the liquid secreted into the intestine passes through the gut so quickly that very little Na^+ is reabsorbed, leading to very low Na^+ levels in the body.

E4 Membrane transport: macromolecules

Key Notes

Exocytosis

Exocytosis is the movement of macromolecules (proteins, hormones, neurotransmitters) out of the cell across the plasma membrane into the extracellular space. All cells continuously secrete proteins via the constitutive secretory pathway, although many cells (e.g. of the pancreas, nerve cells) also secrete proteins via the regulated secretory pathway in response to certain stimuli. Proteins destined to be secreted are synthesized on ribosomes bound to the RER membrane and are then transported in membrane-bound vesicles to the Golgi apparatus where they are sorted and packaged up into secretory vesicles. Coated vesicles transport material around the cell and the protein coats are made of clathrin, COPI or COPII. The SNARE proteins mediate the fusion of vesicles with their target membranes.

Endocytosis

Endocytosis is the uptake of macromolecules from the extracellular space into the cell across the plasma membrane via the formation of an intracellular vesicle pinching off from the plasma membrane.

Phagocytosis

Phagocytosis is the uptake of large particles (e.g. bacteria and cell debris). The particle binds to the surface of the phagocytic cell and the plasma membrane then engulfs the particle and ingests it via the formation of a large endocytic vesicle, a phagosome. Most protozoa utilize phagocytosis as a form of feeding, whereas in multicellular organisms only a few specialized cells (e.g. macrophages and neutrophils) can undergo phagocytosis.

Receptor-mediated endocytosis

Receptor-mediated endocytosis, a form of pinocytosis, is the selective uptake of extracellular macromolecules through their binding to specific cell surface receptors. The receptor–macromolecule complex then accumulates in clathrin-coated pits and is endocytosed via a clathrin-coated vesicle. The polypeptides of clathrin form a three-legged structure known as a triskelion, which assembles into a basket-like convex framework that causes the membrane to invaginate at that point and eventually to pinch off and form a vesicle (endosome). These clathrin-coated vesicles migrate into the cell where the clathrin coats are lost to form early endosomes. As the endosomes mature and eventually fuse with lysosomes, their contents are acidified and the cell surface receptors recycled to the plasma membrane. Cholesterol, in the form of low-density lipoproteins, is taken up into mammalian cells by receptor-mediated endocytosis.

Related topics	(E2) Membrane structure	(H4) Protein targeting
	(E5) Signal transduction	(K6) Lipoproteins
	(E6) Nerve function	

Exocytosis

A cell often needs to secrete larger molecules than can be accommodated by the transport systems dealt with in Section E3. **Exocytosis** refers to the movement of macromolecules such as proteins, hormones, neurotransmitters, etc., out of the cell across the plasma membrane into the extracellular space. **Membrane-bound vesicles** containing the molecule to be secreted fuse with the plasma membrane to release their contents into the extracellular space. In the **constitutive secretory pathway** (Figure 1), vesicles destined for the plasma membrane leave the Golgi apparatus in a steady stream. The membrane proteins and lipids in these vesicles provide new material for the plasma membrane, while the soluble proteins, neurotransmitters, etc. inside the vesicles are secreted to the extracellular space. All cells have this constitutive secretory pathway.

Proteins destined to be secreted from the cell are translated on ribosomes attached to the RER (Section H4). Vesicles containing these proteins then bud off from the RER, migrate through the cytosol and fuse with the membrane of the **Golgi apparatus** (Figure 1). Further vesicles bud off and move the proteins through the different compartments of the Golgi, where various post-translational modifications to the proteins take place, such

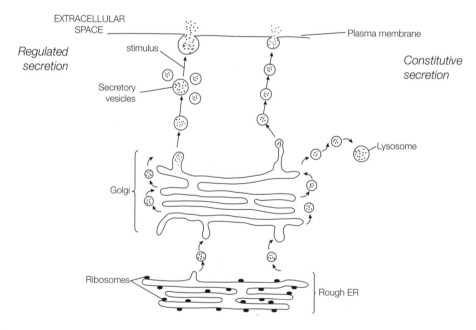

Figure 1. Exocytosis of proteins by the constitutive and regulated secretory pathways.

as O-linked glycosylation (Section H5). The proteins are packaged into discrete vesicles that bud off from the **trans-Golgi network** and are then targeted to the appropriate part of the cell (Section H4). For example, vesicles containing secretory proteins will be targeted to the plasma membrane, those containing lysosomal proteins will be targeted to the lysosomes.

In certain cells, however, an additional secretory pathway exists, the **regulated secretory pathway** (Figure 1). This pathway is found mainly in cells that are specialized for secreting products rapidly on demand in response to a particular stimulus. For example, the hormone insulin and digestive enzymes are secreted by the pancreas, while neurotransmitters are secreted by nerve cells (Section E6). In such secretory cells, the secreted substances are initially stored in **secretory vesicles**, which form by budding from the trans-Golgi network.

Coated vesicles and coat proteins

The vesicles transporting material around the cell are known as **coated vesicles** because they have specific proteins covering their outer (cytosolic) surface. These **coat proteins** act as a flexible scaffold, promoting vesicle formation. The three known types of coated vesicles are characterized by their protein coats:

1. **Clathrin** forms a polyhedral framework around vesicles that transport material from the Golgi to the plasma membrane, as well as from the plasma membrane into the cell (see below).

2. **COPI** (for *co*at *p*rotein) coats vesicles that are involved in both forward (**anterograde**) and backward (**retrograde**) transport between successive Golgi compartments.

3. **COPII** coats vesicles that move from the RER to the Golgi.

Proteins mediate vesicle fusion

On arriving at its target membrane, a vesicle fuses with it and releases its contents on the opposite side of the target membrane (Figure 1). This fusion process is mediated by integral or lipid-linked membrane proteins known as **SNAREs**. **R-SNAREs** or **v-SNAREs** (which contain conserved Arg residues and include **syntaxin**) usually associate with the vesicle membrane, while **Q-SNAREs** or **t-SNAREs** (which contain conserved Gln residues and include **synaptobrevin** and **SNAP-25**) associate with target membranes. Different SNAREs are proteolytically cleaved by the **tetanus and botulinum toxins** that are responsible for the fatal infectious diseases tetanus and botulism, respectively.

Endocytosis

Endocytosis is the uptake of extracellular macromolecules across the plasma membrane into the cell. The material to be ingested is progressively enclosed by a small portion of the plasma membrane, which first invaginates and then pinches off to form an intracellular vesicle containing the ingested molecule.

Phagocytosis

Phagocytosis is the endocytosis of large particles such as bacteria, cell debris or even intact cells via large endocytic vesicles called **phagosomes**. In order to be ingested the particle must first bind to the surface of the phagocyte, usually through specialized cell surface receptors. Once bound to the cell surface, the phagocyte is stimulated to begin engulfing the particle with its plasma membrane, thereby enclosing it within a

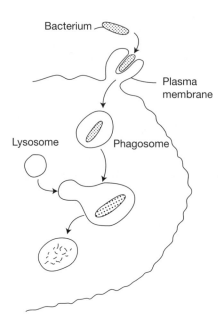

Figure 2. Phagocytosis.

phagosome (Figure 2). The phagosome then fuses with a lysosome (Section A2) and the ingested particle is broken down. Utilizable material will be transported into the cytosol, while indigestible substances will remain in the lysosomes, forming residual bodies. In protozoa, phagocytosis is a form of feeding, where the ingested material is broken down in the lysosomes and utilized as food. In multicellular organisms, only a few specialized cells are capable of phagocytosis. **Macrophages** and **neutrophils** (white blood cells) use phagocytosis to protect the organism against infection by ingesting invading microorganisms. Macrophages are also involved in scavenging dead and damaged cells and cell debris.

Receptor-mediated endocytosis

Receptor-mediated endocytosis, a form of **pinocytosis** (cell-drinking), is the selective uptake of macromolecules from the extracellular fluid, which requires specific cell surface receptors and usually involves clathrin-coated pits and vesicles. This process, which takes place in most animal cells, involves the macromolecule binding specifically to a **cell surface receptor** (usually an integral membrane protein; Sections E2 and E5). Once bound to the receptor, the receptor–macromolecule complex accumulates in a **clathrin-coated pit**. With the aid of the membrane-associated GTP-binding protein, **dynamin**, these pits bud from the membrane to form small **clathrin-coated vesicles** containing the receptors and their bound macromolecules (ligands) (Figure 3).

Receptor-mediated endocytosis provides a way of selectively concentrating particular macromolecules that are at low concentrations in the extracellular fluid, thereby increasing the efficiency of their uptake without having to take in large quantities of extracellular fluid. One of the best-studied and understood receptor-mediated endocytic processes is the uptake into mammalian cells of **cholesterol** in the form of **low-density lipoprotein (LDL) particles** by the **LDL-receptor** (Section K6).

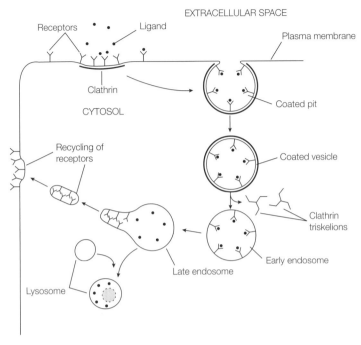

Figure 3. Receptor-mediated endocytosis involves clathrin-coated pits and vesicles.

Many **viruses** and other **toxins** gain entry to animal cells by receptor-mediated endocytosis. Although the cells do not purposely have cell surface receptors that recognize the viral particle, the virus has evolved to express a protein on its surface that mimics the correct ligand recognized by the receptor, thus allowing the virus to bind and be internalized.

Clathrin-coated pits and vesicles

Electron microscopy (Section A4) has revealed that clathrin-coated pits are invaginations of the plasma membrane that are coated on their cytosolic surface with a densely packed material made up predominantly of the protein **clathrin** (Figure 3). This protein, which has been highly conserved throughout evolution, consists of three large and three small polypeptide chains that together form a three-legged structure called a **triskelion**. These clathrin triskelions assemble into a basket-like convex framework of hexagons and pentagons to form the coated pits. The assembly of the clathrin coat is thought to drive the membrane to invaginate at that point. As further clathrin triskelions are added to the structure, a complete cage forms, pinching off a region of the membrane and forming a clathrin-coated vesicle. In endocytosis these vesicles then migrate into the cell and shed their clathrin coat to form **early endosomes**, which in turn mature to **late endosomes** (Figure 3). The pH in the interior of the endosomes decreases due to proton pumps in their membrane acidifying the vesicle lumen. This change in pH often promotes the dissociation of the molecule (ligand) from the receptor, allowing the receptors to be recycled back to the plasma membrane. The endosomes then fuse with **lysosomes** where the contents are degraded and the component parts, such as amino acids and sugars, may be taken up into the cytosol to be used by the cell.

E5 Signal transduction

Key Notes

Cell signaling

Cells communicate with one another in multicellular organisms using extracellular signaling molecules or hormones. The hormone is secreted by the signaling cell and then binds to a receptor on the target cell, initiating a response in that cell. In endocrine signaling the hormone acts at a distant site in the body from where it was produced, in paracrine signaling the hormone acts on nearby cells, and in autocrine signaling the hormone acts on the same cell from which it was secreted.

Signaling molecules with intracellular receptors

Some lipophilic hormones (e.g. the steroid hormones) diffuse across the plasma membrane and interact with intracellular receptors of the nuclear receptor family in the cytosol or nucleus. The gas nitric oxide (NO) is a signaling molecule that crosses the plasma membrane and stimulates the intracellular enzyme guanylyl cyclase to produce cyclic guanosine monophosphate (cGMP).

Signaling molecules with cell surface receptors

Other lipophilic hormones (e.g. the prostaglandins) and hydrophilic hormones (e.g. the peptide hormones insulin and glucagon and the biogenic amines epinephrine and histamine) bind to cell surface receptors. These are integral membrane proteins located in the plasma membrane that bind the hormone (ligand) with high affinity and specificity. On binding the ligand, the receptor may undergo a conformational change or dimerize and transmit the information into the cell (signal transduction).

Enzyme-linked receptors

Enzyme-linked receptors (e.g. the insulin receptor) have an intrinsic enzyme activity. Ligand binding causes autophosphorylation of tyrosine residues in the cytoplasmic domain of these receptor tyrosine kinases. These modified tyrosines are then recognized by other proteins in the cytosol. Other receptors have serine/threonine kinase activity, while some lack intrinsic kinase activity and associate with cytoplasmic tyrosine kinases (e.g. Src family kinases). Many enzyme-linked receptors interact through small binding domains (e.g. SH2) with scaffold proteins inside the cell, which organize groups of proteins into signaling complexes.

Ion channel-linked receptors

Ion channel-linked receptors change conformation to allow ions to flow across the membrane thereby altering the membrane potential.

G protein-coupled receptors	G protein-coupled receptors (GPCRs) contain seven transmembrane α-helices and activate trimeric G [guanosine triphosphate (GTP)-binding] proteins that in turn lead to the production of an intracellular second messenger. Trimeric G proteins contain three subunits: α, β and γ. Different GPCRs interact with different trimeric G proteins leading to different intracellular signaling events. There are two main classes of GTPase switch proteins: the Gα subunit of the trimeric G proteins and the monomeric G proteins (e.g. Ras). They are attached to the cytosolic face of the plasma membrane by a covalently attached lipid and cycle through an inactive GDP bound form and an active GTP bound form.
Second messengers	Intracellular signaling molecules (second messengers) are produced in response to the activation of various cell surface receptors. The second messengers cAMP and cGMP are produced by adenylyl cyclase and guanylyl cyclase, respectively. Activation of phospholipase C leads to the production of the second messengers inositol 1,4,5-trisphosphate (IP_3) and 1,2-diacylglycerol, which in turn cause the release of Ca^{2+} from intracellular stores and activate protein kinase C, respectively.
Calcium ions	Many extracellular signals induce an increase in the level of Ca^{2+} ions in the cytosol. When triggered Ca^{2+} channels, such as voltage-dependent Ca^{2+} channels in the plasma membrane, IP_3-gated Ca^{2+} channels and ryanodine receptors in the ER and sarcoplasmic reticulum, allow Ca^{2+} ions to flow into the cytosol. The rise in cytosolic Ca^{2+} results in altered cellular events through Ca^{2+}-binding proteins, such as troponin C in muscle and calmodulin.
Regulated proteolysis	Some cell surface receptors undergo regulated intramembrane proteolysis upon binding ligand, with the release of the cytoplasmic domain, which translocates to the nucleus.
Related topics	(E2) Membrane structure (J7) Control of glycogen (E6) Nerve function metabolism

Cell signaling

In multicellular organisms, there is a need for the cells to **communicate** with one another in order to **coordinate** their growth and metabolism. The principal way by which cells communicate with each other is by means of **extracellular signaling molecules** or **hormones**. These molecules are synthesized and secreted by **signaling cells** and produce a specific response in **target cells** that have **specific receptors** for the signaling molecule. Different cells can respond differently to the same signaling molecule depending on the type of receptor and the intracellular reactions initiated (see below). Cell signaling can be classified into three distinct types based on the distance over which the signaling molecule acts. In **endocrine signaling**, the signaling molecule (e.g. insulin) acts on target

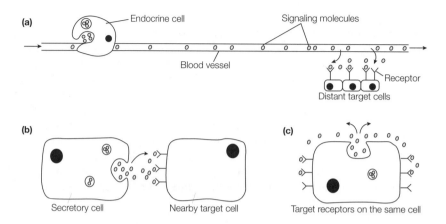

Figure 1. Cellular signaling: (a) endocrine; (b) paracrine; (c) autocrine.

cells distant from its site of synthesis in cells of an endocrine organ (e.g. the pancreas; Figure 1a). The endocrine cells secrete the signaling molecule into the bloodstream (if an animal) or the sap (if a plant), which carries it to the target cells elsewhere in the organism. In **paracrine signaling**, the signaling molecule affects only target cells close to the cell from which it was secreted (Figure 1b). The communication from one nerve cell to another by chemical neurotransmitters is an example of paracrine signaling (Section E6). The third type of cell signaling is **autocrine signaling**, where a cell responds to a molecule that it has produced itself (Figure 1c).

Signaling molecules with intracellular receptors

Small lipophilic (lipid-soluble) hormones diffuse across the plasma membrane and then interact with **intracellular receptors** in the cytosol or nucleus. The receptors are all structurally related, being part of the **nuclear receptor superfamily**. The resulting hormone–receptor complex often binds to regions of the DNA and affects the transcription of certain genes (Section G6). Small lipophilic hormones with intracellular receptors include the **steroid hormones**, which are synthesized from cholesterol (Section K5) (e.g. the female sex hormones estrogen and progesterone), **thyroxine**, which is produced by thyroid cells and is the principal iodinated compound in animals, **retinoic acid**, which is derived from vitamin A, and **vitamin D**, which is synthesized in the skin in response to sunlight (Section K5).

An important and remarkable example of a small signaling molecule that passes readily across the target cell plasma membrane is the gas **nitric oxide** (NO). This is used in both animals and plants. NO is synthesized by the deamination of arginine, catalyzed by the enzyme **NO synthase** (Section M3). NO rapidly diffuses out of the cell where it is produced and into neighboring cells. It only acts locally as it has a short half-life, about 5–10 s. In many target cells, NO binds to the active site of **guanylyl cyclase**, stimulating the enzyme to produce the small intracellular mediator cGMP (see below). **Nitroglycerine**, which is used to treat patients with **angina** (pain resulting from inadequate blood flow to the heart muscle), is converted to NO, which relaxes the blood vessels thereby reducing the workload on the heart. **Carbon monoxide** (CO) is another gas that is used as a signaling molecule, again by stimulating guanylyl cyclase.

Signaling molecules with cell surface receptors

Hydrophilic (water-soluble) molecules (which cannot diffuse across the hydrophobic interior of the lipid bilayer) bind to cell surface receptors. These include the **peptide hormones**, such as **insulin** and **glucagon**, and the small charged molecules, often **biogenic amines**, such as **epinephrine** (**adrenaline**) and **histamine**, that are derived from amino acids and function as hormones and neurotransmitters (Section E6). Some lipophilic (lipid-soluble) hormones also bind to cell surface receptors. These include the **prostaglandins**, a family of structurally similar compounds that are found in both vertebrates and invertebrates. Prostaglandins are synthesized from **arachidonic acid** (a 20-carbon fatty acid with four unsaturated double bonds) (Section K1) and act as paracrine signaling molecules. **Aspirin** and other **anti-inflammatory agents** inhibit the synthesis of prostaglandins (Section K1).

Cell surface receptors are **integral membrane proteins** situated in the plasma membrane (Section E2) that bind the signaling molecule (**ligand**) with high affinity. The ligand binds to a specific site on the receptor in much the same way as a substrate binds to an enzyme (Section D1). Binding of the ligand to the receptor may cause a **conformational change** in the receptor or promote **dimerization** of two receptors that initiates a sequence of reactions in the target cell (often referred to as **signal transduction**) leading to a change in cellular function. The distribution of receptors varies on different cells, and there is often more than one type of receptor for a particular ligand, allowing different target cells to respond differently to the same signaling molecule. Cell surface receptors can be classified into three major classes depending on how they transfer the information from the ligand to the interior of the cell: enzyme-linked receptors, ion channel-linked receptors and G protein-coupled receptors.

Enzyme-linked receptors

Numerous receptors have intrinsic or tightly associated **enzyme activity**. On binding of the ligand to their extracellular face, such receptors undergo a conformational change and activate an **enzyme activity**. In the case of the **insulin receptor**, which is a complex of two α- and two β-subunits held together by disulfide bonds, the polypeptide hormone insulin (the ligand) binds to the extracellular face of the α-subunits (Figure 2). The receptor then undergoes a conformational change leading to the **autophosphorylation** (self-phosphorylation) of the cytosolic domain of the β-subunit. Specifically the hydroxyl groups in the side-chains of certain **tyrosine** residues are phosphorylated, with **ATP** being the phosphate donor. The phosphorylated receptor is then recognized by other proteins in the cytosol that in turn modulate various intracellular events, allowing the cell to respond to the hormone appropriately (Section J7). In addition, the β-subunit can directly phosphorylate other target proteins within the cell.

The insulin receptor is an example of a **receptor tyrosine kinase**, while the **transforming growth factor-β** (TGF-β) family of receptors have **serine/threonine kinase activity** in their cytosolic domain. Other receptors with tightly associated enzyme activity include various **cytokine receptors** that bind interferons, growth hormone, some interleukins and other cytokines. For many enzyme-linked receptors, ligand binding induces the **oligomerization** (formation of dimers or higher oligomers) and it is this rearrangement of the cytosolic domains that enables the neighboring kinase domains of the receptor chains to cross-phosphorylate each other in the process of **autophosphorylation**.

Some enzyme-linked receptors, as well as other receptors, interact with **scaffold proteins** inside the cell, which organize groups of proteins into **signaling complexes** (Figure 3).

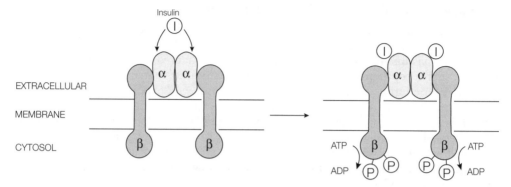

Figure 2. Signal transduction through an enzyme-linked receptor such as the insulin receptor.

The proteins within these signaling complexes assemble through the interactions of a variety of highly conserved, small binding domains such as the **Src homology 2 (SH2) domains** and **phosphotyrosine-binding (PTB) domains** that bind to phosphorylated tyrosine residues, **Src homology 3 (SH3) domains** that bind to short proline-rich amino acid sequences, and **pleckstrin homology (PH) domains** that bind to the headgroups of inositol phospholipids that have been additionally phosphorylated by **phosphatidylinositol 3-kinase (PI 3-kinase)**. Some scaffold proteins contain multiple **PDZ domains**, each of which binds to a specific motif on a receptor or signaling protein. The binding of these scaffold proteins to the activated receptor may help to relay the signal onward or may decrease the signaling process, providing negative feedback.

Some cell surface receptors require tyrosine phosphorylation for their activity and yet lack a tyrosine kinase domain. These receptors act through **cytoplasmic tyrosine kinases** (or nonreceptor tyrosine kinases), which associate with the receptor and phosphorylate

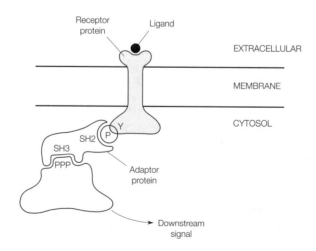

Figure 3. A hypothetical signaling complex. Binding of ligand to the extracellular face of the cell surface receptor results in phosphorylation of a tyrosine residue in its cytosolic domain. The phosphorylated tyrosine residue is recognized by an SH2 domain in the adaptor protein. Elsewhere in the adaptor protein is an SH3 domain that binds to a proline-rich sequence (PPP) in another signaling protein, such that the signal is relayed into the cell.

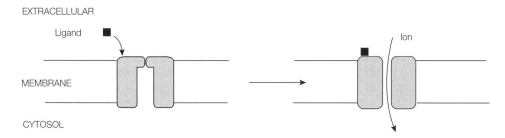

Figure 4. Signal transduction through an ion channel-linked receptor.

various target proteins. The largest family of cytoplasmic tyrosine kinases is the **Src family** that includes Src, Yes, Fyn and Lck, which all contain SH2 and SH3 domains and are located on the cytoplasmic surface of the plasma membrane.

Ion channel-linked receptors

Ion channel-linked receptors (transmitter-gated ion channels or ionotrophic receptors) are involved in the rapid synaptic signaling between electrically excitable cells. Here, binding of the ligand causes a conformational change in the protein such that a specific **ion channel** is opened (Figure 4). This allows a certain ion to flow through that subsequently alters the **electric potential** across the membrane. For example, at the nerve–muscle junction the **neurotransmitter acetylcholine** binds to specific receptors that allow Na^+ ions to flow into, and K^+ ions out of, the target cell (Section E6 for more detail).

G protein-coupled receptors

G protein-coupled receptors (**GPCRs**) form a very large group of cell surface receptors that are coupled to signal-transducing trimeric G proteins. All GPCRs contain **seven membrane-spanning** α-helical regions with their N-terminus on the extracellular face of the plasma membrane and their C-terminus on the cytoplasmic face (Figure 5). The GPCR family includes receptors for numerous hormones and neurotransmitters, light-activated receptors (rhodopsins) in the eye, and thousands of odorant receptors in the mammalian nose. On binding its ligand, a GPCR activates the signal-transducing **trimeric G proteins** [**guanosine triphosphate (GTP)-binding proteins**], which in turn activate or inhibit an effector protein. The trimeric G proteins consist of three subunits: α, β and γ. The Gα subunit is a **GTPase switch protein** that alternates between an active (on) state and an inactive (off) state (see below). In the human genome there are multiple copies of each of the α, β and γ subunits, providing diversity in the signaling through GPCRs.

Figure 5. GPCRs contain seven transmembrane α-helical regions (cylinders 1–7), with their N-terminus on the extracellular side of the membrane and their C-terminus in the cytosol.

The Gα and Gγ subunits are linked to the cytosolic surface of the plasma membrane by covalently attached lipids (Section E2). In the resting state, when no ligand is bound to the GPCR, the Gα subunit is bound to GDP and complexed with Gβγ (Figure 6). Binding of the ligand to the GPCR changes its conformation, causing it to bind to the Gα subunit in such a way that the GDP is displaced and GTP becomes bound. The Gα subunit then dissociates from Gβγ, but both remain anchored to the membrane. The Gα subunit with GTP bound then interacts with and activates an associated **effector protein**, such as adenylyl cyclase, or in some cases regulates the opening of an ion channel causing a change in the membrane potential. However, this activation is short-lived, as the GTP is rapidly, within seconds, hydrolyzed to GDP by the intrinsic GTPase activity in the Gα subunit (see below). The Gα subunit, now with GDP bound, dissociates from the effector protein, deactivating it, and reassociates with Gβγ, ready for another round of activation and nucleotide exchange (Figure 6).

The hormone **epinephrine** binds to several different GPCRs. On binding to β-**adrenergic receptors** on the surface of liver and adipose cells, epinephrine promotes glycogenolysis and lipolysis, respectively (Sections J7 and K4). On smooth muscle cells lining the blood vessels in the intestine, skin and kidneys, epinephrine binds to the α_2-**adrenergic receptor** causing the arteries to constrict. The β-adrenergic receptors are coupled to a **stimulatory G protein** (G_s) that activates the membrane-bound adenylyl cyclase, whereas the α_2-adrenergic receptor is coupled to an **inhibitory G protein** (G_i) that inhibits adenylyl cyclase. Thus through binding to different receptors and activating different G proteins one ligand can trigger a variety of different actions in different target cells.

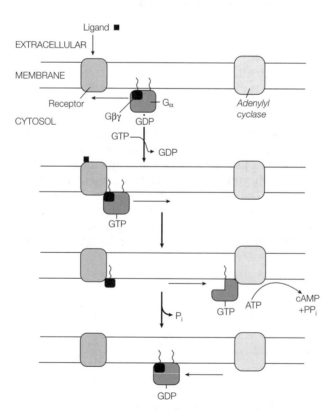

Figure 6. Signal transduction through a G protein-coupled receptor (see text for details).

GTPase switch proteins

The **GTPase family** of proteins is a group of **intracellular switch proteins** (or on–off molecular switches) of which there are two main classes: the Gα subunit of the **trimeric G proteins** (see above) and the **monomeric G proteins** such as the **Ras, Rho and Rab families**. Ras, Rho and Rab proteins that couple to the activated receptor through adaptor scaffold proteins (see above) act as transducers and bifurcation signaling proteins, changing the nature of the signal and sending it along multiple downstream pathways. In the case of Ras, a downstream serine/threonine phosphorylation cascade is activated, that includes the **mitogen-activated protein kinase (MAP-kinase)**. Like the Gα subunits, the Ras proteins are attached to the cytosolic face of the plasma membrane by a covalently attached lipid group (Section E2). When the GTPase has guanosine diphosphate (GDP) bound, it is in the 'off' state. Activation, via a cell surface receptor or a **guanine nucleotide-exchange factor** (GEF), leads to the GDP being exchanged for GTP, converting the GTPase to the 'on' state (Figure 7). The activated GTPase with its bound GTP then dissociates from the receptor and binds to and activates an effector enzyme (e.g. adenylyl cyclase), which in turn catalyzes the formation of a second messenger (e.g. cAMP). The GTPase then hydrolyzes the bound GTP, causing it to revert to the 'off' state (Figure 7). **Cholera toxin** acts by inhibiting the intrinsic GTPase activity (Figure 7), with the result that once activated to the GTP-bound state the GTPase cannot be turned off again.

Second messengers

The binding of ligands to many receptors leads to a short-lived increase in the concentration of certain **intracellular signaling molecules** called **second messengers**. (The hormone/ligand can be considered as the first messenger.) The major second messengers are 3′,5′-cyclic AMP (**cAMP**), 3′,5′-cyclic GMP (**cGMP**), **inositol 1,4,5-trisphosphate** (IP_3), **1,2-diacylglycerol** (DAG) and **Ca^{2+}**. The elevation in the level of one or other of these second messengers then leads to a rapid alteration in cellular function.

cAMP and cGMP are derived from ATP and GTP by the actions of **adenylyl cyclase** and **guanylyl cyclase**, respectively. For example, the action of epinephrine and glucagon on glycogen metabolism is mediated through the second messenger cAMP, which in turn activates the **cAMP-dependent protein kinase**, **protein kinase A** (Figure 3 in Section J7). cAMP and cGMP are short-lived as they are rapidly broken down by **phosphodiesterases**.

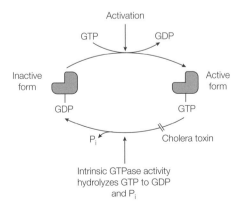

Figure 7. Cycling of the GTPase switch proteins between the active and inactive forms.

IP$_3$ and DAG are derived from the membrane lipid **phosphatidylinositol 4,5-bisphosphate** (which is a phosphorylated derivative of phosphatidylinositol; Section E1) by the action of **phospholipase C**, which is also located in the plasma membrane and, like adenylyl cyclase, is activated via G proteins by GPCRs (Figure 8). One of the main actions of the polar IP$_3$ is to diffuse through the cytosol and interact with Ca^{2+} channels in the membrane of the ER (Figure 8), causing the release of stored Ca^{2+} ions, which in turn mediate various cellular responses. The DAG produced by the hydrolysis of phosphatidylinositol 4,5-bisphosphate, along with Ca^{2+} ions released from the ER, activates **protein kinase C**, a membrane-bound serine/threonine protein kinase, that phosphorylates various target proteins, again leading to alterations in a variety of cellular processes (Figure 8).

Calcium ions

Many extracellular signals induce an increase in the level of **Ca^{2+} ions** in the cytosol. For example, in muscle cells, Ca^{2+} triggers contraction. The concentration of Ca^{2+} in the cytosol is usually kept very low (approximately 0.1 µM), whereas its concentration in the extracellular fluid and in the lumen of the ER is high (approximately 1 mM). Thus there is a **gradient of Ca^{2+} ions** across the plasma membrane and ER membrane, such that when **Ca^{2+} channels** in these membranes are triggered to open, Ca^{2+} ions rapidly flow into the cytosol, raising the Ca^{2+} concentration by 10–20-fold and triggering Ca^{2+}-responsive proteins inside the cell. There are three main types of Ca^{2+} channel: **voltage-dependent Ca^{2+} channels** in the plasma membrane that open in response to membrane depolarization, for example in nerve cells (Section E6); **IP$_3$-gated Ca^{2+} channels** in the ER (see above); and **ryanodine receptors** (so called because they are inhibited by the plant alkaloid ryanodine) that release Ca^{2+} from the sarcoplasmic reticulum in muscle cells (Section B5) or the ER of other cells. Ca^{2+} pumps, such as the Ca^{2+}-ATPase, in the ER and plasma membrane help to maintain the low concentration of Ca^{2+} ions in the cytosol of resting cells. **Ca^{2+}-binding proteins** serve as transducers of the cytosolic Ca^{2+} signal. These Ca^{2+}-binding proteins include **troponin C** in skeletal muscle (Section B5) and **calmodulin**, a ubiquitous protein found in all eukaryotic cells. Calmodulin

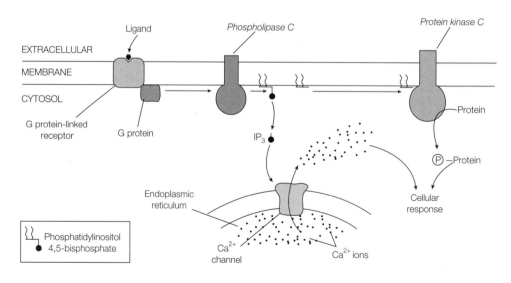

Figure 8. Generation of the intracellular second messengers IP$_3$, DAG and Ca^{2+}.

functions as a multipurpose intracellular Ca^{2+} receptor, mediating many Ca^{2+}-regulated processes, and undergoes a conformational change upon binding Ca^{2+}. The activated calmodulin can then bind to a number of different target proteins, including a family of **Ca^{2+}/calmodulin-dependent protein kinases (CaM kinases)**, which then phosphorylate serines or threonines on other proteins.

Regulated proteolysis

There are several unusual signaling pathways for relaying signals from cell surface receptors to the interior of the cell that involve **regulated proteolysis**. Amongst these pathways is that mediated by the receptor protein **Notch** and the pathway activated by secreted **Hedgehog** proteins that have been highly conserved across evolution and play crucial roles in animal development. In the case of the transmembrane Notch receptor, binding of its **ligand Delta** to the extracellular face leads to a proteolytic cleavage first in the region adjacent to the membrane and then a second cleavage within the hydrophobic transmembrane region (Figure 9). The released cytoplasmic domain then migrates to the nucleus where it activates the transcription of various target genes (Figure 9). Proteolytic cleavage within the hydrophobic transmembrane region is unusual but more and more proteins are being identified that are subject to this **regulated intramembrane proteolysis** (RIP). As peptide bonds in the receptor protein are cleaved in this process, the receptor cannot be re-used.

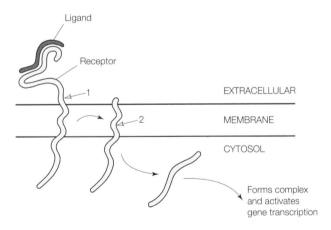

Figure 9. Cell signaling through the Notch receptor. Binding of ligand results in proteolytic cleavage of the receptor (1) on the extracellular face of the membrane. The resulting membrane-bound stub is then cleaved within the transmembrane domain (2), releasing the cytosolic tail, which forms a complex with other proteins and activates gene transcription in the nucleus.

E6 Nerve function

Key Notes

Nerve cells	Nerve cells, or neurons, consist of a cell body from which the dendrites and axon extend. The dendrites receive information from other cells; the axon passes this information on to another cell, the post-synaptic cell. The axon is covered in a myelin membranous sheath except at the nodes of Ranvier. The axon ends at the nerve terminal where chemical neurotransmitters are stored in synaptic vesicles for release into the synaptic cleft.
The action potential	An electric membrane potential exists across the plasma membrane due to the unequal distribution of Na^+ and K^+ ions, which is generated by the Na^+/K^+-ATPase. Upon stimulation, neurons depolarize their membrane potential from the resting state (–60 mV) to +40 mV, generating an action potential. The action potential is caused by Na^+ ions flowing into the cell through voltage-sensitive Na^+ channels. The resting membrane potential is restored by K^+ ions flowing out of the cell through voltage-sensitive K^+ channels. The poison tetrodotoxin acts by blocking the Na^+ channel.
Neurotransmitters	Chemical neurotransmitters, such as acetylcholine, the biogenic amines and small peptides, are stored in the pre-synaptic nerve terminal in synaptic vesicles. When the action potential reaches the nerve terminal, it causes the synaptic vesicles to fuse with the plasma membrane in a Ca^{2+}-dependent manner and to release their contents by exocytosis. The neurotransmitter then diffuses across the synaptic cleft, binds to specific receptors on the post-synaptic cell membrane and initiates a response in that cell.
Related topics	(E2) Membrane structure (E4) Membrane transport: (E3) Membrane transport: macromolecules small molecules (E5) Signal transduction

---

Nerve cells

In eukaryotes, probably the most rapid and complex signaling is mediated by **nerve impulses**. Nerve cells (**neurons**) consist of a cell body with numerous projections of the plasma membrane, called **dendrites** (Figure 1). These interact with other cells and receive information from them in the form of nerve impulses. The cell body then assimilates the information derived from a number of dendritic contacts and passes on the information as another nerve impulse down the large **axon** (Figure 1). The axon ends at the **synapse** where it makes contact with the **post-synaptic** (target) **cell**. The axon is covered in places by a membranous **myelin** sheath, made up mainly of the lipid sphingomyelin (Section E1), which acts as an electrical insulator, enabling the nerve impulses to be transmitted

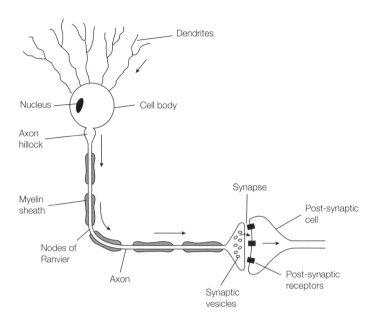

Figure 1. Schematic diagram of a typical nerve cell.

over long distances, sometimes more than 1 m in larger animals. Every millimeter or so along the axon, the myelin sheath is interrupted by unmyelinated regions called the **nodes of Ranvier** (Figure 1). The end of the axon, the nerve terminal, is full of **synaptic vesicles** that store the chemical **neurotransmitters**, such as acetylcholine. When a nerve impulse reaches the nerve terminal, the synaptic vesicles release their contents into the **synaptic cleft**, the space between the pre- and post-synaptic cells. The neurotransmitter then diffuses across the space and interacts with receptors on the surface of the post-synaptic cell, causing a signal to be transduced in that cell.

The action potential

An electric potential (the **membrane potential**) exists across the plasma membrane of all cells. Most cells are electrically inactive, as this membrane potential does not vary with time. However, neurons and muscle cells are **electrically active** as their membrane potential *can* vary with time. In all cells, the membrane potential is generated through the action of the **Na⁺/K⁺-ATPase** (Section E3), with a high concentration of K^+ inside the cell and a high concentration of Na^+ outside. Neurons vary their electric potential by controlled changes in the **permeability** of the plasma membrane to Na^+ and K^+ ions. Upon stimulation, the membrane potential of a neuron rises rapidly from the **resting potential** of –60 mV (millivolts) to approximately +40 mV (Figure 2a); the membrane is said to **depolarize** and an **action potential** is generated. In order for this to occur, the membrane potential has to be depolarized beyond a critical **threshold level** (approximately –40 mV). With time, the membrane potential returns to the resting potential. The action potential is propagated along the axon starting from the axon hillock (Figure 1).

The action potential arises from large, transient changes in the permeability of the plasma membrane of the neuron to Na^+ and K^+ ions. Two types of **voltage-sensitive ion channels** are present in the membrane: one is selectively permeable to Na^+ ions,

Figure 2. The action potential: (a) depolarization of the membrane potential; (b) changes in the permeability of the plasma membrane to Na+ and K+.

the other to K+ ions (Figure 3a). These **integral membrane proteins** (Section E2) are sensitive to the membrane potential, undergoing **conformational changes** as the potential alters. First, the conductance of the membrane to Na+ changes. Depolarization of the membrane beyond the threshold level causes a conformational change in the Na+ channel, allowing Na+ ions to flow down their concentration gradient from the outside of the cell into the interior (Figure 3b). The entry of Na+ further depolarizes the membrane, causing more Na+ channels to open, resulting in a rapid influx of Na+ and a change in the membrane potential from –60 mV to +40 mV in a millisecond. The Na+ channels then spontaneously close, and the K+ channels open, allowing K+ ions to flow out of the cell and restore the negative resting potential within a few milliseconds (Figure 3c). The wave of depolarization is propagated along the axon by the opening of Na+ channels on the

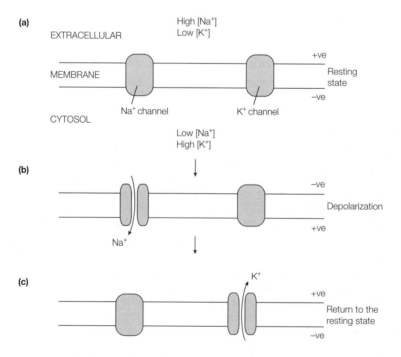

Figure 3. Mechanism of depolarization of the nerve membrane by the opening and closing of selective Na+ and K+ ion channels.

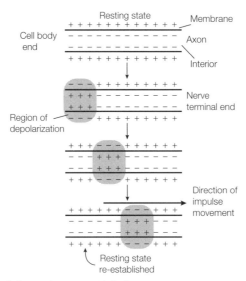

Figure 4. Propagation of the action potential along an axon.

nerve terminal side of the initial depolarized region (Figure 4). The action potential can only move in that direction, as the Na^+ channels have a **refractory period** when they are insensitive to further stimulation. Only approximately one in a million of the Na^+ and K^+ ions in a neuron flow across the plasma membrane during the action potential. Thus, this is a very efficient way of signaling over long distances. The activity of a single channel can be measured using the **patch-clamp technique** in which a clean glass pipette with a tip diameter of approximately 1 µm is pressed against an intact cell to form a seal.

The neurotoxin, **tetrodotoxin**, a highly potent poison from the puffer fish, blocks the conduction of nerve impulses along axons and so leads to respiratory paralysis by binding very tightly to the Na^+ channel and blocking its action.

Neurotransmitters

When the action potential reaches the nerve terminal, it causes the release of a **chemical neurotransmitter** from the synaptic vesicles. The mammalian nervous system employs numerous substances as neurotransmitters. These include the **amino acids** glutamate and glycine, **acetylcholine**, the **biogenic amines** such as **epinephrine** and **dopamine**, and a variety of **small peptides** such as the **enkephalins**. For example, acetylcholine is stored in **synaptic vesicles**, a specialized form of secretory vesicle, and is released into the **synaptic cleft** by **exocytosis** (Section E4) in a Ca^{2+}-dependent manner (Figure 5). The acetylcholine molecules then diffuse across to the plasma membrane of the **post-synaptic cell** where they bind to specific **receptors**. The acetylcholine receptor is a 250 kDa complex of four polypeptide chains that forms a **gated channel** through the membrane (Section E5). On binding of two acetylcholine molecules, the channel opens, allowing Na^+ and K^+ ions to flow in and out of the cell, respectively. The resulting depolarization of the post-synaptic membrane initiates a new action potential in that cell. The acetylcholine in the synaptic cleft is rapidly broken down by the enzyme **acetylcholinesterase**, which is the target of compounds such as **diisopropylphosphofluoridate** (see Section D4 for its structure and mechanism of action), used as a component of some nerve gases.

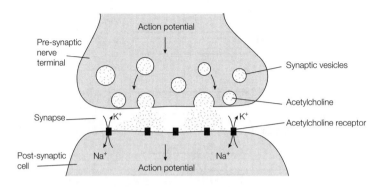

Figure 5. Release of a neurotransmitter into the synaptic cleft.

F1 An introduction to DNA

Key Notes

Bases	In DNA, there are four bases: adenine (abbreviated A), guanine (G), thymine (T) and cytosine (C). Adenine and guanine are purines; thymine and cytosine are pyrimidines.
Nucleosides	A nucleoside is a pyrimidine or purine base covalently bonded to a sugar. In DNA, the sugar is deoxyribose and so this is a deoxyribonucleoside. There are four types of deoxyribonucleoside in DNA; deoxyadenosine, deoxyguanosine, deoxythymidine and deoxycytidine.
Nucleotides	A nucleotide is base+sugar+phosphate covalently bonded together. In DNA, where the sugar is deoxyribose, this unit is a deoxyribonucleotide.
3′5′ Phosphodiester bonds	In DNA, the nucleotides are covalently joined together by 3′5′ phosphodiester bonds to form a repetitive sugar–phosphate chain, which is the backbone to which the bases are attached.
DNA sequence	The DNA sequence is the sequence of A, C, G and T along the DNA molecule, which carries the genetic information.
DNA double helix	In a DNA double helix, the two strands of DNA are wound round each other with the bases on the inside and the sugar–phosphate backbones on the outside. The two DNA chains are held together by hydrogen bonds between pairs of bases; adenine (A) always pairs with thymine (T), and guanine (G) always pairs with cytosine (C).
Related topics	(F3) DNA replication in bacteria (F4) DNA replication in eukaryotes (G1) An introduction to RNA (G2) Transcription in prokaryotes (G4) Transcription in eukaryotes: an overview

Bases

The bases in DNA have carbon–nitrogen ring structures; because of the nitrogen atoms, they are called nitrogenous bases. There are two types of ring structure. **Adenine** and **guanine** are **purines** (Figure 1a), each having two joined carbon–nitrogen rings but with different side-chains. **Thymine** and **cytosine** are **pyrimidines** (Figure 1a); each has only one carbon–nitrogen ring and again they differ in their side-chains.

(a)

Adenine (A)　　　　　　Guanine (G)　　　　　　Cytosine (C)　　　　　　Thymine (T)

(b)

Deoxyribose　　　　　　Deoxycytidine　　　　　　　　Deoxyadenosine

(c)

Deoxyadenosine 5'-triphosphate; dATP

Figure 1. (a) The purines, adenine and guanine, and the pyrimidines, thymine and cytosine; (b) deoxyribose and two deoxyribonucleosides, deoxycytidine and deoxyadenosine; (c) a deoxyribonucleotide, deoxyadenosine 5' triphosphate (dATP).

Nucleosides

In RNA, the nucleosides have ribose as the sugar component (Section G1) and so are ribonucleosides. In DNA the sugar is deoxyribose (Figure 1b) (i.e. the 2'-OH group in ribose is replaced by a hydrogen atom; hence 'deoxy') and so the nucleosides are **deoxyribonucleosides**. For DNA, these are **deoxyadenosine, deoxyguanosine, deoxythymidine** and **deoxycytidine**. In each case, the C-1 of the sugar is joined to the base via one of its nitrogen atoms. If the base is a pyrimidine, the nitrogen at the 1 position (i.e. N-1) is involved in bonding to the sugar. If the base is a purine, the bonding is to the N-9 position of the base (Figure 1b).

Nucleotides

A nucleotide is a phosphate ester of a nucleoside. It consists of a phosphate group joined to a nucleoside at the hydroxyl group attached to the C-5 of the sugar, that is it is a **nucleoside 5'-phosphate** or a **5'-nucleotide**. The primed number denotes the atom of the sugar to which the phosphate is bonded. In DNA, the nucleotides have deoxyribose as the sugar and hence are called **deoxyribonucleotides**. Deoxyribonucleotides may have a single phosphate group (**deoxyribonucleoside 5′-monophosphates**, dNMPs), two phosphate groups (**deoxyribonucleoside 5′-diphosphates**, dNDPs) or three phosphate groups (**deoxyribonucleoside 5′-triphosphates**, dNTPs). Deoxyribonucleoside triphosphates are the precursors for DNA synthesis. These are deoxyadenosine 5′-triphosphate (dATP) (Figure 1c), deoxyguanosine 5′-triphosphate (dGTP), deoxycytidine 5′-triphosphate (dCTP) and deoxythymidine 5′-triphosphate (dTTP). In each case the 'd' in the abbreviation (for example in dATP) indicates that the sugar in the nucleotide is deoxyribose. During DNA synthesis (Sections F3 and F4), two of the phosphates of each deoxyribonucleotide are split off (as pyrophosphate) so that only a single phosphate (the α phosphate) is incorporated into DNA.

3′5′ Phosphodiester bonds

In a DNA molecule, the different nucleotides are covalently joined to form a long polymer chain by covalent bonding between the phosphates and sugars. For any one nucleotide, the phosphate attached to the hydroxyl group at the 5′ position of the sugar is in turn bonded to the hydroxyl group on the 3′ carbon of the sugar of the next nucleotide. Since each phosphate–hydroxyl bond is an ester bond, the linkage between the two deoxyribonucleotides is a **3′5′ phosphodiester bond** (Figure 2). Thus, in a DNA chain, all of the 3′ and 5′ hydroxyl groups are involved in phosphodiester bonds except for the first and the last nucleotide in the chain. The first nucleotide has a 5′ phosphate not bonded to any other nucleotide and the last nucleotide has a free 3′ hydroxyl. Thus each DNA chain has **polarity**; it has a **5′ end** and a **3′ end**.

Figure 2. 3′5′ Phosphodiester bonds formed between nucleotides in a DNA molecule.

DNA sequence

Each nucleotide can be thought of as a single letter in an alphabet that has only four letters, A, G, C and T. Different genes have different sequences of these four nucleotides and so code for different biological messages. Since the deoxyribonucleotides in DNA differ only in the bases they carry, this sequence of deoxyribonucleotides can be recorded simply as a **base sequence**. For example, ACTTTCAGACC is part of the base sequence of one gene and codes for part of one protein whereas TGGAACCGTCA is part of the base sequence of a different gene coding for a different protein. Traditionally the base sequence is written in the order from the 5′ end of the DNA strand to the 3′ end, i.e. it is **written in the 5′ → 3′ direction**. Given that there are four types of nucleotide, the number of different possible sequences (or messages) in a DNA strand n nucleotides long is 4^n. DNA molecules are typically many thousands of nucleotides long so that the number of possible messages is enormous.

DNA double helix

In 1953, Watson and Crick worked out the three-dimensional structure of DNA, starting from X-ray diffraction photographs taken by Franklin and Wilkins. They deduced that DNA is composed of two strands wound round each other to form a double helix, with the bases on the inside and the sugar–phosphate backbones on the outside. In the double helix (Figure 3), the two DNA strands are organized in an **antiparallel** arrangement (i.e. the two strands run in opposite directions, one strand is orientated 5′ → 3′ and the other is orientated 3′ → 5′). The bases of the two strands form hydrogen bonds to each other; A pairs with T, and G pairs with C. This is called **complementary base pairing** (Figure 4). Thus a large two-ringed purine is paired with a smaller single-ringed pyrimidine and the

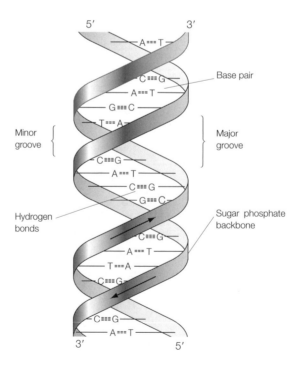

Figure 3. The DNA double helix.

two bases fit neatly in the gap between the sugar–phosphate strands and maintain the correct spacing. There would be insufficient space for two large purines to pair and too much space for two pyrimidines to pair, which would be too far apart to bond. The G:C and A:T base pairing also maximizes the number of effective hydrogen bonds that can form between the bases; there are three hydrogen bonds between each G:C base pair and two hydrogen bonds between each A:T base pair. Thus A:T and G:C base pairs form the most stable conformation both from steric considerations and from the point of view of maximizing hydrogen bond formation.

Adenine : thymine   Guanine : cytosine

Figure 4. The DNA base pairs. Hydrogen bonds are shown as dashed lines. dR, deoxyribose.

F2 Genes and chromosomes

Key Notes

Concept of the gene
The original concept of a gene was a region of DNA that encoded a single polypeptide (or RNA) product. However, post-transcriptional mechanisms in eukaryotes are now known that can generate multiple (sequence-related) polypeptides from a single RNA transcript so the working definition needs to be modified to accommodate this new knowledge.

Prokaryotic chromosomes
The DNA in a bacterium is a supercoiled double-stranded circular molecule that is packaged in the nucleoid region of the cell. The DNA is negatively supercoiled, complexed to several histone-like proteins (mainly proteins HU, HLP-1 and H-NS) and organized into about 50 domains bound to a protein scaffold.

Eukaryotic chromosomes
Eukaryotic cells contain much more DNA than prokaryotes. In the nucleus, the DNA is packaged into chromosomes that consist mainly of DNA and proteins called histones although other nonhistone proteins (NHP) are also present. Each chromosome contains a single linear double-stranded DNA molecule.

Nucleosomes
The chromosomal DNA is complexed with five types of histone (H1, H2A, H2B, H3 and H4). These are very basic proteins, rich in arginine and lysine. The amino acid sequences of histones are highly conserved in evolution. The DNA is wound round a histone octamer (two molecules each of H2A, H2B, H3 and H4) to form a nucleosome. The DNA between neighboring nucleosomes (linker DNA) binds histone H1. The packing ratio of nucleosomes is about 7.

The 30 nm fiber
Nucleosomes are organized into a 30 nm fiber. The exact arrangement of nucleosomes in the 30 nm fiber is unclear; possibilities are a higher order helix (called a solenoid) or a zigzag arrangement. The overall packing ratio is about 40.

Higher order structures
During interphase, the chromosomes are dispersed and presumably present mainly as 30 nm fiber. During cell division, the chromosomes become compacted and highly condensed but little is known as to how this occurs. However, one higher order structure is that the 30 nm fiber is attached to a central protein scaffold in each chromosome in a series of radial loops.

Organelle genomes
Most mitochondrial and chloroplast genomes are double-stranded circular DNA molecules but the genome may be

spread over several circular molecules, rather than a single molecule. In some unicellular eukaryotes, the mitochondrial genome is linear whereas in other eukaryotes the organelle has both circular and linear forms. Organelles may also have multiple copies of the genome.

Related topics	(F1) An introduction to DNA	(F4) DNA replication in eukaryotes

Concept of the gene

By the 1960s, the gene was clearly defined as the region of DNA that gives rise to a single polypeptide (or to a single RNA for genes whose final product is RNA not protein, e.g. ribosomal RNA genes). The existence of operons in prokaryotes (Section G3) did not challenge this concept since, although several clustered genes produced a single polycistronic mRNA, one could still identify single DNA regions as genes based on the distinct polypeptides they encoded. The concept even accommodated the discovery that many protein-coding genes in eukaryotes comprise coding regions (exons) separated by noncoding sequences (introns; see Section G5) since, again, only one polypeptide was encoded by this region of DNA. More recently, however, other mechanisms have come to light in eukaryotic cells that can lead to a variety of polypeptides being produced from a single DNA sequence; for example, alternative RNA splicing, alternative polyadenylation sites and RNA editing (Section G7). Nevertheless, in each of these cases, the protein products are closely related by sequence and all are derived from the same single region of DNA. Thus the original definition perhaps needs amending to indicate that a protein-coding gene is a region of DNA that encodes a single polypeptide or a set of closely related polypeptides, but otherwise the definition is intact. The alternative scenario, to regard a single DNA sequence that gives rise to, say, 10 closely related polypeptides by post-transcriptional processing as representing 10 genes, would not be reasonable.

The genome of an organism encompasses all of the genes of that organism. Even in a bacterial cell such as *Escherichia coli* (*E. coli*), the amount of DNA required is substantial (4.6 million base pairs) and so this DNA must be packaged. In a eukaryotic cell, the problem is even greater. A typical human cell, for example, contains about 1000 times more DNA than an *E. coli* cell. The rest of this section describes how DNA is packaged in both prokaryotes and eukaryotes.

Prokaryotic chromosomes

The classical view based on studies of *Escherichia coli* is that the bacterial DNA is a circular double-stranded molecule (often referred to as the **bacterial chromosome**). If it were an open circle it would measure about 1.6 mm in circumference but in reality it is **negatively supercoiled**, that is, it is twisted upon itself (Figure 1b). Supercoiled DNA is more compact than open circular DNA so supercoiling is a good approach to packaging the bacterial chromosome into the small bacterial cell, which is only 1×2 μm in size.

The supercoiled DNA is located in a region of the cell called the **nucleoid** (Section A1) where it is organized into 50 or so **loops** or **domains** that are bound to a central **protein core** or **scaffold**, attached to the cell membrane. Figure 1c illustrates this organization, although only six loops are shown for clarity. Within this structure, the DNA is complexed

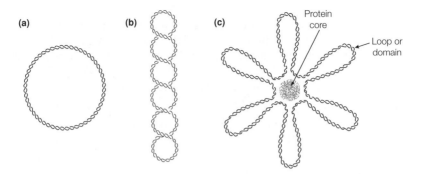

Figure 1. (a) A circular double-stranded DNA molecule; (b) supercoiled DNA; (c) the association of circular bacterial DNA with a protein core.

with at least four **DNA-binding proteins** that seem to have a role in packaging it, including proteins **HU, HLP-1** and **H-NS**. The most abundant of these is HU. Its structure is very different from histones (which are involved in packaging eukaryotic DNA; see below) but it packages bacterial DNA in a similar way by forming a tetramer around which about 60 base pairs (bp) of DNA is wound.

In recent years, it has become increasingly clear that not all prokaryotes have their DNA organized in the classical manner of *E. coli*.

● A second group of prokaryotes called the **archaea** do not contain protein HU but have proteins more similar to histones, which form a tetramer around which about 80 bp of DNA is wound.

● Most bacterial and archaeal genomes are circular DNA molecules like *E. coli* described above but some bacteria with linear genomes are now known including *Streptomyces coelicolor*, *Agrobacterium tumefaciens* and *Borrelia burgdorferi*.

● Some prokaryotes have genomes that are divided into several molecules, i.e. they have **multipartite genomes**. For example, the genome of *Vibrio cholerae*, the bacterium that causes cholera, has its essential genes distributed across two circular DNA molecules.

Eukaryotic chromosomes

The large amount of genomic DNA in a eukaryotic cell is tightly packaged in chromosomes contained within a specialized organelle, the nucleus. With the exception of the sex chromosomes, diploid eukaryotic organisms such as humans have two copies of each chromosome, one inherited from the father and one from the mother.

Chromosomes contain both DNA and protein. Most of the protein on a weight basis is **histones**, but there are also many thousands of other proteins found in far less abundance and these are collectively called **nonhistone proteins** (NHP). This nuclear DNA–protein complex is called **chromatin**.

In the nucleus, each chromosome contains a single linear double-stranded DNA molecule. The length of the packaged DNA molecule varies. In humans, the shortest DNA molecule in a chromosome is about 1.6 cm and the longest is about 8.4 cm. During the **metaphase** stage of mitosis, when the chromosomes align on the mitotic spindle ready for segregation, they are at their most condensed and range in size from only 1.3 μm to 10 μm long. Thus the **packing ratio**, that is the ratio of the length of the linear DNA molecule to the length of the metaphase chromosome, is about 10^4. In the time period between the end of one mitosis and the start of the next (i.e. **interphase**), the chromatin is more disperse. Here the packing ratio is in the range 10^2–10^3. Overall, the extensive packaging of DNA in chromosomes results from three levels of folding involving nucleosomes, 30 nm filaments and radial loops.

Nucleosomes

The first level of packaging involves the binding of the chromosomal DNA to histones. Overall, in chromosomes, the ratio of DNA to histones on a weight basis is approximately 1:1. There are five main types of histones called H1, H2A, H2B, H3 and H4. Histones are very basic proteins; about 25% of their amino acids are lysine or arginine so histones have a large number of positively charged amino acid side-chains. These positively charged groups therefore bind to the negatively charged phosphate groups of DNA. Not surprisingly given their importance in packaging DNA, the amino acid sequences of histones have been highly conserved in evolution. The most conserved are histones H3 and H4; for example, H3 and H4 from peas and cows differ in only four and two amino acids respectively! Histone H1 is the least conserved histone, which reflects its somewhat different role in packaging DNA compared with the other histones (see below). In sperm heads, DNA is particularly highly condensed and here the histones are replaced with small basic proteins called **protamines**.

When chromosomes are gently 'decondensed', they have the appearance under the electron microscope of 'beads on a string' (Figure 2). The 'beads' are called **nucleosomes** and consist of DNA complexed with histones. The 'string' is linear double-stranded DNA between adjacent nucleosomes and is called **linker DNA** (Figure 2). The distance between nucleosomes, that is the length of the linker DNA, varies from about 50–70 base pairs (bp) from organism to organism. Even in a single nucleus, the distance between adjacent nucleosomes varies depending on, for example, the presence of other sequence-specific DNA-binding proteins.

If a chromatin preparation is incubated with micrococcal nuclease, an enzyme that degrades DNA, the linker DNA is destroyed leaving **nucleosome core particles** in which the histones protect the associated DNA from digestion. Each nucleosome core particle contains a double-stranded DNA fragment 140–150 bp long (depending on the species)

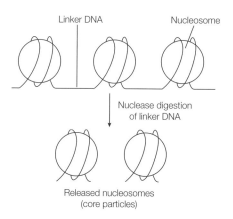

Figure 2. 'Beads-on-a-string' structure of chromatin.

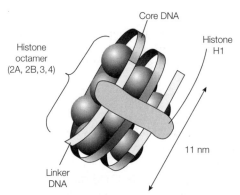

Figure 3. Schematic diagram of a nucleosome consisting of the DNA double helix wound round a histone octamer (two molecules each of histones H2A, H2B, H3 and H4).

bound to a complex of eight histones, the **histone octamer**, consisting of two molecules each of histones H2A, H2B, H3 and H4 (Figure 3). The DNA is wound round the outside of the histone octamer in about 1.65 turns of a left-handed supercoil. DNA–histone contacts are made along the inside face of this superhelix. Overall the packing ratio is about 7, that is the DNA length is shortened about seven-fold by winding around the nucleosome.

Histone H1 is called a **linker histone**. One linker histone is attached to each nucleosome but its exact position is not known. It was thought to bind to the outside of the histone octamer in the region of the linker DNA and act to hold the coiled DNA on the nucleosomes (Figure 3), perhaps even hold adjacent nucleosomes together in close proximity. However, some experimental data suggest that the linker histone may be located between the histone octamer and the DNA rather than on the surface.

The 30 nm fiber

If nuclei are lysed very gently, the chromatin is seen to exist as a 30 nm diameter fiber. This diameter is much larger than a single nucleosome (which is about 11 nm) and suggests that the nucleosomes are organized into a higher order structure. Exactly how the nucleosomes are organized to form the 30 nm fiber is not known; one possibility is that the nucleosomes wind up into a higher order helix with six nucleosomes per turn to form a **solenoid** (Figure 4). This would give a fiber three nucleosomes wide, which is indeed the diameter observed. In such a solenoid, the linear length of the DNA has been reduced by a further factor of 6 (equivalent to six nucleosomes per turn of the solenoid). Coupled with the packing ratio of 7 for the nucleosome itself (see above), this gives a packing ratio for the solenoid of approximately 6×7 (i.e. about 40). An alternative model is the **zigzag model** (Figure 4b). Other models also exist, including one called a helical ribbon.

It is important to realize that not all of the protein in the histone octamer is contained within the body of the nucleosome. Each histone molecule has a flexible N-terminal tail that extends out of the octamer. These tails play an important role in gene control in that when they are chemically modified, the 30 nm fiber can become more accessible for the transcription of genes in the associated nucleosomes.

Higher order structures

During interphase (i.e. between nuclear divisions), the chromosomes are dispersed in the nucleus and presumably present mainly as 30 nm fiber. However, during mitosis or

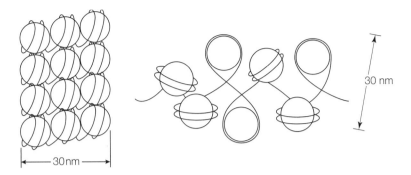

Figure 4. (a) The proposed solenoid model of chromatin to yield a 30 nm fiber. The structure consists of six nucleosomes per turn of the helix and hence would be three nucleosomes wide. In the diagram, only three nucleosomes of each turn are visible; the other three nucleosomes per turn are hidden from view. (b) Zigzag model. For each nucleosome, the DNA is wound round the outside (1.65 turns of a left-handed supercoil). However, in this simple diagram, because of the orientation of some nucleosomes, not all of the DNA supercoil is shown.

meiosis, the chromosomes become compacted and highly condensed. Little is known as to how this condensation is organized. When chromosomes are depleted of histones, they are seen to have a central fibrous '**protein scaffold**' (or **nuclear matrix**) to which the DNA is attached in loops (Figure 5). Therefore, *in vivo* it seems that the next order of packaging involves the attachment of the 30 nm fiber to multiple locations on this central protein scaffold in a series of radial loops.

Figure 5. Attachment of the 30 nm fiber to a central protein scaffold with the loops arranged radially around the scaffold. The diagram on the right shows a representational cross-section through a chromosome.

Organelle genomes

The mitochondria and chloroplasts of eukaryotic cells also contain DNA but, unlike the nuclear DNA, most mitochondrial and chloroplast genomes are double-stranded circular DNA molecules resembling bacterial chromosomes. However, the situation is more complex than this.

- The chloroplast genome may be spread over several circular DNA molecules; in the dinoflagellates (marine algae), the chloroplast genome is present as multiple circles each apparently containing just one gene.

- The mitochondrial genomes of *Paramecium*, *Chlamydomonas* and some yeasts (i.e. all unicellular eukaryotes) are linear molecules.

- In some eukaryotes there are linear versions of the organelle genome co-existing alongside the circular form(s).

Added to this complexity is the fact that organelle genomes may contain multiple copies of the genome; for example human mitochondria contain about 10 identical copies.

F3 DNA replication in bacteria

Key Notes

DNA polymerases
E. coli DNA polymerase I requires all four deoxyribonucleoside 5′-triphosphates (dNTPs) as precursors, Mg^{2+}, a DNA template and a primer with a 3′-OH end. DNA synthesis occurs in a 5′ → 3′ direction. DNA polymerase I also has a 3′ → 5′ exonuclease (proof-reading) activity and a 5′ → 3′ exonuclease activity. *E. coli* DNA polymerase II and III lack the 5′ → 3′ exonuclease activity. *E. coli* DNA polymerase III is the main enzyme that synthesizes DNA during replication whereas DNA polymerase I fills the gaps between segments of DNA made by DNA polymerase III. DNA polymerase II is involved mainly in DNA repair.

Replication forks
Replication starts at a single origin, is bi-directional and semi-conservative. Each replication bubble (or eye) consists of two replication forks.

Okazaki fragments
DNA synthesis proceeds in a 5′ → 3′ direction on each strand of the parental DNA. On the strand with 3′ → 5′ orientation (the leading strand) the new DNA is synthesized continuously. On the strand that has 5′ → 3′ orientation (the lagging strand), the DNA is synthesized discontinuously as a series of short Okazaki fragments that are then joined together.

RNA primer
DNA replication requires an RNA primer that is synthesized by an RNA polymerase called primase. This is extended by DNA polymerase III, which makes the DNA for both the leading and lagging strands. DNA polymerase I degrades the primer and replaces it with DNA. DNA ligase then joins DNA ends.

Accessory proteins
A helicase unwinds the DNA double helix and single-stranded DNA-binding (SSB) protein stabilizes the single-stranded regions during replication. DNA topoisomerase I is needed to allow the helix to unwind without causing extensive rotation of the chromosome. DNA topoisomerase II separates the two daughter DNA circles following replication.

Related topics
(F1) An introduction to DNA (F4) DNA replication in eukaryotes

DNA polymerases

DNA polymerase I from *E. coli* catalyzes the stepwise addition of deoxyribonucleotides to the 3'-OH end of a DNA chain:

$$(DNA)_{n\ residues} + dNTP \rightarrow (DNA)_{n+1} + PP_i$$

The enzyme has the following requirements:

● All four dNTPs (dATP, dGTP, dTTP and dCTP) must be present to be used as precursors; Mg^{2+} is also required

● A DNA **template** is essential, to be copied by the DNA polymerase

● A **primer** with a free 3'-OH that the enzyme can extend

DNA polymerase I is a template-directed enzyme, that is it recognizes the next nucleotide on the DNA template and then adds a complementary nucleotide to the 3'-OH of the primer, creating a 3'5' phosphodiester bond, and releasing pyrophosphate. The reaction is shown in Figure 1. It involves nucleophilic attack of the 3'-OH of the primer on the α-phosphate group of the incoming nucleotide. The primer is extended in a 5' → 3' direction.

DNA polymerase I also corrects mistakes in DNA by removing mismatched nucleotides (i.e. it has **proof-reading activity**). Thus, during polymerization, if the nucleotide that has just been incorporated is incorrect (mismatched), it is removed using a 3' → 5' exonuclease activity. This gives very high fidelity; an error rate of less than 10^{-8} per base pair. DNA polymerase I also has a 5' → 3' exonuclease activity, i.e. it can hydrolyze nucleic acid starting from the 5' end of a chain. This activity plays a key role in removing the RNA primer used during replication (see below). Thus, overall, DNA polymerase I has three different active sites on its single polypeptide chain; 5' → 3' polymerase, 3' → 5' exonuclease and 5' → 3' exonuclease.

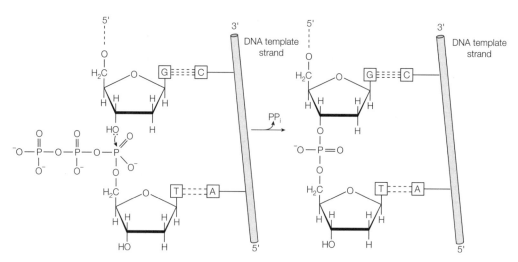

Figure 1. DNA synthesis. In this schematic diagram, the incoming dTTP hydrogen bonds with the adenine on the template DNA strand and a 3'5' phosphodiester bond is formed, releasing pyrophosphate.

E. coli also contains two other DNA polymerases, **DNA polymerase II** and **DNA polymerase III**. As with DNA polymerase I, these enzymes also catalyze the template-directed synthesis of DNA from deoxyribonucleotidyl 5′-triphosphates, need a primer with a free 3′-OH group, synthesize DNA in the 5′ → 3′ direction, and have 3′ → 5′ exonuclease activity. Neither enzyme has 5′ → 3′ exonuclease activity.

Although DNA polymerase I was the first bacterial DNA polymerase isolated (in 1957 by Arthur Kornberg), it is in fact DNA polymerase III (isolated in 1972) that is the main replicating enzyme responsible for copying the DNA template. However, DNA polymerase I has an important role in replication; as described below, it fills the gaps between DNA regions synthesized by DNA polymerase III. DNA polymerase II is mainly involved in **DNA repair** not genome replication.

Replication forks

When the bacterial circular chromosome is replicated, replication starts at a **single origin**. In *E. coli* this is called the **oriC locus** and contains five copies of a DNA sequence that acts as a binding site for the **DnaA** origin-recognition protein. Several other proteins now bind including **DnaB**, a DNA helicase. Local regions of the oriC unwind ready for both strands to serve as template for the synthesis of new DNA. DNA synthesis then proceeds outward in both directions from the single origin (i.e. it is **bi-directional**; Figure 2). The products of the reaction are two daughter double-stranded DNA molecules each of which has one original template strand and one strand of newly synthesized DNA. Thus, replication is **semi-conservative**. The region of replicating DNA associated with the single origin is called a **replication bubble** or **replication eye** and consists of two **replication forks** moving in opposite directions around the DNA circle (Figure 2).

Okazaki fragments

Double-stranded DNA is **antiparallel** (Section F1); one strand runs 5′ → 3′ and the complementary strand runs 3′ → 5′. As the original double-stranded DNA opens up at a replication fork, new DNA is made against each template strand. Superficially, therefore, one might expect new DNA to be made 5′ → 3′ for one daughter strand and 3′ → 5′ for the other daughter strand. However, all DNA polymerases make DNA only in the 5′ → 3′ direction and never in the 3′ → 5′ direction. What actually happens is that on the template strand with 3′ → 5′ orientation, new DNA is made in a continuous piece in the correct

(a) Replication bubble   **(b)** Replication forks

Figure 2. Replication of the bacterial circular chromosome. Replication starts from a single origin and proceeds bi-directionally (a), moving around the chromosome with time (b). The two replication forks eventually meet and fuse. The two circular daughter DNA molecules produced each have one original template DNA strand (thin line) and one new strand (thick line).

Figure 3. Synthesis of DNA at a replication fork. As the parental DNA (thin line) opens up, each of the two parental strands acts as a template for new DNA synthesis (thick lines). The leading strand is synthesized continuously but the lagging strand is synthesized as short (Okazaki) DNA fragments that are then joined together.

$5' \rightarrow 3'$ direction. This new DNA is called the **leading strand** (Figure 3). On the other template strand (that has a $5' \rightarrow 3'$ orientation), DNA polymerase synthesizes short pieces of new DNA (about 1000–2000 nucleotides long) in the $5' \rightarrow 3'$ direction (Figure 3) and then joins these pieces together. The small fragments are called **Okazaki fragments** after their discoverer. The new DNA strand, which is made by this discontinuous method is called the **lagging strand**.

RNA primer

DNA polymerase cannot start DNA synthesis without a primer. Even on the lagging strand, each Okazaki fragment requires an RNA primer before DNA synthesis can start. The primer used in each case is a short piece of RNA [about 4–15 nucleotides (nt) long] and is synthesized by an RNA polymerase called **primase** (Figure 4a). Primase can make RNA directly on the single-stranded DNA template because, like all RNA polymerases, it does not require a primer to begin synthesis. The RNA primer made by primase (Figure 4b) is then extended by DNA polymerase III (Figure 4c). DNA polymerase III synthesizes DNA for both the leading and lagging strand (Figure 4d). After DNA synthesis by DNA polymerase III, DNA polymerase I uses its $5' \rightarrow 3'$ exonuclease activity to remove the RNA primer and then fills the gap with new DNA (Figures 4e and 4f). DNA polymerase III cannot carry out this task because it lacks the $5' \rightarrow 3'$ exonuclease activity of DNA polymerase I. Finally, DNA ligase joins the ends of the DNA fragments together (Figure 4g).

Accessory proteins

DNA polymerases I and III, primase and DNA ligase are not the only proteins needed for replication of the bacterial chromosome. The DNA template is a double helix with each strand wound tightly around the other and hence the two strands must be unwound during replication. How is this **unwinding problem** solved? **DnaB helicase** is used to unwind the double helix (using ATP as energy source) and **SSB (single-stranded DNA-binding) protein** prevents the single-stranded regions from base pairing again so that each of the two DNA strands is accessible for replication. In principle, for a replication fork to move along a piece of DNA, the DNA helix would need to unwind ahead of it, causing the DNA to rotate rapidly. However, the bacterial chromosome is circular and so there are no ends to rotate. The solution to the problem is that an enzyme called **topoisomerase I** breaks a phosphodiester bond in one DNA strand (a single-strand break) a small distance ahead of the fork, allowing the DNA to rotate freely (swivel) around the other (intact) strand. The phosphodiester bond is then re-formed by the topoisomerase.

Figure 4. Details of DNA replication: (a) primase binds to the DNA template strand (thin line) and (b) synthesizes a short RNA primer (dotted line); (c) DNA polymerase III now extends the RNA primer by synthesizing new DNA (thick line); (d) during synthesis of the lagging strand, adjacent Okazaki fragments are separated by the RNA primers; (e) the RNA primers are now removed and the gaps filled with DNA by DNA polymerase I (f) generating adjacent DNA fragments that are then (g) joined by DNA ligase.

After the bacterial circular DNA has been replicated, the result is two double-stranded circular DNA molecules that are interlocked. **Topoisomerase II** separates them as follows. This enzyme works in a similar manner to topoisomerase I but causes a transient break in each strand (a double-strand break) of a double-stranded DNA molecule. Thus topoisomerase II binds to one double-stranded DNA circle and causes a transient double-strand break that acts as a 'gate' through which the other DNA circle can pass (Figure 5). Topoisomerase II then re-seals the strand breaks.

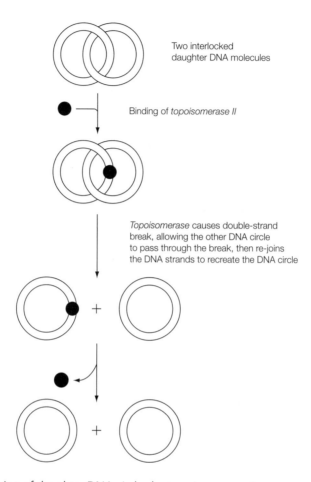

Two interlocked
daughter DNA molecules

Binding of *topoisomerase II*

Topoisomerase causes double-strand
break, allowing the other DNA circle
to pass through the break, then re-joins
the DNA strands to recreate the DNA circle

Figure 5. Separation of daughter DNA circles by topoisomerase II.

F4 DNA replication in eukaryotes

Key Notes

Multiple replicons   DNA replication occurs only in the S phase. It occurs at many chromosomal origins, is bi-directional and semi-conservative. Sets of 20–80 replicons act as replication units that are activated in sequence. Unlike bacterial origins of replication, those in mammals are less well defined.

At least nine DNA polymerases   DNA polymerase δ is the main enzyme responsible for replicating chromosomal DNA. DNA polymerase α primes the process. DNA polymerases β and ε repair DNA, and DNA polymerase γ replicates mitochondrial DNA.

Leading and lagging strands   DNA polymerase α and δ synthesize the lagging strand, via Okazaki fragments. The RNA primers are synthesized by DNA polymerase α, which carries a primase subunit. DNA polymerase δ is the main replicating polymerase and synthesizes DNA using the RNA primer made by DNA polymerase α.

Telomere replication   The regions at the ends of chromosomes (telomeres) contain multiple copies of a repeated G-rich hexanucleotide repeat. Telomerase is used to replicate telomere DNA. It is a reverse transcriptase with an integral RNA that contains a sequence that is the reverse complement of the telomere repeat. It extends the telomere end by copying the RNA template in multiple rounds of synthesis and then translocation.

Replication of chromatin   Nucleosomes do not dissociate from the DNA during DNA replication; rather they must open up to allow the replication apparatus to pass. Both daughter DNA molecules have old histones bound to them but new histones must also be synthesized to allow all the DNA to be packaged correctly in nucleosomes.

Related topics   (A3) Cell growth   (F3) DNA replication in (F1) An introduction to DNA   bacteria

---

Multiple replicons

In eukaryotes, replication of chromosomal DNA occurs only in the S phase of the cell cycle (Section A3). As for bacterial DNA (Section F3), eukaryotic DNA is replicated **semi-conservatively**. Replication of each linear DNA molecule in a chromosome starts at **many origins**, one every 30–300 kilobases (kb) of DNA depending on the species and tissue, and proceeds **bi-directionally** from each origin.

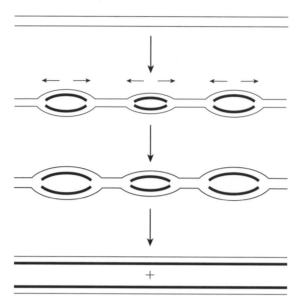

Figure 1. Replication of eukaryotic chromosomal DNA. Replication begins at many origins and proceeds bi-directionally at each location. Eventually the replication eyes merge together to produce two daughter DNA molecules, each of which consists of one parental DNA strand (thin line) and one newly synthesized DNA strand (thick line).

The use of multiple origins is essential in order to ensure that the large amount of chromosomal DNA in a eukaryotic cell is replicated within the necessary time period; a human diploid cell must copy about 6 billion base pairs compared to only about 4.6 million base pairs that *E. coli* needs to replicate. Not surprisingly, every chromosome in a human cell has several hundred origins of replication. Unlike *E. coli*, mammalian replication origins do not appear to have sharply defined sequences for binding proteins that initiate replication but more general AT-rich regions at which **origin of replication complexes** (**ORCs**) are formed. Each ORC contains six different proteins, each of which is homologous to the *E. coli* DnaA protein.

At each origin, a **replication bubble** forms consisting of two **replication forks** moving in opposite directions. The DNA replicated under the control of a single origin is called a **replicon**. DNA synthesis proceeds until replication bubbles merge together (Figure 1).

All of the regions of a chromosome are not replicated simultaneously. Rather, many replication eyes will be found in one part of the chromosome and none in another section. Thus replication origins are activated in clusters, called **replication units**, consisting of 20–80 origins. During S phase, the different replication units are activated in a set order until eventually the whole chromosome has been replicated. Transcriptionally active DNA appears to be replicated early in S phase, whilst chromatin that is condensed and not transcriptionally active is replicated later.

At least nine DNA polymerases

Eukaryotic cells contain at least nine different DNA polymerases denoted by Greek suffixes; α, β, γ, δ, ε, etc. The main enzyme involved in replicating chromosomal DNA is DNA polymerase δ. DNA polymerase α also plays an important role in replication since it primes the process (see below).

DNA polymerase γ is coded by a nuclear gene but the enzyme is targeted to mitochondria where it replicates mitochondrial DNA.

DNA polymerases β and ε are involved mainly in DNA repair, as are the other DNA polymerases not considered here.

Leading and lagging strands

The basic scheme of replication of double-stranded chromosomal DNA in eukaryotes follows that for bacterial DNA replication (Section F3); a leading strand and a lagging strand are synthesized, the latter involving discontinuous synthesis via Okazaki fragments. The RNA primers required (8–12 nt long) are made by DNA polymerase α, which carries a **primase subunit**. DNA polymerase α initiates synthesis of the lagging strand, making first the RNA primer and then extending it with a short region of DNA. DNA polymerase δ then synthesizes the rest of the Okazaki fragment. Similarly the leading strand is also synthesized by DNA polymerase δ. The δ enzyme has 3′ → 5′ exonuclease activity and so can proof-read the DNA made, but DNA polymerase α has no such activity.

Telomere replication

The replication of a linear DNA molecule in a eukaryotic chromosome creates a problem that does not exist for the replication of bacterial circular DNA molecules. The normal mechanism of DNA synthesis (see above) means that the 3′ end of the lagging strand is not replicated. This creates a gap at the end of the chromosome and therefore a shortening of the double-stranded replicated portion. The effect is that the chromosomal DNA would become shorter and shorter after each replication. Various mechanisms have evolved to solve this problem. In many organisms, the solution is to use an enzyme called **telomerase** to replicate the chromosome ends (**telomeres**).

Each telomere contains many copies of a repeated hexanucleotide sequence that is G-rich; in humans it is 5′-TTAGGG-3′. Telomerase contains both a protein and, as an integral part of its structure, an RNA molecule about 450 nt long. Near the 5′ end of this RNA is the sequence 5′-CUAACCCUAAC-3′, which is the reverse complement of the human telomere repeat (5′-TTAGGG-3′).

The mechanism of action of telomerase is shown in Figure 2. The RNA molecule of telomerase hydrogen-bonds to the telomere end. Then, using the RNA as a template, telomerase (which is a reverse transcriptase; Section I4) copies the RNA template adding deoxyribonucleotides to the telomere DNA end. Telomerase then translocates further along the strand towards the new telomere end and repeats the extension process. It can do this hundreds of times before finally dissociating. That this is indeed the way telomerase works is confirmed by finding that in other species the telomere repeat differs in sequence from the human (e.g. in *Tetrahymena*, it is 5′-TTGGGG-3′) but the corresponding telomerase always contains an RNA sequence that is the reverse complement of that species' repeat sequence.

How the other strand (the C-rich strand) at the telomere end is extended is still not perfectly clear but probably the newly extended DNA strand can act as a template for normal DNA replication (lagging strand synthesis by DNA polymerase α) to form double-stranded chromosomal DNA.

Amazingly, in mammals, telomerase is active in the early embryo but after birth is active only in stem cells and reproductive cells so in other mammalian cells, the chromosomal DNA shortens every time the cell divides.

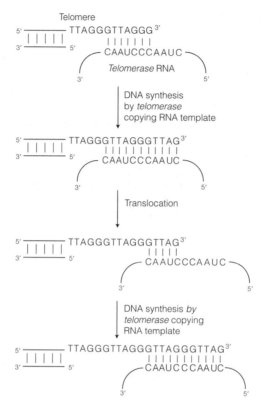

Figure 2. Replication of telomeric DNA. Telomerase is a reverse transcriptase with an integral RNA molecule it uses as template to direct DNA synthesis and hence extension of the ends of chromosomal DNA.

Replication of chromatin

When a chromosome is replicated, the replication machinery passes through the nucleosomes without removing the histones from the DNA. How this occurs is still not fully understood. Both of the daughter DNA molecules that result from replication have old histones bound to them but overall, since the amount of DNA has now doubled, more histones are needed to package the DNA correctly into nucleosomes. Not surprisingly, therefore, the S phase of the cell cycle is also the time when large amounts of histones are synthesized.

G1 An introduction to RNA

Key Notes

Covalent structure	RNA is a polymer chain of ribonucleotides joined by 3′5′ phosphodiester bonds. The covalent structure is very similar to that for DNA except that uracil replaces thymine and ribose replaces deoxyribose.
RNA secondary structure	RNA molecules are largely single-stranded but there are regions of self-complementarity where the RNA chain forms internal double-stranded regions.
Related topics	(F1) An introduction to DNA (G2) Transcription in prokaryotes (G3) Operons (G4) Transcription in eukaryotes: an overview (G5) Transcription of protein-coding genes in eukaryotes (G6) Regulation of transcription by RNA Pol II

Covalent structure

Like DNA (Section F1), RNA is a long polymer consisting of nucleotides joined by 3′5′ phosphodiester bonds. However, there are some differences.

- The bases in RNA are adenine (abbreviated A), guanine (G), uracil (U) and cytosine (C). Thus thymine in DNA is replaced by **uracil** in RNA, a different pyrimidine (Figure 1a). However, like thymine (Section F1), uracil can form base pairs with adenine.

- The sugar in RNA is **ribose** rather than deoxyribose as in DNA (Figure 1b).

The corresponding ribonucleosides are adenosine, guanosine, cytidine and uridine. The corresponding ribonucleotides are adenosine 5′-triphosphate (ATP), guanosine 5′-triphosphate (GTP), cytidine 5′-triphosphate (CTP) and uridine 5′-triphosphate (UTP).

As with DNA, the nucleotide sequence of RNA is also written as a base sequence in the 5′ → 3′ direction. Thus GUCAAGCCGGAC is the sequence of one short RNA molecule.

Figure 1. (a) Uracil; (b) ribose.

RNA secondary structure

Most RNA molecules are single-stranded but an RNA molecule may contain regions which can form complementary base pairing where the RNA strand loops back on itself (Figure 2). If so, the RNA will have some double-stranded regions. Ribosomal RNAs (rRNAs) and transfer RNAs (tRNAs) (Sections G8 and G9, respectively) exhibit substantial secondary structure, as do some messenger RNAs (mRNAs).

Figure 2. An example of self-complementarity in RNA, forming an internal double-stranded region; hydrogen bonding between bases is shown by the symbol •.

G2 Transcription in prokaryotes

Key Notes

Three phases of transcription	Transcription by *E. coli* RNA polymerase occurs in three phases: initiation, elongation and termination. Initiation involves binding of the enzyme to a promoter upstream of the gene. During elongation, the antisense DNA strand is used as the template so that the RNA made has the same base sequence as the sense (coding) strand, except that U replaces T. A termination signal is eventually encountered that halts synthesis and causes release of the completed RNA.
Promoters and initiation	RNA polymerase holoenzyme (containing $\alpha_2\beta\beta'\omega\sigma$ subunits) initiates transcription by binding to a 40–60 bp region that contains two conserved promoter elements, the –10 sequence (Pribnow box) with the consensus TATAAT and the –35 sequence with the consensus TTGACA. The σ factor is essential for initiation. No primer is required. Promoters vary up to 1000-fold in their efficiency of initiation, which depends on the exact sequence of the key promoter elements as well as flanking sequences. Some highly expressed genes also have an upstream element located 40–60 nucleotides (nt) upstream of the transcriptional start site that increases the efficiency of initiation by acting as an additional binding site for RNA polymerase.
Elongation	Following initiation, the σ subunit dissociates from RNA polymerase to leave the core enzyme ($\alpha_2\beta\beta'\omega$) that continues RNA synthesis in a $5' \rightarrow 3'$ direction using the four ribonucleoside 5'-triphosphates as precursors. The DNA double helix is unwound for transcription, forming a transcription bubble, and is then rewound after the transcription complex has passed.
Termination	A common termination signal is a hairpin structure formed by a palindromic GC-rich region, followed by an AT-rich sequence. Other signals are also used, which require the assistance of rho (ρ) protein for effective termination.
RNA processing	Messenger RNA transcripts of protein-coding genes in prokaryotes require little or no modification before translation. Ribosomal RNAs (rRNAs) and transfer RNAs (tRNAs) are synthesized as precursor molecules that require processing by specific ribonucleases to release the mature RNA molecules.

Related topics	(F1) An introduction to DNA	(G6) Regulation of
	(G1) An introduction to RNA	transcription by RNA Pol II
	(G3) Operons	(G7) Processing of eukaryotic
	(G4) Transcription in	pre-mRNA
	eukaryotes: an overview	(G8) Transcription and
	(G5) Transcription of	processing of ribosomal RNA
	protein-coding genes in	(G9) Transcription and
	eukaryotes	processing of transfer RNA

Three phases of transcription

Gene transcription by *E. coli* RNA polymerase takes place in three phases: **initiation**, **elongation** and **termination**. During initiation, RNA polymerase recognizes a specific site on the DNA, upstream from the gene that will be transcribed, called a **promoter site** and binds to it. This complex of bound RNA polymerase to the double-stranded promoter is called a **closed promoter complex**. The DNA then unwinds locally, creating an **open promoter complex**, and RNA polymerase starts transcribing the single-stranded DNA template. During elongation, the RNA polymerase uses the **antisense (–) strand** of DNA as template and synthesizes a complementary RNA molecule using ribonucleoside 5′-triphosphates as precursors. The RNA produced has the same sequence as the nontemplate strand, called the **sense (+) strand** (or **coding strand**) except that the RNA contains U instead of T. At different locations on the bacterial chromosome, sometimes one strand is used as template, sometimes the other, depending on which strand is the coding strand for the gene in question. The correct strand to be used as template is identified for the RNA polymerase by the presence of the promoter site. Finally, the RNA polymerase encounters a termination signal and ceases transcription, releasing the RNA transcript and dissociating from the DNA.

Promoters and initiation

In *E. coli*, all genes are transcribed by a single large RNA polymerase with the subunit structure $\alpha_2\beta\beta'\omega\sigma$. This complete enzyme, called the **holoenzyme**, is needed to initiate transcription since the σ factor is essential for recognition of the promoter; it decreases the affinity of the core enzyme for nonspecific DNA binding sites and increases its affinity for the promoter.

The holoenzyme binds to a promoter region about 40–60 bp in size and then initiates transcription a short distance downstream (i.e. 3′ to the promoter). Within the promoter lie two common 6 bp sequences that are particularly important for promoter function and which are therefore highly conserved between species. Using the convention of calling the first nucleotide of a transcribed sequence as +1, these two **promoter elements** lie at positions –10 and –35, that is about 10 and 35 bp, respectively, upstream of where transcription will begin (Figure 1).

- The **–10 sequence** has the **consensus sequence** (i.e. average sequence comparing many promoters) of TATAAT. Because this element was discovered by Pribnow, it is also known as the **Pribnow box**. It is an important recognition site that interacts with the σ factor of RNA polymerase.

- The **–35 sequence** has the consensus sequence TTGACA and is important in DNA unwinding during transcriptional initiation.

Figure 1. Prokaryotic promoter showing the –10 sequence and –35 sequence. By convention, the first nucleotide of the template DNA that is transcribed into RNA is denoted +1, the transcriptional start site.

The sequence between the –10 sequence and the –35 sequence is not conserved (i.e. it varies from promoter to promoter) but the distance between these two sites is extremely important for correct functioning of the promoter because it places the two elements on the same side of the double helix where they are accessible to be bound by RNA polymerase.

Promoters differ by up to 1000-fold in their efficiency of initiation of transcription so that genes with strong promoters are transcribed very frequently whereas genes with weak promoters are transcribed far less often. The –10 and –35 sequences of strong promoters correspond well with the consensus sequences shown in Figure 1, whereas weaker promoters tend to differ from the consensus at several nucleotides. The nature of the sequences around the transcriptional start site can also influence the efficiency of initiation.

It is common for prokaryotes to have several σ factors that recognize different types of promoter. In *E. coli*, the most common σ factor is σ^{70} (so called because it has a mass of 70 kDa), which recognizes the promoter consensus sequences described above. However, under other environmental conditions, such as nitrogen starvation or elevated temperatures (heat-shock), other σ factors are expressed, which bind to promoters with different consensus sequences. This leads to the preferential transcription of those genes bearing the different consensus promoter. Therefore, the σ factor has an important role in directing which genes are transcribed by RNA polymerase.

As well as the core promoter, some *E. coli* genes that are highly expressed also have another sequence (called the **UP element** for **upstream element**) located 40–60 nucleotides (nt) upstream of the transcriptional start site. This increases the efficiency of initiation by acting as an additional binding site for RNA polymerase; it binds the α subunit of RNA polymerase.

Once RNA polymerase has bound to a promoter, it does not need a primer to begin transcription (cf. DNA polymerases, Sections F3 and F4); the RNA polymerase begins transcription directly.

Elongation

After transcription initiation, the σ factor is released from the transcriptional complex to leave the **core enzyme** ($\alpha_2 \beta\beta'\omega$), which continues elongation of the RNA transcript. Thus the core enzyme contains the catalytic site for polymerization. The first nucleotide in the RNA transcript is usually pppG or pppA. The RNA polymerase then synthesizes RNA in the $5' \rightarrow 3'$ direction, using the four ribonucleoside 5'-triphosphates (ATP, CTP, GTP, UTP) as precursors. The 3'-OH at the end of the growing RNA chain attacks the α phosphate group of the incoming ribonucleoside 5'-triphosphate to form a 3'5' phosphodiester bond (Figure 2). The region of unwound DNA that is undergoing transcription is called the **transcription bubble** (Figure 3). The RNA transcript forms a transient RNA–DNA hybrid helix with its template strand but then peels away from the DNA as transcription proceeds. The DNA is unwound ahead of the transcription bubble and after the transcription complex has passed, the DNA rewinds.

Figure 2. Transcription by RNA polymerase. In each step the incoming ribonucleotide selected is that which can base pair with the next base of the DNA template strand. In the diagram, the incoming nucleotide is rUTP to base pair with the A residue of the template DNA. A 3′5′ phosphodiester bond is formed, extending the RNA chain by one nucleotide, and pyrophosphate is released. Overall the RNA molecule grows in a 5′ to 3′ direction.

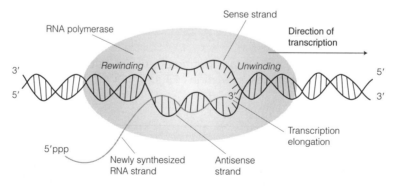

Figure 3. A transcription bubble. The DNA double helix is unwound and RNA polymerase then synthesizes an RNA copy of the DNA template strand. The nascent RNA transiently forms an RNA–DNA hybrid helix but then peels away from the DNA, which is subsequently rewound into a helix once more.

Termination

Transcription continues until a termination sequence is reached. At this point, transcription ceases, the RNA transcript and the RNA polymerase dissociate from the DNA, and the single-stranded region of DNA rewinds into a double helix.

The most common termination signal is a GC-rich region that is a **palindrome**, followed by an AT-rich sequence. The RNA made from the DNA palindrome is self-complementary and so base pairs internally form a **hairpin structure** rich in GC base pairs followed by four or more U residues (Figure 4). However, not all termination sites have this hairpin structure. Those that lack such a structure require an additional protein, called **rho** (ρ), to help recognize the termination site and stop transcription.

RNA processing

In prokaryotes, RNA transcribed from protein-coding genes (**messenger RNA, mRNA**) requires little or no modification prior to translation. In fact, many mRNA molecules begin to be translated even before RNA synthesis has finished. However, **ribosomal RNA** (**rRNA**) and **transfer RNA** (**tRNA**) are synthesized as precursor molecules that do require post-transcriptional processing (Sections G8 and G9, respectively).

Figure 4. A typical hairpin structure formed by the 3′ end of an RNA molecule during termination of transcription.

G3 Operons

Key Notes

An overview

Operons are clusters of structural genes under the control of a single operator site and regulator gene, which ensures that expression of the structural genes is coordinately controlled.

The *lac* operon

The *lac* operon contains *lacZ*, *lacY* and *lacA* genes encoding β-galactosidase, galactoside permease, and thiogalactoside transacetylase, respectively, preceded by an operator site (O_{lac}) and a promoter (P_{lac}). The operon is transcribed by RNA polymerase to produce a single polycistronic mRNA that is translated to produce all three enzymes. When lactose is present, the background level of β-galactosidase converts some lactose to allolactose, which induces transcription of the *lac* operon. Isopropylthiogalactoside (IPTG) can also act as an inducer. Transcription of the operon is controlled by the lac repressor protein encoded by the *lacI* gene.

The lac repressor

In the absence of an inducer, RNA polymerase binds to the *lacI* gene promoter (P_{lacI}) and transcribes the gene, producing lac repressor protein, which binds to the *lac* operator site, O_{lac}, and prevents transcription of the *lac* operon. In the presence of an inducer (e.g. allolactose or IPTG), the inducer binds to the repressor and changes its conformation, reducing its affinity for the *lac* operator. The repressor dissociates and allows RNA polymerase to transcribe the *lac* operon.

CRP/CAP

When glucose is present, transcription of the *lac* operon is inhibited (catabolite repression). Catabolite activator protein, CAP (also called cAMP response protein, CRP) is required for high-level transcription of the *lac* operon. It associates with 3′, 5′-cAMP to form a CRP–cAMP complex which binds to the *lac* promoter and increases RNA polymerase binding, stimulating *lac* operon transcription. When glucose is present, the intracellular level of cAMP falls, CRP alone cannot bind to the *lac* promoter and the *lac* operon is only weakly transcribed. When glucose is absent, the level of intracellular cAMP rises, and the CRP–cAMP complex is formed and stimulates *lac* operon transcription, allowing lactose to be used as an alternative carbon source.

Positive and negative regulation

In negative regulation of prokaryotic gene expression, bound repressor prevents transcription of the structural genes. In positive regulation of gene expression, an activator binds to DNA and increases the rate of transcription. The *lac* operon is subject to both negative and positive control.

The *trp* operon	The *trp* operon contains five structural genes encoding enzymes for tryptophan biosynthesis, a *trp* promoter (P_{trp}) and a *trp* operator sequence (O_{trp}). The operon is transcribed only when tryptophan is scarce.
The trp repressor	When tryptophan is lacking, a trp repressor protein (encoded by the *trpR* operon) is synthesized but cannot bind to the *trp* operator and so the *trp* operon is transcribed to produce the tryptophan synthetic enzymes. When tryptophan is present, it binds to the repressor and activates it so that the repressor binds to the *trp* operator and stops *trp* operon transcription.
Attenuation	A leader sequence in the polycistronic mRNA transcribed from the *trp* operon can form several possible stem-loop secondary structures, one of which can act as a transcription terminator. A different stem-loop can act as an anti-terminator. In the presence of tryptophan, ribosomes bind to the *trp* polycistronic mRNA that is being transcribed, following closely behind the RNA polymerase, and begin to translate the leader sequence. The position of the bound ribosomes prevents formation of the anti-terminator stem-loop but allows the terminator loop to form, which then inhibits further *trp* operon transcription. If tryptophan is scarce, the ribosome pauses when attempting to translate the two trp codons in the leader sequence, which leaves the leader sequence available to form the anti-terminator stem-loop and transcription of the *trp* operon continues.
Attenuation vs. repression	The *trp* operon is regulated by both repression (which determines whether transcription will occur or not) and attenuation (which fine-tunes transcription). Other operons for amino acid biosynthetic pathways are regulated by both repression and attenuation or by attenuation only.
Related topics	

An overview

Many protein-coding genes in bacteria are clustered together in **operons**, which serve as transcriptional units that are coordinately regulated. It was Jacob and Monod in 1961 who proposed the operon model for the regulation of transcription. The operon model proposes three elements:

- A set of **structural genes** (i.e. genes encoding the proteins to be regulated)

- An **operator site**, which is a DNA sequence that regulates transcription of the structural genes

- A **regulator gene**, which encodes a protein that recognizes the operator sequence

The *lac* operon

One of the most studied operons is the ***lac* operon** in *E. coli*. This codes for key enzymes involved in lactose metabolism: **galactoside permease** (also known as **lactose permease**; it transports lactose into the cell across the cell membrane) and **β-galactosidase** (which hydrolyzes lactose to glucose and galactose). It also codes for a third enzyme, **thiogalactoside transacetylase**. Normally *E. coli* cells make very little of any of these three proteins but when lactose is available, it causes a large and coordinated increase in the amount of each enzyme to be synthesized. Thus each enzyme is an **inducible enzyme** and the process is called **induction**. The mechanism is that the few molecules of β-galactosidase in the cell before induction convert the lactose to allolactose, which then turns on transcription of these three genes in the *lac* operon. Thus allolactose is an **inducer**. Another inducer of the *lac* operon is **isopropylthiogalactoside** (IPTG). Unlike allolactose, this inducer is not metabolized by *E. coli* and so is useful for experimental studies of induction.

In the *lac* operon (Figure 1), the **structural genes** are the *lacZ*, *lacY* and *lacA* genes encoding β-galactosidase, the permease and the transacetylase, respectively. They are transcribed to yield a single **polycistronic mRNA** that is then translated to produce all three enzymes (Figure 1*)*. The existence of a polycistronic mRNA ensures that the amounts of all three gene products are regulated coordinately. Transcription occurs from a single promoter (P_{lac}) that lies upstream of these structural genes (Figure 1) and binds RNA polymerase (Section G2). However, also present are an operator site (O_{lac}) between the promoter and the structural genes, and a *lac*I gene that codes for the **lac repressor** protein.

The lac repressor

The *lac*I gene has its own promoter (P_{lac}) that binds RNA polymerase and leads to transcription of lac repressor mRNA and hence production of lac repressor protein monomers. Four identical repressor monomers come together to form the active tetramer, which can bind tightly to the *lac* operator site, O_{lac}. The O_{lac} sequence is **palindromic**, that is it has the same DNA sequence when one strand is read 5′ to 3′ and the complementary strand is read 5′ to 3′. This symmetry of the operator site is matched by the symmetry of the repressor tetramer.

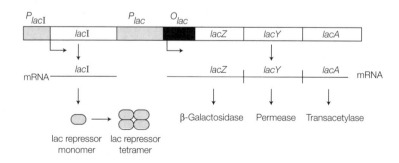

Figure 1. Structure of the *lac* operon.

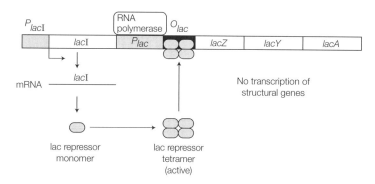

Figure 2. Repression of transcription by the lac repressor in the absence of inducer.

Figure 3. Inducer inactivates the lac repressor and so allows transcription of the structural genes.

In the absence of an inducer such as allolactose or IPTG, the *lac*I gene is transcribed and the resulting repressor protein binds to the operator site of the *lac* operon, O_{lac}, and prevents transcription of the *lacZ*, *lacY* and *lacA* genes (Figure 2). During induction, the inducer binds to the repressor. This causes a change in conformation of the repressor that greatly reduces its affinity for the *lac* operator site. The lac repressor now dissociates from the operator site and allows the RNA polymerase (already in place on the adjacent promoter site) to begin transcribing the *lacZ*, *lacY* and *lacA* genes (Figure 3). This yields many copies of the polycistronic mRNA and, after translation, large amounts of all three enzymes.

If inducer is removed, the lac repressor rapidly binds to the *lac* operator site and transcription is inhibited almost immediately. The *lacZYA* RNA transcript is very unstable and so degrades quickly such that further synthesis of the β-galactosidase, permease and transacetylase ceases.

CRP/CAP

When *E. coli* is growing on glucose, it has no need for enzymes needed to metabolize other sugars such as lactose. Indeed to synthesize such enzymes under such conditions would be energetically wasteful. So when glucose is present, transcription of the *lac* operon is inhibited, a phenomenon called **catabolite repression**.

How does catabolite repression work? High level transcription of the *lac* operon requires the presence of a specific activator protein called **catabolite activator protein** (**CAP**), also called **cAMP response protein** (**CRP**). This protein, which is a dimer, cannot bind to DNA unless it is complexed with 3′, 5′-cAMP (cAMP). The CRP–cAMP complex binds to the *lac* promoter just upstream from the binding site for RNA polymerase. It increases the binding of RNA polymerase and so stimulates transcription of the *lac* operon.

Glucose inhibits **adenylate cyclase**, the enzyme that synthesizes cAMP from ATP. Thus, in the presence of glucose the intracellular level of cAMP falls, so CRP cannot bind to the *lac* promoter, and the *lac* operon is only weakly active (even in the presence of lactose). When glucose is absent, adenylate cyclase is not inhibited, the level of intracellular cAMP rises and binds to CRP. When glucose is absent but lactose is present, the CRP–cAMP complex then stimulates transcription of the *lac* operon and allows the lactose to be used as an alternative carbon source. In the absence of lactose, the lac repressor of course ensures that the *lac* operon remains inactive. These combined controls ensure that the *lacZ, lacY* and *lacA* genes are transcribed strongly only if glucose is absent and lactose is present.

Positive and negative regulation

The *lac* operon is a good example of **negative control** (**negative regulation**) of gene expression in that bound repressor prevents transcription of the structural genes. **Positive control** (**positive regulation**) of gene expression is when the regulatory protein binds to DNA and increases the rate of transcription. In this case, the regulatory protein is called an activator. The CAP/CRP involved in regulating the *lac* operon is a good example of an activator. Thus the *lac* operon is subject to both negative and positive control.

The *trp* operon

The tryptophan (*trp*) operon (Figure 4) contains five structural genes encoding enzymes for tryptophan biosynthesis with an upstream *trp* promoter (P_{trp}) and *trp* operator sequence (O_{trp}). The *trp* operator region partly overlaps the *trp* promoter. The operon is regulated such that transcription occurs when tryptophan in the cell is in short supply.

The trp repressor

In the absence of tryptophan (Figure 4a), a trp repressor protein encoded by a separate operon, *trpR*, is synthesized and forms a dimer. However, this is inactive and so is unable to bind to the *trp* operator and the structural genes of the *trp* operon are transcribed. When tryptophan is present (Figure 4b), the enzymes for tryptophan biosynthesis are not needed and so expression of these genes is turned off. This is achieved by tryptophan binding to the repressor to activate it so that it now binds to the operator and stops transcription of the structural genes. In this role, tryptophan is said to be a **co-repressor**. This is negative control, because the bound repressor prevents transcription, but note that the *lac* operon and *trp* operon show two ways in which negative control can be achieved; either (as in the *lac* operon) by having an active bound repressor that is inactivated by a bound ligand (the inducer) or (as in the *trp* operon) by having a repressor that is inactive normally but activated by binding the ligand. As in the case of the *lac* operator, the core binding site for the trp repressor in the *trp* operator is palindromic.

Attenuation

A second mechanism, called **attenuation**, is also used to control expression of the *trp* operon. The 5′ end of the polycistronic mRNA transcribed from the *trp* operon has a

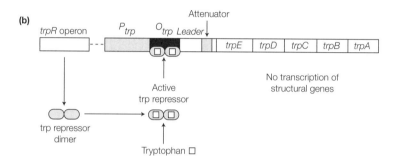

Figure 4. Regulation of the *trp* operon: (a) transcription in the absence of tryptophan; (b) no transcription in the presence of tryptophan.

leader sequence upstream of the coding region of the *trpE* structural gene (Figure 4). This leader sequence encodes a 14 amino acid **leader peptide** containing two tryptophan residues. The function of the leader sequence is to fine-tune expression of the *trp* operon based on the availability of tryptophan inside the cell. It does this as follows.

The leader sequence contains four regions (Figure 5, numbered 1–4) that can form a variety of base paired **stem-loop** ('**hairpin**') secondary structures. Now consider the two extreme situations: the presence or absence of tryptophan. Attenuation depends on the fact that, in bacteria, ribosomes attach to mRNA as it is being synthesized and so translation starts even before transcription of the whole mRNA is complete. When tryptophan is abundant (Figure 5a), ribosomes bind to the *trp* polycistronic mRNA that is being transcribed and begin to translate the leader sequence. Now, the two trp codons for the leader peptide lie within sequence 1, and the translational Stop codon (Section H1) lies between sequence 1 and 2.

During translation, the ribosomes follow very closely behind the RNA polymerase and synthesize the leader peptide, with translation stopping eventually between sequences 1 and 2. At this point, the position of the ribosome prevents sequence 2 from interacting with sequence 3. Instead, sequence 3 base pairs with sequence 4 to form a 3:4 stem-loop, which acts as a **transcription terminator**. Therefore, when tryptophan is present, further transcription of the *trp* operon is prevented.

If, however, tryptophan is in short supply (Figure 5b), the ribosome pauses at the two trp codons contained within sequence 1, leaving sequence 2 unbound by the ribosome and hence free to base pair with sequence 3 to form a 2:3 structure (also called the

Figure 5. Attenuation of the *trp* operon: (a) when tryptophan is plentiful, sequences 3 and 4 base pair to form a 3:4 structure that stops transcription; (b) when tryptophan is in short supply, the ribosome stalls at the trp codons in sequence 1, leaving sequence 2 available to interact with sequence 3. Thus a 3:4 transcription terminator structure cannot form and transcription continues.

anti-terminator). Because the 2:3 stem-loop has formed, sequence 3 is unavailable for 3:4 stem-loop formation and transcription continues to the end of the *trp* operon. Hence the availability of tryptophan controls whether transcription of this operon will stop early (attenuation) or continue to synthesize a complete polycistronic mRNA.

Historically, attenuation was discovered when it was noticed that deletion of a short sequence of DNA between the operator and the first structural gene, *trpE*, increased the level of transcription. This region was named the **attenuator** (Figure 4) and is the DNA that encodes that part of the leader sequence that forms the transcription terminator stem-loop.

Attenuation vs. repression

Overall, for the *trp* operon, repression via the trp repressor determines whether transcription will occur or not and attenuation then fine-tunes transcription. Attenuation occurs in at least six other operons that encode enzymes for amino acid biosynthetic pathways. In some cases, such as the *trp* operon, both repression and attenuation operate to regulate expression. In contrast, for some other operons such as the *his*, *thr* and *leu* operons, transcription is regulated only by attenuation.

G4 Transcription in eukaryotes: an overview

<table>
<tr><td colspan="2">Key Notes</td></tr>
<tr><td>Three RNA polymerases</td><td>In eukaryotes, RNA is synthesized by three RNA polymerases: RNA Pol I is a nucleolar enzyme that transcribes rRNAs; RNA Pol II is located in the nucleoplasm and transcribes mRNAs, most snRNAs, snoRNAs and miRNAs; RNA Pol III is also nucleoplasmic and transcribes tRNA, 5S rRNA, U6 snRNA, and the 7S RNA of the signal recognition particle (SRP).</td></tr>
<tr><td>RNA synthesis</td><td>Each RNA polymerase transcribes only one strand, the antisense (–) strand, of a double-stranded DNA template, directed by a promoter. Synthesis occurs $5' \rightarrow 3'$ and does not require a primer.</td></tr>
<tr><td>RNA polymerase subunits</td><td>Each of the three RNA polymerases contains 12 or more subunits, some of which are similar to those of E. coli RNA polymerase. However, four to seven subunits in each enzyme are unique to that enzyme.</td></tr>
<tr><td>Organelles</td><td>Chloroplasts and mitochondria each contain a single type of RNA polymerase. The chloroplast enzyme is similar to a bacterial RNA polymerase while the mitochondrial enzyme has similarity to some bacteriophage RNA polymerases.</td></tr>
<tr><td>Related topics</td><td>(F1) An introduction to DNA
(G1) An introduction to RNA
(G2) Transcription in prokaryotes
(G3) Operons
(G5) Transcription of protein-coding genes in eukaryotes
(G6) Regulation of transcription by RNA Pol II
(G7) Processing of eukaryotic pre-mRNA
(G8) Transcription and processing of ribosomal RNA
(G9) Transcription and processing of transfer RNA</td></tr>
</table>

Three RNA polymerases

Unlike prokaryotes where all RNA is synthesized by a single RNA polymerase, the nucleus of a eukaryotic cell has three RNA polymerases responsible for transcribing different types of RNA.

- **RNA polymerase I** (**RNA Pol I**) is located in the nucleolus and transcribes the **28S, 18S** and **5.8S rRNA** genes.

- **RNA polymerase II** (**RNA Pol II**) is located in the nucleoplasm and transcribes **protein-coding genes** to yield pre-mRNA and genes for **small nuclear RNAs** (**snRNAs**) involved in mRNA processing (Section G7) (except for U6 snRNA). It also synthesizes

small nucleolar RNAs (**snoRNAs**) involved in rRNA processing (Section G8), and **microRNA** (**miRNA**) (Section G7).

- **RNA polymerase III** (**RNA Pol III**) is also located in the nucleoplasm. It transcribes the genes for **tRNA, 5S rRNA,** and **U6 snRNA** and the **7S RNA** associated with the signal recognition particle (SRP) involved in the translocation of proteins across the endoplasmic reticulum membrane (Section H4).

RNA synthesis

The basic mechanism of RNA synthesis by the three eukaryotic RNA polymerases is the same as for the prokaryotic enzyme (Section G2), that is:

- The initiation of RNA synthesis by RNA polymerase is directed by the presence of a promoter site on the 5′ side of the transcriptional start site

- The RNA polymerase transcribes one strand, the **antisense** (–) **strand**, of the DNA template

- RNA synthesis does not require a primer

- RNA synthesis occurs in the 5′ → 3′ direction with the RNA polymerase catalyzing a nucleophilic attack by the 3′-OH of the growing RNA chain on the α phosphorus atom on an incoming ribonucleoside 5′-triphosphate

RNA polymerase subunits

Each of the three eukaryotic RNA polymerases contains 12 or more subunits and so these are large complex enzymes. The genes encoding some of the subunits of each eukaryotic enzyme show DNA sequence similarities to genes encoding subunits of the core enzyme ($\alpha_2\beta\beta'\omega$) of *E. coli* RNA polymerase (Section G2). However, four to seven other subunits of each eukaryotic RNA polymerase are unique in that they show no similarity either with bacterial RNA polymerase subunits or with the subunits of other eukaryotic RNA polymerases.

Organelles

It is generally accepted that mitochondria and chloroplasts were originally bacteria, which, during early evolution, formed a symbiotic relationship with a primitive eukaryotic cell (the **endosymbiont theory**). In support of this, many of the features of gene organization and expression are similar in these organelles to what occurs in bacteria. For example, chloroplasts contain a single type of RNA polymerase similar in structure to a bacterial RNA polymerase. Mitochondria also contain a single type of RNA polymerase, which has similarities to some bacteriophage RNA polymerases.

G5 Transcription of protein-coding genes in eukaryotes

Key Notes

Gene organization

Most protein-coding genes in eukaryotes consist of coding sequences called exons interrupted by noncoding sequences called introns. The number of introns and their size varies from gene to gene. The primary transcript (pre-mRNA) undergoes processing reactions to yield mature mRNA, which typically has a 5′ cap and 3′ poly(A) tail with a central coding region flanked by a 5′ untranslated region (UTR) and a 3′ UTR.

Promoters for RNA polymerase II

The transcription of eukaryotic protein-coding genes is controlled by a combination of sequence elements in the core promoter near the transcriptional start site and by upstream control elements. The two main consensus sequences in the core promoter are the TATA box and the initiator sequence (Inr). Genes transcribed by RNA polymerase II may have both of these elements, just one or neither. Genes lacking a TATA box often have a downstream promoter element (DPE). Genes lacking both a TATA box and an Inr element tend to be transcribed at low rates and not at a defined transcriptional start site.

Initiation of transcription

RNA polymerase binding to the promoter requires the formation of a transcription initiation complex involving several general (basal) transcription factors that assemble in a strict order. For genes that have a TATA box in their core promoter, the key initial event is the binding of the transcription factor IID (TFIID) protein complex to the TATA box.

Elongation and termination

After TFIIH phosphorylates the C-terminal domain (CTD) of RNA polymerase II, this enzyme starts moving along the DNA template synthesizing RNA. Elongation continues until transcription comes to a halt soon after a poly(A) signal sequence has been synthesized.

Related topics

(F1) An introduction to DNA
(G1) An introduction to RNA
(G2) Transcription in prokaryotes
(G3) Operons
(G4) Transcription in eukaryotes: an overview

(G6) Regulation of transcription by RNA Pol II
(G7) Processing of eukaryotic pre-mRNA
(G8) Transcription and processing of ribosomal RNA
(G9) Transcription and processing of transfer RNA

Gene organization

In marked contrast to prokaryotic genes where proteins are encoded by a continuous sequence of triplet codons, the vast majority of protein-coding genes in eukaryotes are **discontinuous**. The coding sections of the gene (called **exons**) are interrupted by non-coding sections of DNA (called **introns**; Figure 1). Nevertheless, the triplet codons within the exons and the order of exons themselves in the gene is still **colinear** with the amino acid sequence of the encoded polypeptide. The number of introns in a protein-coding gene varies and they range in size from about 80 bp to over 10 000 bp.

The primary transcript of a eukaryotic protein-coding gene is a **pre-mRNA** molecule that must be processed to yield mature mRNA ready for translation; the pre-mRNA receives a **5′ cap** and (usually but not always) a **poly(A) tail** of about 200 A residues, and the intron sequences are removed by **RNA splicing**. These RNA processing reactions are covered in detail in Section G7 but it must be stressed that they are closely linked to the transcription process itself. For example, cap addition occurs very soon after the 5′ end of the pre-mRNA has been synthesized and while transcription elongation is continuing. Likewise, polyadenylation is an integral part of the transcription termination process for most protein-coding genes.

The mature mRNA molecule typically has a 5′ cap and a 3′ poly(A) tail with the various exons (e.g. exons 1, 2 and 3 in Figure 1) joined together to form a central **coding region**. During protein synthesis (Section H3), the coding region is translated to yield the

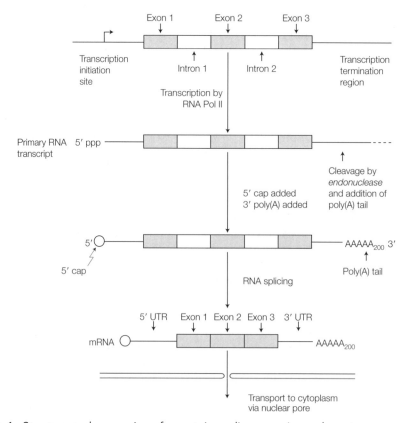

Figure 1. Structure and expression of a protein-coding gene in a eukaryote.

corresponding polypeptide for this gene. The regions between the 5′ cap and the coding region and between the coding region and the 3′ poly(A) tail are not translated and so are called the **5′ untranslated region (5′ UTR)** and **3′ untranslated region (3′ UTR)**, respectively (Figure 1).

Promoters for RNA polymerase II

The transcription of eukaryotic protein-coding genes is controlled by a variety of sequence elements. Some of these are close to the transcriptional start site and have conserved sequences, and are called the **core promoter**. However, alone they are often only weakly efficient and interaction with **upstream control elements** is needed to drive efficient initiation of transcription. For these genes therefore, the promoter is a combination of sequences near to the start site of transcription and sequences located further upstream.

Two main consensus sequences are found in the core promoter (Figure 2).

- Most promoter sites for RNA polymerase II include a highly conserved sequence located about 25–35 bp upstream (i.e. to the 5′ side) of the start site (although it can occur as far upstream as –100). This sequence has the consensus 5′-TATA(A/T)AA(A/G)-3′ and is called the **TATA box** (Figure 2). Since the start site is denoted as position +1, the TATA box position is said to be located at about position –25 to –35. The TATA box sequence resembles the –10 sequence (Section G2) in prokaryotes (TATAAT) but is located further upstream. Both elements have essentially the same function, namely recognition by the RNA polymerase in order to position the enzyme at the correct location to initiate transcription. The sequence around the TATA box is also important in that it influences the efficiency of initiation.

- An **initiator sequence (Inr)** is often found at the transcriptional start site in the region –3 to +5. It has the consensus (in mammalian genes) of 5′-YCANTYY-3′ where N is any of the four standard nucleotides and Y represents C or T.

Genes transcribed by RNA polymerase II often have both a TATA box and an Inr sequence but others have only one element, not both.

Genes that lack a TATA box often have a third sequence called a **downstream promoter element (DPE)** located downstream of the transcriptional start site in the region +28 to +32 (Figure 2). The DPE does not have a defined consensus sequence (i.e. its sequence

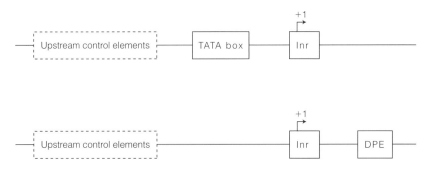

Figure 2. Example core promoters for RNA Pol II. When present, the TATA box is located approximately 25–35 bp upstream of the transcriptional start site (denoted as +1) whereas the Inr sequence spans the transcriptional start site since it is located at between –3 and +5. The DPE is located downstream of the start site in the region +28 to +32.

varies) but it is known to be involved in transcription initiation by binding to TFIID, one of the key proteins involved (see below).

Some genes transcribed by RNA polymerase II have neither a TATA box nor an Inr element; these genes tend to be transcribed at low rates and initiate transcription somewhere within a broad region of DNA (about 200 bp or so) rather than at a defined transcriptional start site.

As mentioned above, transcription is also regulated by upstream control elements; these are considered in Section G6.

Initiation of transcription

In order to initiate transcription, RNA polymerase II requires the assistance of several other proteins or protein complexes, called **general** (or **basal**) **transcription factors**, which must assemble into a complex on the promoter in order for RNA polymerase to bind and start transcription (Figure 3). These all have the generic name of **TFII** (transcription factor for RNA polymerase II).

For genes that have a TATA box, the first event in initiation is the binding of the **transcription factor IID** (**TFIID**) protein complex to the TATA box via one of its subunits called **TBP** (**TATA box binding protein**). As soon as the TFIID complex has bound, **TFIIA** binds and stabilizes the TFIID–TATA box interaction. Next, **TFIIB** binds to TFIID. However, TFIIB can also bind to RNA polymerase II and so acts as a bridging protein. Thus, RNA polymerase II, which has already complexed with **TFIIF**, now binds. This is followed by the binding of **TFIIE** and **TFIIH**. This final protein complex contains at least 40 polypeptides and is called the **transcription initiation complex**.

Elongation and termination

TFIIH has two functions. It is a helicase, which means that it can use ATP to unwind the DNA helix, converting the closed promoter complex to an open promoter complex and allowing transcription to begin. In addition, it phosphorylates RNA polymerase II, which causes this enzyme to change its conformation and dissociate from other proteins in the initiation complex. RNA polymerase II now starts moving along the DNA template, synthesizing RNA, that is, the process enters the **elongation phase**. The key phosphorylation occurs on a long C-terminal tail called the **C-terminal domain** (**CTD**) of the RNA polymerase II molecule. This is a substantial event; the CTD in mammals consists of 52 repeats of the sequence Tyr-Ser-Pro-Thr-Ser-Pro-Ser and two of the three serines in each repeat can be phosphorylated. Interestingly, only RNA polymerase II that has a nonphosphorylated CTD can initiate transcription but only an RNA polymerase II with a phosphorylated CTD can elongate RNA.

Elongation of the RNA chain continues until termination occurs. Most eukaryotic mRNAs have a **poly(A) tail** at their 3′ end, i.e. up to 250 A residues. This stretch of A residues is not coded by the DNA but is added to the RNA by an enzyme called **poly(A) polymerase** in a process called **polyadenylation**. Polyadenylation requires the existence of a **polyadenylation signal sequence** near the 3′ end of the pre-mRNA. Until recently, polyadenylation was regarded as an event that happened after transcription had terminated, i.e. a post-transcriptional event. However, it is now clear that transcription stops soon after the poly(A) signal sequence has been synthesized and that polyadenylation is a key part of the transcription termination process. This is described in more detail in Section G7.

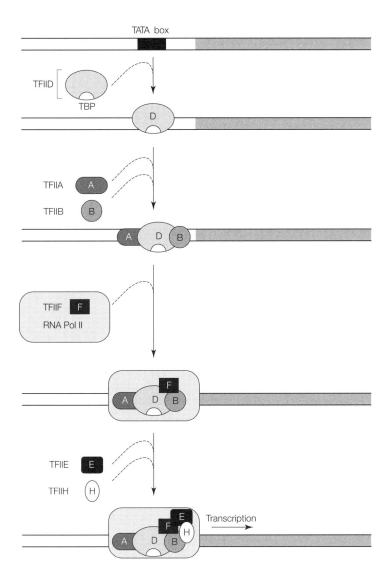

Figure 3. Initiation of transcription by RNA polymerase II. TFIID binds to the TATA box followed in order by the binding of TFIIA, TFIIB and a preformed complex of TFIIF–RNA polymerase II. Subsequently TFIIE and TFIIH bind in order and transcription then starts about 25 bp downstream from the TATA box. Note that the placement of the various factors in this diagram is arbitrary; their exact positions in the complex are not known.

G6 Regulation of transcription by RNA Pol II

Key Notes

Mechanism of regulation

Many genes are active in all cells but some are transcribed only in specific cell types, at specific times and/or only in response to specific external stimuli. Transcriptional regulation occurs via transcription factors that bind to short control elements associated with the target genes and then interact with each other and with the transcription initiation complex to increase or decrease the rate of transcription of the target gene.

Regulatory elements

Many transcription factors bind to control elements located upstream within a few hundred base pairs of the protein-coding gene. Some regulatory elements (e.g. GC box and CAAT box) are found upstream of most protein-coding genes, but other elements are associated with only a few genes and are responsible for gene-specific transcriptional regulation (e.g. cAMP response element, hormone response elements).

Enhancers

Enhancers are positive transcriptional control elements typically 100–200 bp long that can be located either upstream or downstream of the target gene, are active in either orientation, and can activate transcription from the target gene even when located a long distance away (sometimes 10–50 kb). The transcription factors bound to these long-distance elements interact with the transcription initiation complex by looping out of the DNA.

Transcription factors have multiple domains

Transcription factors that increase the rate of transcription usually have at least two domains of protein structure, a DNA-binding domain that recognizes the specific DNA control element to bind to, and an activation domain that interacts with other transcription factors or the RNA polymerase. Many transcription factors operate as dimers (homodimers or heterodimers) held together via dimerization domains. Some transcription factors interact with small ligands via a ligand-binding domain.

DNA-binding domains

DNA-binding domains contain characteristic protein motifs. The helix-turn-helix (HTH) motif contains two α-helices separated by a short β-turn. When the transcription factor binds to DNA, the recognition helix lies in the major groove of the DNA double helix. The zinc finger motif consists of a peptide loop with either two cysteines and two histidines (the C_2H_2 finger) or four cysteines (the C_4 finger) at the base of the loop that tetrahedrally coordinate a zinc ion. The zinc finger secondary structure is two β-strands and one α-helix. Transcription factors often contain several zinc fingers; in each case the α-helix binds in the major groove of the DNA double helix. Some transcription factors (e.g. bZIP proteins, basic HLH proteins) contain basic domains that interact with the target DNA.

Dimerization domains

A leucine zipper has a leucine every seventh amino acid and forms an α-helix with the leucines presented on the same side of the helix every second turn, giving a hydrophobic surface. Two transcription factor monomers can interact via the hydrophobic faces of their leucine zipper motifs to form a dimer. The helix-loop-helix (HLH) motif contains two α-helices separated by a nonhelical loop. The C-terminal α-helix has a hydrophobic face; two transcription factor monomers, each with an HLH motif, can dimerize by interaction between the hydrophobic faces of the two C-terminal α-helices.

Activation domains

No common structural motifs are known for the activation domains of transcription factors. Activation domains that are rich in acidic amino acids, glutamines or prolines have been reported.

Repressors

Repressor proteins that inhibit the transcription of specific genes in eukaryotes may bind either to control elements near to the target gene or to silencers that may be located a long distance away. The repressor may inhibit transcription of the target gene directly or may do so by interfering with the function of an activator protein required for efficient gene transcription.

Related topics

Mechanism of regulation

A number of protein-coding genes are active in all cells and are required for so-called 'house-keeping' functions, such as the enzymes of glycolysis (Section J3) and the citric acid cycle (Section L1) and the proteins of the electron transport chain (Section L2). However, some genes are active only in specific cell types and are responsible for defining the specific characteristics and function of those cells; for example, immunoglobulin genes in lymphocytes, myosin in muscle cells. In addition, the proteins expressed by any given cell may change over time (for example during early development) or in response to external stimuli, such as hormones. Eukaryotic cells can regulate the expression of protein-coding genes at a number of levels but a prime site of regulation is transcription.

Transcriptional regulation in a eukaryotic cell (i.e. which genes are transcribed and at what rate) is mediated by **transcription factors** (other than the general transcription factors; Section G5), which recognize and bind to short regulatory DNA sequences (**control elements**) associated with the gene. These sequences are also called *cis*-**acting elements** (or simply *cis*-**elements**) since they are on the same DNA molecule as the gene being controlled (*cis* is Latin for 'on this side'). The protein transcription factors that bind to these elements are also known as *trans*-**acting factors** (or simply *trans*-**factors**) in that the genes encoding them can be on different DNA molecules (i.e. on different chromosomes). The transcription factors which regulate specific gene transcription do so by interacting with the proteins of the transcription initiation complex and may either increase (activate) or decrease (repress) the rate of transcription of the target gene.

Typically each protein-coding gene in a eukaryotic cell has several control elements in its promoter (Figure 1) and hence is under the control of several transcription factors, which interact with each other and with the transcription initiation complex by protein:protein interaction to determine the rate of transcription of that gene.

Regulatory elements

Many transcription factors bind to control sequences (**regulatory elements**) within a few hundred base pairs of the protein-coding gene being regulated. Positive control elements that lie upstream of the gene, usually within 200 bp of the transcriptional start site (Figure 1), function to increase the transcriptional activity of the gene well above that of the basal promoter.

Some of these elements, for example the **GC box** and the **CAAT box**, are found in the promoters of many eukaryotic protein-coding genes; indeed genes often have several copies of one or both elements. The CAAT box has the consensus sequence 5′-GGCCAATCT-3′ and binds transcription factors NF-1 and NF-Y, whereas the GC box has the consensus sequence 5′-GGGCGG-3′ and binds transcription factor SP1, which then interacts with general transcription factor TFIID (Section G5).

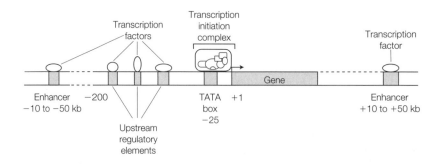

Figure 1. Control regions that regulate transcription of a typical eukaryotic protein-coding gene. Although shown as distinct entities here for clarity, *in vivo* the different regulatory proteins bound to the control elements and distant enhancers interact with each other and with the general transcription factors of the transcription initiation complex to modulate the rate of transcriptional initiation.

In contrast, some upstream regulatory elements are associated only with a few specific genes and are responsible for limiting the transcription of those genes to certain tissues or in response to certain stimuli. For example:

- The cyclic AMP (cAMP) response element (CRE) has the consensus sequence 5′-WCGTCA-3′ (where W is A or T) and is involved in cAMP modulation of gene transcription, binding the CREB transcription factor which activates transcription.

- The pituitary cell element (consensus sequence 5′-ATATTCAT-3′) binds transcription factor Pit-1 and activates the transcription of specific genes in pituitary cells.

- Steroid hormones also control gene transcription. They enter the target cell and bind to specific **steroid hormone receptors** in the cytoplasm. The binding of the hormone releases the receptor from an inhibitor protein that normally keeps the receptor in the cytoplasm. The hormone–receptor complex, now free of inhibitor, dimerizes and travels to the nucleus where it binds to a transcriptional control element, called a **hormone response element**, in the promoters of target genes. Then, like other transcription factors, the bound hormone–receptor complex interacts with the transcription initiation complex to increase the rate of transcription of the gene. The result is hormone-specific transcription of a subset of genes in target cells that contain the appropriate steroid hormone receptor. Here, the hormone receptor is itself a transcription factor that is activated by binding the hormone ligand.

The multiple regulatory elements associated with eukaryotic genes means that multiple transcription factors bind and interact with each other and with the transcription initiation complex to activate or repress transcription. So the transcriptional activity of that gene depends on the *combination* of factors that bind to it at that particular time.

Enhancers

Although many positive control elements lie close to the gene they regulate, others can be located long distances away (sometimes 10–50 kb) either upstream or downstream of the gene (Figure 1). A long-distance positive control sequence of this kind is called an **enhancer** if the transcription factor(s) that binds to it increases the rate of transcription. An enhancer is typically 100–200 bp long and contains several sequence elements that act together to give the overall enhancer activity. Enhancers have the following characteristics; they:

- Can activate transcription over long distances

- Can be located upstream or downstream of the gene being controlled

- Are active in either orientation with respect to the gene.

For enhancers located a long distance away from the gene being controlled, interaction between transcription factors bound to the enhancer and to promoter elements near the gene may occur by looping out of the DNA between the two sets of elements (Figure 2).

Transcription factors have multiple domains

In most cases, the transcription factors in eukaryotes that bind to enhancer or promoter sequences are activator proteins that induce transcription. These proteins usually have at least two distinct domains of protein structure, a **DNA-binding domain** that recognizes the specific DNA sequence to bind to, and an **activation domain** responsible for bringing about the transcriptional activation by interaction with other transcription factors

Figure 2. Looping out of DNA allowing the interaction of enhancer-bound factor(s) with the transcription initiation complex.

and/or the RNA polymerase molecule. Many transcription factors operate as dimers, either **homodimers** (identical subunits) or **heterodimers** (dissimilar subunits) with the subunits held together via **dimerization domains**. DNA-binding domains and dimerization domains contain characteristic protein structures (**motifs**) that are described below. Finally, some transcription factors (e.g. steroid hormone receptors) are responsive to specific small molecules (**ligands**), which regulate the activity of the transcription factor. In these cases, the ligand binds at a **ligand-binding domain**.

DNA-binding domains

Several DNA-binding domains are known but cannot all be covered here. Some of the more prominent ones are as follows.

Helix-turn-helix (HTH motif)

The HTH motif consists of two α-helices separated by a short (four amino acid) peptide sequence that forms a β-**turn** (Figure 3a). When the transcription factor binds to DNA, one of the helices, called the **recognition helix**, lies in the major groove of the DNA double helix (Figure 3b). The HTH motif in eukaryotes was originally discovered in certain transcription factors that play major roles in *Drosophila* early development. These proteins each contain a 60-amino acid DNA-binding region called a **homeodomain** (encoded by a DNA sequence called a **homeobox**). The homeodomain has four α-helices in which helices II and III are the classic HTH motif. Since the original discovery, the HTH motif has been found in a wide range of eukaryotic transcription factors, including many that have no role in development. It is also found in many prokaryotic regulatory proteins, including the lac repressor that controls the *lac* operon (Section G3).

Zinc finger

At least six types of zinc finger have been reported, two of which are the C_2H_2 **finger** and the C_4 **finger**. The C_2H_2 zinc finger is a loop of 12 or so amino acids with two cysteines and two histidines at the base of the loop that tetrahedrally coordinate a zinc ion (Figure 4a). This forms a compact structure of two β-strands and one α-helix (Figure 4b). The α-helix contains a number of conserved basic amino acids and interacts directly with the DNA, binding in the major groove of the double helix. Transcription factors that contain zinc fingers often contain several such motifs, arranged such that the α-helix of each contacts the DNA. Indeed RNA polymerase III transcription factor A (**TFIIIA**; Section G8) contains nine zinc fingers! The SP1 transcription factor, which binds to the GC box, has three zinc fingers.

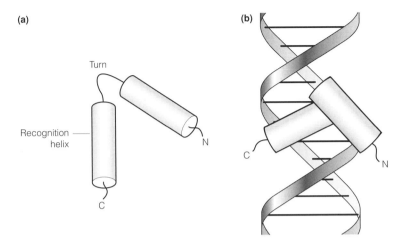

Figure 3. (a) A helix-turn-helix (HTH) motif of a DNA-binding protein; (b) binding of the HTH motif to target DNA showing the recognition helix lying in the major groove of the DNA.

The C_4 zinc finger is also found in a number of transcription factors, including steroid hormone receptor proteins. This motif forms a similar structure to that of a C_2H_2 zinc finger but has four cysteines coordinated to the zinc ion instead of two cysteines and two histidines (Figure 4c).

Figure 4. (a) A C_2H_2 zinc finger; (b) C_2H_2 zinc finger secondary structure. Adapted from Travers, A. (1993) *DNA–Protein Interactions*, Chapman & Hall; (c) a C_4 zinc finger.

Basic domains

DNA-binding domains called **basic domains** (an α-helix rich in basic amino acids) sometimes occur in transcription factors in combination with leucine zipper or helix-loop-helix (HLH) dimerization domains (see below). The combination of basic domain and dimerization domain gives these proteins their names of **basic leucine zipper proteins (bZIP)** or **basic HLH proteins**, respectively. In each case, the dimerization means that two basic domains (one from each monomer) interact with the target DNA.

Dimerization domains

Leucine zippers

The leucine zipper motif contains a leucine every seventh amino acid in the primary sequence and forms an α-helix with the leucines presented on the same side of the helix every second turn, giving a hydrophobic surface. The transcription factor dimer is formed by the two monomers interacting via the hydrophobic faces of their leucine zipper motifs (Figure 5a). In the case of bZIP proteins, each monomer also has a basic DNA-binding domain located N-terminal to the leucine zipper. Thus the bZIP protein dimer has two basic domains. These actually face in opposite directions, which allows them to bind to DNA sequences that have inverted symmetry. They bind in the major groove of the target DNA (Figure 5b). The leucine zipper domain also acts as the dimerization domain in transcription factors that use DNA-binding domains other than the basic domain. For example, some homeodomain proteins, containing the HTH motif for DNA binding, have leucine zipper dimerization domains. In all cases, the dimers that form may be homodimers or heterodimers.

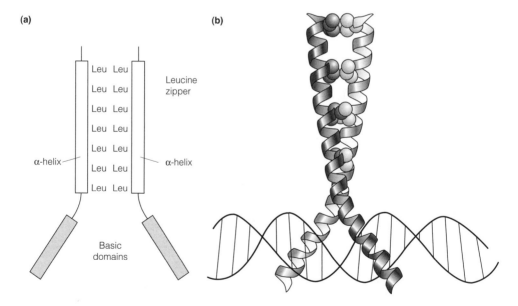

Figure 5. (a) A bZIP protein dimer showing the leucine zipper dimerization domain and the two basic domains; (b) folded structure of a bZIP protein showing the basic domains binding in the major groove of the target DNA. Adapted from Travers, A. (1993) *DNA–Protein Interactions*, Chapman & Hall.

Helix-loop-helix (HLH motif)

The HLH dimerization domain is quite distinct from the HTH motif described above (which is involved in DNA binding *not* dimerization) and must not be confused with it. The HLH domain consists of two α-helices separated by a nonhelical loop. The C-terminal α-helix has hydrophobic amino acids on one face. Thus two transcription factor monomers, each with an HLH motif, can dimerize by interaction between the hydrophobic faces of the two C-terminal α-helices. Like the leucine zipper (see above), the HLH motif is often found in transcription factors that contain basic DNA-binding domains. Again, like the leucine zipper, the HLH motif can dimerize transcription factor monomers to form either homodimers or heterodimers. This ability to form heterodimers markedly increases the variety of active transcription factors that are possible and so increases the potential for gene regulation.

Activation domains

Unlike DNA-binding domains and dimerization domains, no common structural motifs have yet been identified in the activation domains of diverse transcription factors. Some types of activation domain are as follows:

- **Acidic activation domains** are rich in acidic amino acids (aspartic and glutamic acids). For example, mammalian glucocorticoid receptor proteins contain this type of activation domain
- **Glutamine-rich domains** (e.g. as in SP1 transcription factor)
- **Proline-rich domains** (e.g. as in *c-jun* transcription factor)

Repressors

Gene repressor proteins that inhibit the transcription of specific genes in eukaryotes also exist. They may act by binding either to control elements within the promoter region near the gene or at sites located a long distance away from the gene, called **silencers**. The repressor protein may inhibit transcription directly. One example is the **mammalian thyroid hormone receptor**, which, in the absence of thyroid hormone, represses transcription of the target genes. However, other repressors inhibit transcription by blocking activation. This can be achieved in one of several ways: by blocking the DNA binding site for an activator protein, by binding to and masking the activation domain of the activator factor, or by forming a non-DNA binding complex with the activator protein. Several examples of each mode of action are known.

G7 Processing of eukaryotic pre-mRNA

Key Notes

Overview	The primary RNA transcript from a protein-coding gene in a eukaryotic cell is modified by RNA processing reactions in order to become functional mRNA. The 5′ end is modified with a 5′ cap structure. Most pre-mRNAs are cleaved near the 3′ end and a poly(A) tail is added. Intron sequences are removed by RNA splicing. Transcription by RNA polymerase II and pre-mRNA processing are tightly coupled events; key components involved in the processing steps are associated with the C-terminal domain (CTD) of RNA polymerase II and transfer to the growing RNA molecule to carry out the processing steps.
5′ Processing: capping	Immediately after transcription, the 5′ phosphate is removed, guanylyl transferase adds a G residue linked via a 5′–5′ covalent bond, and this is methylated to form a 7-methylguanosine (m^7G) cap. The ribose residues of either the adjacent one or two nucleotides may also be methylated. The cap protects the 5′ end of the mRNA against ribonuclease degradation and also functions in the initiation of protein synthesis.
RNA splicing	Intron sequences are removed by RNA splicing that cleaves the RNA at exon–intron boundaries and ligates the ends of the exon sequences together. The cleavage sites are marked by conserved consensus sequences; in most cases the intron starts with GU and ends with AG, a polypyrimidine tract lies upstream of the AG, and a conserved branchpoint sequence is located about 20–50 nt upstream of the 3′ splice site. The splicing reaction involves two transesterification steps, which release the intron as a branched lariat structure containing a 2′5′ bond with a conserved A residue in the branchpoint sequence. The RNA splicing reactions require snRNPs and accessory proteins that assemble into a spliceosome. The RNA components of the snRNPs are complementary to the conserved sequences and so can base pair with them. AT-AC introns start with AU and end with AC, instead of GU and AG, respectively, and their splicing requires a different set of snRNPs than those used for splicing of the major form of intron, except both use U5 snRNP. A few organisms can splice together exons from two different RNA molecules: *trans*-splicing. Some introns are self-splicing; the intron RNA sequence catalyzes its own excision without the involvement of a spliceosome.

3′ Processing: cleavage and polyadenylation

Most pre-mRNA transcripts are cleaved near the 3′ end between a polyadenylation signal sequence (5′-AAUAAA-3′) and a GU-rich (or U-rich) sequence downstream. Specific proteins bind to these sequences to form a complex. One of the bound proteins, poly(A) polymerase, adds a poly(A) tail of about 200–250 A residues to the new 3′ end of the RNA molecule and poly(A) binding protein molecules bind to this. Termination of transcription by RNA polymerase II is coupled to polyadenylation; transcription stops soon after RNA polymerase II synthesizes the polyadenylation signal sequence. The exact role of the poly(A) tail is unclear but it may protect the 3′ end of the final mRNA against nuclease degradation and it increases translational efficiency of the mRNA. Some pre-mRNAs (e.g. histone pre-mRNAs) are cleaved near the 3′ end but no poly(A) tail is added.

Alternative processing

Some pre-mRNAs contain more than one set of sites for 3′ end cleavage and polyadenylation. The use of alternative sites can lead to mRNAs that contain different 3′ noncoding regions or have different coding capacities. Some alternative splice pathways affect which exons are retained in the final mRNA, allowing several different proteins to be synthesized from a single gene.

RNA editing

RNA editing can change the sequence of an mRNA molecule after synthesis and processing; nucleotides may be substituted, added or deleted. In human liver, apolipoprotein B pre-mRNA does not undergo editing and translation yields apolipoprotein B100. In the small intestine, RNA editing converts a single C residue in apolipoprotein B pre-mRNA to U, changing a codon for glutamine (CAA) to a termination codon (UAA). Translation of the edited mRNA yields the shorter protein, apolipoprotein B48, that lacks a protein domain for receptor binding. Many other examples of editing occur, including trypanosome mitochondrial mRNAs, where over half of the uridines in the final mRNA are acquired through the editing process.

mRNA silencing

The translation of specific mRNAs can be inhibited by microRNAs (miRNAs), a process called mRNA silencing. Most miRNAs are transcribed by RNA Pol II with a few transcribed by RNA Pol III. The miRNA is derived either by processing a large RNA precursor transcribed from an miRNA gene or by processing an intron containing an miRNA sequence within a protein-coding gene. The mature miRNA forms an miRNP complex with proteins (called the miRISC) and binds to complementary sequences (usually in the 3′ UTR) in the target mRNA, inhibiting translation.

Overview

In eukaryotes, the product of transcription of a protein-coding gene is the **primary RNA transcript pre-mRNA** (Section G5), which requires processing to generate functional mRNA. As described earlier (Section G5), transcription and RNA processing in eukaryotes are tightly coupled events. A number of the enzymes involved in the processing steps are associated with the C-terminal domain (CTD) of RNA polymerase II and then transfer to the growing RNA molecule as they are required for processing steps.

Several processing reactions occur. Immediately after it has been synthesized by RNA polymerase II, the 5′ end of the pre-mRNA is modified by the addition of a **5′ cap** (a process known as **capping**). The primary RNA transcript includes both coding (exon) and noncoding (intron) regions (Section G5, Figure 1). The latter need to be removed and the exon sequences joined together by RNA splicing (Section G5, Figure 1) to generate a continuous RNA coding sequence ready for translation. The RNA splicing reactions are catalyzed by a large RNA–protein complex called a **spliceosome** (see below) that assembles on the primary RNA transcript as it is being synthesized, so RNA splicing occurs soon after synthesis of the relevant RNA. The 3′ ends of most (but not all) pre-mRNAs are also modified by cleavage and the addition of about 200–250 A residues to form a **poly(A) tail**. This **polyadenylation** reaction occurs during transcription termination. The following sections describe each RNA processing reaction in more detail.

5′ Processing: capping

Capping of the pre-mRNA occurs very soon after synthesis of the 5′ end of the molecule and usually before this RNA is more than 30 nt long. It involves the addition of **7-methylguanosine** (m^7G) to the 5′ end (Figure 1). To achieve this, the terminal 5′ phosphate is first removed by a phosphatase. **Guanylyl transferase** then catalyzes a reaction whereby the resulting diphosphate 5′ end attacks the α phosphorus atom of a GTP molecule to add a G residue in an unusual 5′5′ triphosphate link (Figure 1). The G residue is then methylated by **guanine methyltransferase** adding a methyl group to the N-7 position of the guanine ring, using **S-adenosyl methionine** as methyl donor. This structure, with just the m^7G in position, is called a **cap 0 structure**. The ribose of the adjacent nucleotide (nucleotide 2 in the RNA chain) or the riboses of both nucleotides 2 and 3 may also be methylated to give **cap 1** or **cap 2** structures, respectively. In these cases, the methyl groups are added to the 2′-OH groups of the ribose sugars (Figure 1).

The cap protects the 5′ end of the primary transcript against attack by ribonucleases that have specificity for 3′5′ phosphodiester bonds and so cannot hydrolyze the 5′5′ bond in

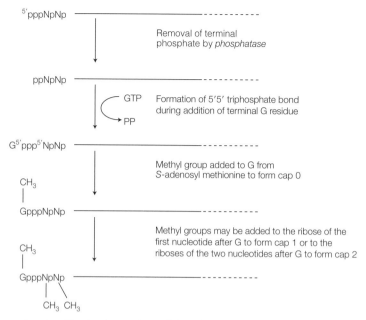

Figure 1. Steps involved in the formation of the 5′ cap.

the cap structure. In addition, the cap plays a role in the initiation step of protein synthesis in eukaryotes. Only RNA transcripts from eukaryotic protein-coding genes become capped; prokaryotic mRNA and eukaryotic rRNA and tRNAs are uncapped.

RNA splicing

A key step in RNA processing is the precise removal of intron sequences and joining the ends of neighboring exons to produce a functional mRNA molecule, a process called **RNA splicing**. The exon–intron boundaries are marked by specific sequences (Figure 2). In most cases, at the 5′ boundary between the exon and the intron (**the 5′ splice site**), the intron starts with the sequence GU and at the 3′ exon–intron boundary (**the 3′ splice site**) the intron ends with the sequence AG. Each of these two sequences lies within a longer consensus sequence. For example, in vertebrates, the consensus 5′ splice site sequence is 5′-AG<u>GU</u>AAGU-3′ (the invariant GU is underlined). A **polypyrimidine tract** (a conserved stretch of about 11 pyrimidines) lies upstream of the AG at the 3′ splice site (Figure 2).

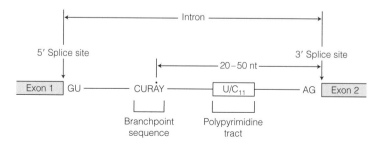

Figure 2. Conserved sequences for RNA splicing. The residue marked as Ȧ in the branchpoint sequence is the site of formation of the 2′5′ branch.

A key signal sequence is the **branchpoint sequence**, which is located about 20–50 nt upstream of the 3′ splice site. In vertebrates this sequence is 5′-CURAY-3′ where R=purine and Y=pyrimidine (in yeast this sequence is 5′-UACUAAC-3′).

RNA splicing occurs in two steps (Figure 3). In the first step, the 2′-OH of the A residue at the branch site (indicated as Å in Figure 2) attacks the 3′5′ phosphodiester bond at the 5′ splice site causing that bond to break and the 5′ end of the intron to loop round and form an unusual 2′5′ bond with the A residue in the branchpoint sequence. Because this A residue already has 3′5′ bonds with its neighbors in the RNA chain, the intron becomes branched at this point to form what is known as a **lariat** intermediate (named as such since it resembles a cowboy's lasso). The new 3′-OH end of exon 1 now attacks the phosphodiester bond at the 3′ splice site causing the two exons to join and release the intron, still as a lariat. In each of the two splicing reactions, one phosphate–ester bond is exchanged for another (i.e. these are two **transesterification reactions**). Since the number of phosphate–ester bonds is unchanged, no energy (ATP) is consumed.

RNA splicing requires the involvement of several **small nuclear RNAs** (**snRNAs**) each of which is associated with several proteins to form a small nuclear ribonucleoprotein particle or **snRNP** (pronounced 'snurp'). Because snRNAs are rich in U residues, they are named U1, U2, etc. The RNA components of the snRNPs have regions that are complementary to the 5′ and 3′ splice site sequences and to other conserved sequences in the intron and so can base pair with them. The U1 snRNP binds to the 5′ splice site and U2 snRNP binds to the branchpoint sequence (Figure 4). A **tri-snRNP complex** of U4, U5 and U6 snRNPs then binds, as do other accessory proteins, so that a multicomponent complex (called a **spliceosome**) is formed at the intron to be removed and causes the intron to be looped out (Figure 4). Thus, through interactions between the snRNAs and the pre-mRNA, the spliceosome brings the upstream and downstream exons together ready for splicing. The spliceosome next catalyzes the two-step splicing reaction to remove the intron and ligate together the two exons. The spliceosome then dissociates and the released snRNPs can take part in further splicing reactions at other sites on the pre-mRNA. Note that as well as snRNPs, hundreds of other proteins called **splicing factors** are also involved in a spliceosome and the functions of most of these have yet to be determined.

Although the vast majority of pre-mRNA introns start with GU at the 5′ splice site and end with AG at the 3′ splice site, some introns (possibly as many as 1%) have different splice

Figure 3. The two steps of RNA splicing.

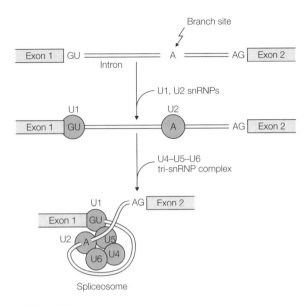

Figure 4. Formation of the spliceosome.

site consensus sequences. In these cases, the intron starts with AU and ends with AC instead of GU and AG, respectively (Figure 5). Since RNA splicing involves recognition of the splice site consensus sequences by key snRNPs (see above), and since these sequences are different in the minor intron class, U1, U2, U4 and U6 snRNPs do not take part in splicing these so-called '**AT-AC introns**' (the AT-AC refers of course to the corresponding DNA sequence). Instead, U11, U12, U4$_{atac}$ and U6$_{atac}$ snRNPs are involved, replacing the roles of U1, U2, U4 and U6, respectively, and assemble to form the '**AT-AC spliceosome**'. U5 snRNP is required for splicing both classes of intron.

A few organisms, such as nematodes and trypanosomes, are able to splice together exons from two different RNA molecules, a process called ***trans*-splicing**. In this context, the more usual splicing together of two exons in the same RNA molecule would be ***cis*-splicing**.

Some **self-splicing introns** are also known, for example, *Tetrahymena* rRNA (Section G8) and some mitochondrial and chloroplast mRNAs. In these cases, the intron RNA sequence catalyzes its own cleavage out of the RNA precursor without the need for a spliceosome. Such catalytic RNA molecules are called **ribozymes** (a name that is clearly fashioned on 'enzymes', i.e. protein catalysts). The chemical similarity of some of these self-splicing reactions with the reactions that occur during spliceosome-mediated splicing has led to a realization of the central role of RNA catalysis in the latter. Spliceosome-mediated

Figure 5. Comparison of the conserved splice site sequences of the majority of introns (top diagram) with those for AT-AC introns (bottom diagram).

splicing probably evolved from self-splicing entities, with snRNAs having roles not only in recognition of splice sites but also in the catalytic reactions of spliceosome-mediated splicing. In particular, the structure formed by U2 snRNA base paired to U6 snRNA probably forms the catalytic center of the spliceosome.

3′ Processing: cleavage and polyadenylation

Most eukaryotic pre-mRNAs undergo polyadenylation, which involves cleavage of the RNA at its 3′ end and the addition of about 200–250 A residues to form a poly(A) tail. The cleavage and polyadenylation reactions require the existence of a **polyadenylation signal sequence** (5′-AAUAAA-3′) located near the 3′ end of the pre-mRNA followed by a sequence **5′-YA-3′** (where Y=a pyrimidine), often 5′-CA-3′, in the next 11–20 nt (Figure 6). A **GU-rich sequence** (or U-rich sequence) is also usually present further downstream. After these RNA sequence elements have been synthesized, two multisubunit proteins called **CPSF (cleavage and polyadenylation specificity factor)** and **CstF (cleavage stimulation factor F)** are transferred from the CTD of RNA polymerase II to the RNA molecule and bind to the sequence elements. A protein complex is formed, which includes at least two additional **cleavage factors** (CFI and CFII), and an enzyme called **poly(A) polymerase (PAP)**. This complex cleaves the RNA between the AAUAAA sequence and the GU-rich sequence (Figure 6). Poly(A) polymerase then adds about 200–250 A residues to the new 3′ end of the RNA molecule using ATP as precursor. As it is made, the poly(A) tail immediately binds multiple copies of a poly(A) binding protein.

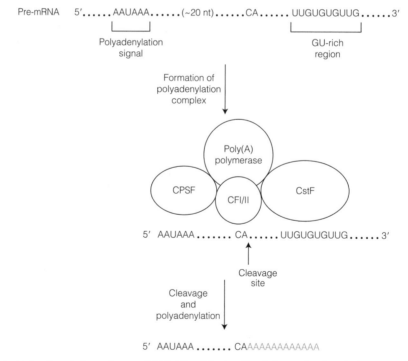

Figure 6. Polyadenylation of pre-mRNA. The full details of the polyadenylation complex are not known but include proteins CPSF and CstF that bind to the consensus sequences indicated, plus poly(A) polymerase and at least two additional cleavage factors (CFI and CFII). After cleavage, the 3′ end of the RNA molecule containing the GU-rich region is lost and the new 3′ end of the pre-mRNA (ending in CA) is extended by the addition of 200–250 A residues, added by poly(A) polymerase, to form a poly(A) tail.

An intact polyadenylation signal is essential for the termination of transcription of protein-coding genes in, for example, human cells, and so we know that transcription termination and polyadenylation are linked.

CPSF binds to TFIID and so becomes incorporated into the transcription initiation complex during initiation (Section G5). It stays associated with the RNA polymerase II during elongation and so is ready to bind to the polyadenylation signal sequence as soon as this has been synthesized. Transcription stops soon after the poly(A) signal sequence has been synthesized. Exactly how this occurs is not fully understood. Two models have been proposed and it looks likely that the actual process is a combination of both models. The first model (the allosteric or anti-terminator model) proposes that transcription through the poly(A) signal sequence (which of course is followed rapidly by the formation of the polyadenylation complex) causes a conformational change in the RNA polymerase II elongation complex that triggers termination rather than further elongation. In this context, it is notable that both CPSF and CstF can interact with the CTD of RNA polymerase II. The second model (the torpedo model) proposes that after cleavage at the poly(A) site, a 5′ to 3′ exonuclease binds to the newly created 5′ end and rapidly degrades the RNA, catching up with the RNA polymerase II and promoting its release from the DNA template.

The exact role of the poly(A) tail is still not fully understood. It has been suggested to protect the 3′ end of the final mRNA against ribonuclease digestion and hence stabilize the mRNA. It also increases the efficiency of translation of the mRNA. However, some mRNAs, notably histone pre-mRNAs, lack a poly(A) tail. (Histone pre-mRNA is still subject to 3′ processing. It is cleaved near the 3′ end by a protein complex that recognizes specific signals, one of which is a stem-loop structure, to generate the 3′ end of the mature mRNA molecule.)

Alternative processing

Alternative polyadenylation sites

Certain pre-mRNAs contain more than one set of signal sequences for 3′ end cleavage and polyadenylation. In some cases, the location of the alternative polyadenylation sites is such that, depending on the site chosen, particular exons may be lost or retained in the subsequent splicing reactions (Figure 7). Here the effect is to change the coding capacity of the final mRNA so that different proteins are produced depending on the polyadenylation site used. In other cases, the alternative sites both lie within the 3′ noncoding region of the pre-mRNA so that the same coding sequences are included in the final mRNA irrespective of which site is used but the 3′ noncoding region can vary. Since the 3′ noncoding sequence may contain signals to control mRNA stability, the choice of polyadenylation site in this situation can affect the lifetime of the resulting mRNA.

Figure 7. Use of alternative polyadenylation sites.

Figure 8. Alternative RNA splicing pathways. In the simple example shown, the transcript can be spliced by alternative pathways leading to two mRNAs with different coding capacities, i.e. exons 1, 2 and 3 or just exons 1 and 3. For genes containing many exons, a substantial number of alternative splice pathways may exist, which are capable of generating many possible mRNAs from the single gene.

Alternative splicing

Many cases are now known where different tissues splice the primary RNA transcript of a single gene by alternative pathways, where the exons that are lost and those that are retained in the final mRNA depend upon the pathway chosen (Figure 8). Presumably, some tissues contain regulatory proteins that promote or suppress the use of certain splice sites to direct the splicing pathway selected. These **alternative splicing pathways** are very important since they allow cells to synthesize a range of functionally distinct proteins from the primary transcript of a single gene.

RNA editing

RNA editing is the name given to several reactions whereby the nucleotide sequence on an mRNA molecule may be changed by mechanisms other than RNA splicing. Individual nucleotides within the mRNA may be changed to other nucleotides, deleted entirely or additional nucleotides inserted. The effect of RNA editing is to change the coding capacity of the mRNA so that it encodes a different polypeptide than that originally encoded by the gene. An example of RNA editing in humans is **apolipoprotein B mRNA**. In liver, the mRNA does not undergo editing and the protein produced after translation is called **apolipoprotein B100** (Figure 9a). In cells of the small intestine, RNA editing (Figure 9b) causes the conversion of a single C residue in the mRNA to U and, in so doing, changes a codon for glutamine (CAA) to a termination codon (UAA). Subsequent translation of the edited mRNA yields the much shorter **apolipoprotein B48** (48% of the size of apolipoprotein B100). This is not a trivial change; apolipoprotein B48 lacks a protein domain needed for receptor binding, which apolipoprotein B100 possesses, and hence the functional activities of the two proteins are different. Many other cases of RNA editing are also known. Trypanosome mitochondrial mRNAs, for example, undergo extensive RNA editing, which results in over half of the uridines in the final mRNA being acquired through the editing process.

mRNA silencing

MicroRNAs (**miRNAs**) are a class of RNAs, found only in eukaryotes, that are only about 21 nt long. The first miRNA was discovered in 1993 but since 2000 many more miRNAs have been identified (over 5000 have so far been reported) and they have been recognized as an important new class of RNA regulatory molecules. In mammals, miRNAs have been predicted to regulate up to 50% of all protein-coding genes. They do so by inhibiting the translation of the target mRNA, i.e. they cause **mRNA silencing**.

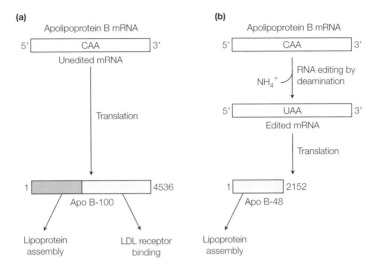

Figure 9. RNA editing: (a) unedited apolipoprotein B mRNA is translated to yield apoB-100, a 4536-amino acid long polypeptide with structural domains for lipoprotein assembly and receptor binding functions; (b) translation of the edited mRNA yields the shorter apoB-48, which lacks the receptor binding domain.

Synthesis of miRNAs

Most miRNAs are transcribed by RNA Pol II with a few transcribed by RNA Pol III. Two routes of miRNA synthesis by RNA Pol II are known (Figure 10). In the first, an miRNA gene with its own promoter is transcribed to yield a long primary transcript called **pri-miRNA** containing up to six miRNA sequences. The pri-miRNA undergoes processing reactions that add a 5′ cap and a 3′ poly(A) tail. Within this molecule, the miRNA sequences form typical internal hairpin loops each containing about 70 nt. These fold-back RNAs are recognized in the nucleus by a protein called **Pasha** (also known as DGCR8), which binds to a protein called **Drosha** to form a **microprocessor complex**. Drosha cuts out the individual hairpin regions from the pri-miRNA to form miRNA precursors called **pre-miRNA** (Figure 10). The pre-miRNAs are exported to the cytoplasm where a ribonuclease called **Dicer** cuts away the loop and trims the ends to leave a double-stranded miRNA molecule about 21 nt long. The two strands are not usually perfectly base-paired (i.e. in some regions, the two sequences are not complementary) and so this is an imperfect RNA duplex. One strand of the duplex is miRNA and the other is called **miRNA*** (also called the **'passenger' strand**).

The second route for miRNA synthesis depends on the fact that, amazingly, some miRNAs are coded by sequences that reside in the introns of protein-coding genes (Figure 10) known as **mirtrons**. The loop of the spliced out intron is removed by a debranching enzyme to release the pre-miRNA.

Mechanism of miRNA action

A mature miRNA exists as a **miRNP** (microribonucleoprotein) complex, which contains only the miRNA strand of the RNA duplex; the miRNA* ('passenger') is degraded. The miRNP complex also contains several proteins including Argonaute protein(s) that are important for miRNP function. Because the function of the miRNP complex is to inhibit

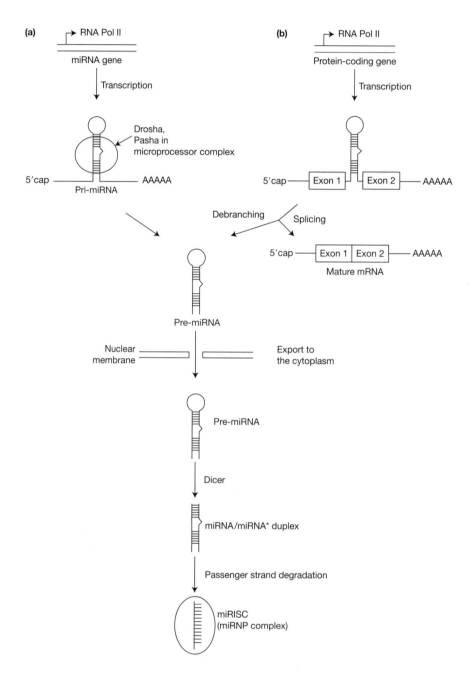

Figure 10. Synthesis of miRNA: (a) an miRNA gene is transcribed by RNA polymerase II to yield a pri-mRNA. This is then cleaved by the microprocessor complex to form pre-miRNA followed by export to the cytoplasm and cleavage by Dicer to yield an miRNA/miRNA* duplex. (b) Transcription of a protein-coding gene, an intron of which contains an miRNA sequence. Splicing generates the mature mRNA and releases the intron, which yields pri-miRNA that is processed as in (a) to give an miRNA/miRNA* duplex. Finally, an miRISC complex is formed containing the single miRNA strand and associated proteins.

the translation of target mRNAs, it is often called the **miRISC (miRNA-induced silencing complex)**.

The miRNA in the miRISC is complementary to a sequence in one or more mRNAs. In animals (Figure 11), most miRISCs bind to sequences in the 3′ untranslated region (3′ UTR; Section G5); indeed multiple miRNAs may bind to different sites in this region on a single mRNA molecule. Some miRNAs have also been reported to bind to coding regions of the mRNA or to sequences in the 5′ UTR (Section G5). The binding of the miRISC to the mRNA causes translational inhibition, i.e. it results in 'silencing' of the mRNA. The exact mechanism is unclear but in some cases it involves cleavage of the mRNA at the base-paired region, and in other cases deadenylation, i.e. removal of A residues from the poly(A) tail.

Figure 11. mRNA silencing. The miRISC complex binds usually to the 3′ untranslated region (3′ UTR) of the target mRNA through base pairing between the miRNA and a complementary sequence in the mRNA. The base pairing may be perfectly matched or, as shown here, may contain one or more regions where the two sequences differ and no base pair forms.

G8 Transcription and processing of ribosomal RNA

Key Notes

Ribosomes

A prokaryotic 70S ribosome comprises two subunits (50S and 30S). The 50S subunit has 23S and 5S rRNAs complexed with 34 polypeptides, whereas the 30S subunit contains 16S rRNA and 21 polypeptides. A eukaryotic 80S ribosome comprises two subunits (60S and 40S). The 60S subunit has 28S, 5.8S and 5S rRNAs complexed with about 49 polypeptides whereas the 40S subunit contains 18S rRNA and about 33 polypeptides. The three-dimensional structure of bacterial ribosomes shows that it is the rRNAs that fold and base pair with each other to form the overall structure of the ribosome, with the proteins located peripherally. In addition, the A, P and E sites, and even the catalytic site for peptide bond formation, are formed by rRNAs, which therefore have a catalytic function as well as structural roles.

Transcription and processing of prokaryotic rRNA

E. coli has seven rRNA transcription units, each containing one copy each of the 23S, 16S and 5S rRNA genes as well as one to four tRNA genes. Transcription produces a 30S pre-rRNA transcript. This folds up to form stem-loop structures, ribosomal proteins bind, and a number of nucleotides become methylated. The modified pre-rRNA transcript is then cleaved at specific sites by RNase III and the ends are trimmed by ribonucleases M5, M6 and M23 to release the mature rRNAs.

Synthesis of eukaryotic 28S, 18S and 5.8S rRNA

The 28S, 18S and 5.8S rRNA genes are present as multiple copies clustered together as tandem repeats. These rRNA transcription units are transcribed, in the nucleolus, by RNA Pol I. The promoter contains a core element that straddles the transcriptional start site and an upstream control element (UCE) about 50–80 bp in size, located about 100 bp further upstream. Four protein complexes, one of which contains the TATA binding protein (TBP) as a component, interact with each other and RNA Pol I to form a transcription initiation complex bound to the promoter elements. Transcription produces a 45S pre-rRNA, which has external transcribed spacers (ETSs) at the 5′ and 3′ ends and internal transcribed spacers (ITSs) internally separating rRNA sequences. The pre-rRNA folds up to form a defined secondary structure with stem-loops, ribosomal proteins bind to selected sequences, and multiple methylation and isomerization reactions (of uridine to pseudouridine) occur

	at specific sites, guided by interaction of the pre-rRNA with snoRNAs (as snoRNPs). The 45S pre-rRNA molecule is then cleaved, releasing 32S and 20S precursor rRNAs that are processed further to generate mature 28S, 18S and 5.8S rRNAs.
Ribozymes	In *Tetrahymena*, the pre-rRNA molecule contains an intron that is removed by self-splicing (in the presence of guanosine, GMP, GDP or GTP) without the need for involvement of any protein. This was the first ribozyme discovered but many have since been reported.
Synthesis of eukaryotic 5S rRNA	Eukaryotic cells contain multiple copies of the 5S rRNA gene. Unlike other eukaryotic rRNA genes, the 5S rRNA genes are transcribed by RNA Pol III. Two control elements, an A box and a C box, lie downstream of the transcriptional start site. The C box binds TFIIIA, which then recruits TFIIIC. TFIIIB now binds and interacts with RNA Pol III to form the transcription initiation complex. Transcription produces a mature 5S rRNA that requires no processing.
Related topics	(F1) An introduction to DNA (G1) An introduction to RNA (G2) Transcription in prokaryotes (G3) Operons (G4) Transcription in eukaryotes: an overview (G5) Transcription of protein-coding genes in eukaryotes (G6) Regulation of transcription by RNA Pol II (G7) Processing of eukaryotic pre-mRNA (G9) Transcription and processing of transfer RNA

Ribosomes

Each ribosome consists of two subunits, a small subunit and a large subunit, each of which is a multicomponent complex of **ribosomal RNAs** (**rRNAs**) and **ribosomal proteins** (Figure 1). One way of distinguishing between particles such as ribosomes and ribosomal subunits is to place the sample in a tube within a centrifuge rotor and spin this at very high speed. This causes the particles to sediment to the tube bottom. Particles that differ in mass, shape and/or density sediment at different velocities (sedimentation velocities). Thus a particle with twice the mass of another will always sediment faster provided both particles have the same shape and density. The sedimentation velocity of any given particle is also directly proportional to the gravitational forces (the centrifugal field) experienced during the centrifugation, which can be increased simply by spinning the rotor at a higher speed. However, it is possible to define a **sedimentation coefficient** that depends solely on the size, shape and density of the particle and is independent of the centrifugal field. Sedimentation coefficients are usually measured in **Svedberg units** (**S**). A prokaryotic ribosome has a sedimentation coefficient of 70S whereas the large and small subunits have sedimentation coefficients of 50S and 30S, respectively (note that S values are not additive). The 50S subunit contains two rRNAs (23S and 5S) complexed with 34 polypeptides whereas the 30S subunit contains 16S rRNA and 21 polypeptides (Figure 1). In eukaryotes the ribosomes are larger and more complex; the ribosome monomer is 80S and consists of 60S and 40S subunits. The 60S subunit contains three

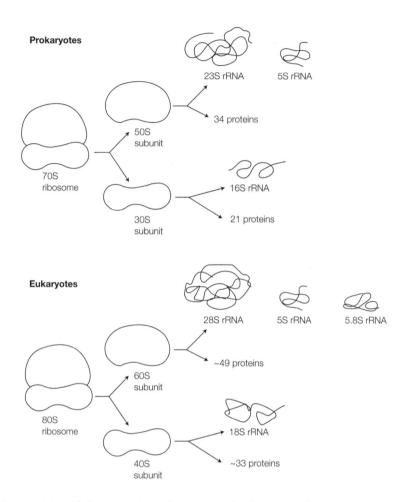

Figure 1. Composition of ribosomes in prokaryotic and eukaryotic cells.

rRNAs (28S, 5.8S and 5S) and about 49 polypeptides and the 40S subunit has 18S rRNA and about 33 polypeptides (Figure 1). However, despite this extra complexity, the overall structure and function of eukaryotic ribosomes are very similar to those from bacteria. In each case, about two-thirds of the structure is rRNA and one-third is protein.

A wide range of studies have built up a detailed picture of the fine structure of ribosomes, mapping the location of the various RNA and protein components and their interactions. The overall shape of a 70S ribosome, gained through electron microscopy studies, is shown in Figure 2. The three-dimensional structure of bacterial ribosomal subunits, determined only a few years ago, shows that the various rRNAs are tightly folded and base pair extensively with each other to form the core of the ribosomal subunits with the proteins restricted to the surface and filling gaps between RNA folds. Not only are the rRNAs largely responsible for the overall structure of the prokaryotic ribosome but they also form the three main binding sites (the A, P and E sites) involved in protein synthesis (Section H2). In addition, the 23S rRNA, rather than a protein, forms the catalytic site for peptide bond formation (Section H2). Therefore, protein synthesis involves rRNAs acting as ribozymes.

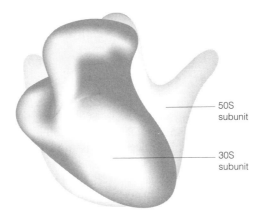

Figure 2. The prokaryotic 70S ribosome.

Transcription and processing of prokaryotic rRNA

In *E. coli* there are seven rRNA transcription units scattered throughout the genome, each of which contains one copy of each of the 23S, 16S and 5S rRNA genes and one to four copies of various tRNA genes (Figure 3). This gene assembly is transcribed by the single prokaryotic RNA polymerase to yield a single **30S pre-rRNA transcript** (about 6000 nt in size). This arrangement ensures that stoichiometric amounts of the various rRNAs are synthesized for ribosome assembly. Following transcription, the 30S pre-rRNA molecule forms internally base paired regions to give a series of stem-loop structures and ribosomal proteins bind to form a **ribonucleoprotein (RNP) complex**. A number of the nucleotides in the folded pre-rRNA molecule are now methylated, on the ribose moieties, using *S*-adenosyl methionine as the methyl donor. Next, the pre-rRNA molecule is cleaved at specific sites by **RNase III** to release precursors of the 23S, 16S and 5S rRNAs.

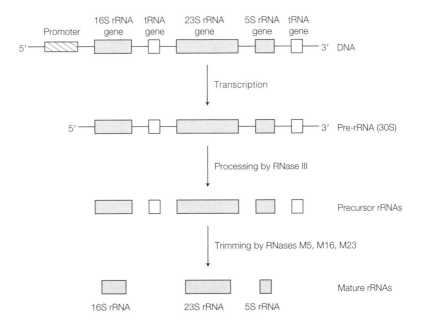

Figure 3. Transcription and processing of prokaryotic rRNA.

The precursors are then trimmed at their 5′ and 3′ ends by ribonucleases **M5**, **M16** and **M23** (which act on the 5S, 16S and 23S precursor rRNAs, respectively) to generate the mature rRNAs.

Synthesis of eukaryotic 28S, 18S and 5.8S rRNA

In eukaryotes, the genes for 28S, 18S and 5.8S rRNA are typically **clustered** together and **tandemly repeated** in that one copy each of the 18S, 5.8S and then 28S genes occur, followed by untranscribed spacer DNA, then another set of 18S, 5.8S and 28S genes occurs and so on (Figure 4a). In humans, there are about 200 copies of this rRNA transcription unit arranged as five clusters of about 40 copies on separate chromosomes. These rRNA transcription units are transcribed by **RNA polymerase I** (**RNA Pol I**) in a region of the nucleus known as the **nucleolus** (Section A2). The nucleolus contains loops of DNA extending from each of the rRNA gene clusters on the various chromosomes and hence each cluster is called a **nucleolar organizer**.

The rRNA promoter consists of a **core element**, which straddles the transcriptional start site (designated as position +1) and is located between residues –45 to +20 plus an **upstream control element** (**UCE**) about 50–80 bp in size and located about 100 bp further upstream (Figure 4b). A transcription factor called **upstream binding factor** (**UBF**) binds both to the UCE as well as to a region next to and overlapping with the core element.

Figure 4. (a) rRNA transcription units; (b) transcription of a single transcription unit by RNA Pol I and processing of pre-rRNA.

Interestingly, **TATA box binding protein** (**TBP**; Section G5) also binds to the rRNA promoter (in fact, TBP is required for initiation by all three eukaryotic RNA polymerases).

In total, four protein complexes interact with RNA polymerase I to initiate ribosomal gene transcription. The first multiprotein complex (called **SL1** in humans) contains TBP, the second complex is UBF and the two other protein complexes involved are **TIFIA** and **TIFIC**. The four complexes interact with each other and with RNA Pol I to form a **transcription initiation complex** bound to the rRNA promoter. The RNA Pol I then transcribes the whole transcription unit of 28S, 18S and 5.8S genes to synthesize a single large pre-rRNA molecule (Figure 4b).

In humans, the product of transcription is a **45S pre-rRNA**, which has non-rRNA **external transcribed spacers** (**ETSs**) at the 5′ and 3′ ends and non-rRNA **internal transcribed spacers** (**ITSs**) internally separating the rRNA sequences (Figure 4b). This 45S molecule folds up to form a defined secondary structure with stem-loops, ribosomal proteins bind to selected sequences, **methylation** of ribose moieties occurs (at over 100 nucleotide sites) and about 95 uridine residues are modified to pseudouridine (Ψ), an isomerization process called **pseudouridylation**. The 45S pre-rRNA molecule is then cleaved, first in the ETSs and then in the ITSs, to release precursor rRNAs, which are cleaved further and trimmed to release the mature 28S, 18S and 5.8S rRNAs (Figure 4b).

In eukaryotes, selection of the sites in pre-rRNA that will be methylated depends upon small RNAs (70–100 nt long) found in the nucleolus called **small nucleolar RNAs** (**snoRNAs**) that exist in ribonucleoprotein complexes called **snoRNPs**. The snoRNAs contain long regions (10–21 nt) that are complementary to specific regions of the pre-rRNA molecule, form base pairs with the pre-rRNA at these sites and then guide where methylation of specific ribosome residues (2′O methylation) and isomerization of uridine to pseudouridine will occur.

There are two distinct families of snoRNAs, called antisense **C/D box** and **H/ACA box snoRNAs**. The naming of these snoRNAs is based on the conserved sequences they contain; C/D box snoRNAs contain two short conserved sequences, C (UGAUGA) and D (CUGA) and likewise the H/ACA box snoRNAs contain other conserved sequences called an H box and an ACA box. In general, the C/D box snoRNPs direct methylation and H/ACA box snoRNPs direct pseudouridylation. Once the snoRNP has bound to the rRNA, it is believed that the protein components of the snoRNP carry out the required chemical modification.

Ribozymes

In at least one eukaryote, *Tetrahymena*, the pre-rRNA molecule contains an intron. Removal of the intron during processing of the pre-rRNA does not require the assistance of any protein! Instead, in the presence of guanosine, GMP, GDP or GTP, the intron excises itself, a phenomenon known as **self-splicing**. This was the first demonstration of **ribozymes**, that is, **catalytic RNA** molecules that catalyze specific reactions. Self-splicing introns have also been discovered in some eukaryotic mRNAs and even peptidyl transferase, a key activity in protein synthesis, is now known to be a ribozyme (Section H2).

Synthesis of eukaryotic 5S rRNA

In eukaryotes, the 5S rRNA gene is also present in multiple copies (2000 copies in human cells, all clustered together at one chromosomal site). Unlike other eukaryotic rRNA genes, the 5S rRNA genes are transcribed by **RNA polymerase III** (**RNA Pol III**), the same enzyme that transcribes tRNA genes.

The promoters for RNA polymerase III vary in structure depending on the gene being transcribed. The promoters of tRNA genes contain control elements called the A box and B box located downstream of the transcriptional start site (Section G9). The promoter for 5S rRNA genes also has two control elements located downstream of the transcriptional start site, but these are an **A box** and a **C box** (Figure 5). The C box binds **transcription factor IIIA** (**TFIIIA**), which then in turn interacts with **TFIIIC** to cause it to bind, a process that probably also involves recognition of the A box. Once TFIIIC has bound, **TFIIIB** binds and interacts with RNA Pol III, causing that to bind also to form the **transcription initiation complex**. One of the three subunits of TFIIIB is **TBP** (Section G5), the transcription factor required for transcription by all three eukaryotic RNA polymerases. Following transcription, the 5S rRNA transcript requires no processing. It migrates to the nucleolus and is recruited into ribosome assembly.

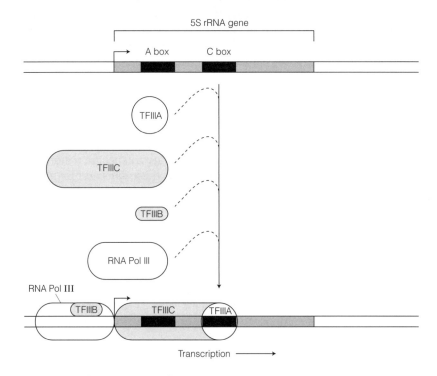

Figure 5. Initiation of transcription of a 5S rRNA gene by RNA Pol III.

G9 Transcription and processing of transfer RNA

Key Notes

tRNA structure

Each tRNA has a cloverleaf secondary structure containing an anticodon arm, a D (or DHU) arm, a T or TΨC arm, and an amino acid acceptor stem to which the relevant amino acid becomes covalently bound, at the 3′-OH group. Some tRNAs also have a variable (or optional) arm. The three-dimensional structure is more complex because of additional interactions between the nucleotides.

Transcription and processing of tRNA in prokaryotes

E. coli contains clusters of up to seven tRNA genes separated by spacer regions, as well as tRNA genes within ribosomal RNA transcription units. Following transcription, the primary RNA transcript folds up into specific stem-loop structures and is then processed by ribonucleases D, E, F and P in an ordered series of reactions to release the individual tRNA molecules.

Transcription and processing of tRNA in eukaryotes

In eukaryotes, tRNA genes are present as multiple copies and are transcribed by RNA Pol III. Several tRNA genes may be transcribed to yield a single pre-tRNA that is then processed to release individual tRNAs. The tRNA promoter includes two control elements, called the A box and the B box, located within the tRNA gene itself and hence downstream of the transcriptional start site. Transcription initiation requires transcription factor IIIC (TFIIIC), which binds to the A and B boxes and TFIIIB that binds upstream of the A box. The primary RNA transcript folds up into stem-loop structures and non-tRNA sequence is removed by ribonuclease action. Unlike prokaryotes, in eukaryotes the CCA sequence at the 3′ end of the tRNA is added after the trimming reactions (by tRNA nucleotidyl transferase). Unlike prokaryotes, pre-tRNA molecules in eukaryotes may also contain a short intron in the loop of the anticodon arm. The intron is removed by tRNA splicing reactions involving endonuclease cleavage at both ends of the intron and then ligation of the cut ends of the tRNA.

Modification of tRNA

Following synthesis, nucleotides in the tRNA molecule may undergo modification to create unusual nucleotides such as 1-methylguanosine (m^1G), pseudouridine (Ψ), dihydrouridine (D), inosine (I) and 4-thiouridine (S^4U).

Related topics	(F1) An introduction to DNA	(G5) Transcription of protein-
	(G1) An introduction to RNA	coding genes in eukaryotes
	(G2) Transcription in	(G6) Regulation of
	prokaryotes	transcription by RNA Pol II
	(G3) Operons	(G7) Processing of eukaryotic
	(G4) Transcription in	pre-mRNA
	eukaryotes: an overview	(G8) Transcription and
		processing of ribosomal RNA

tRNA structure

Transfer RNA (tRNA) molecules play an important role in protein synthesis (Sections H2 and H3). Each tRNA becomes covalently bonded to a specific amino acid to form **aminoacyl-tRNA**, which recognizes the corresponding codon in mRNA and ensures that the correct amino acid is added to the growing polypeptide chain. The tRNAs are small molecules, only 74–95 nt long, which form distinctive **cloverleaf** secondary structures (Figure 1a) by internal base pairing. The stem-loops of the cloverleaf are known as **arms**:

- The **anticodon arm** contains in its **loop** the three nucleotides of the **anticodon**, which will form base pairs with the complementary codon in mRNA during translation

- The **D** or **DHU arm** (with its **D loop**) contains **dihydrouracil**, an unusual pyrimidine

- The **T** or **TΨC arm** (with its **T loop**) contains another unusual base, **pseudouracil** (denoted Ψ) in the sequence TΨC

- Some tRNAs also have a **variable arm** (**optional arm**), which is 3–21 nt in size

Figure 1. (a) Cloverleaf secondary structure of tRNA; (b) tertiary structure of tRNA. Adapted from Freifelder, D. (1986) *Molecular Biology*, 2nd Edn; Jones and Bartlett.

The other notable feature is the **amino acid acceptor stem**. This is where the amino acid becomes attached, at the 3'-OH group of the 3' CCA sequence.

The three-dimensional structure of tRNA (Figure 1b) is even more complex because of additional interactions between the various units of secondary structure.

Transcription and processing of tRNA in prokaryotes

The rRNA transcription units in *E. coli* contain some tRNA genes that are transcribed and processed at the time of rRNA transcription (Section G8). Other tRNA genes occur in clusters of up to seven tRNA sequences separated by spacer regions. Following transcription by the single prokaryotic RNA polymerase, the primary RNA transcript folds up into the characteristic stem-loop structures (Figure 2) and is then processed in an ordered series of cleavages by ribonucleases (RNases), which release and trim the tRNAs to their final lengths. The cleavage and trimming reactions at the 5' and 3' ends of the precursor tRNAs involves **RNases D**, **E**, **F** and **P**. RNases E, F and P are **endonucleases**, cutting the RNA internally, whilst RNase D is an **exonuclease**, trimming the ends of the tRNA molecules.

Transcription and processing of tRNA in eukaryotes

In eukaryotes, the tRNA genes exist as multiple copies and are transcribed by **RNA polymerase III** (**RNA Pol III**). As in prokaryotes, several tRNAs may be transcribed together to yield a single **pre-tRNA molecule** that is then processed to release the mature tRNAs. The promoters of eukaryotic tRNA genes are unusual in that the transcriptional control elements are located downstream (i.e. on the 3' side) of the transcriptional start site (at position +1). In fact, they lie within the gene itself. Two such elements have been identified, called the **A box** and **B box** (Figure 3). Transcription of the tRNA genes by RNA

Figure 2. Transcription and processing of tRNA in prokaryotes.

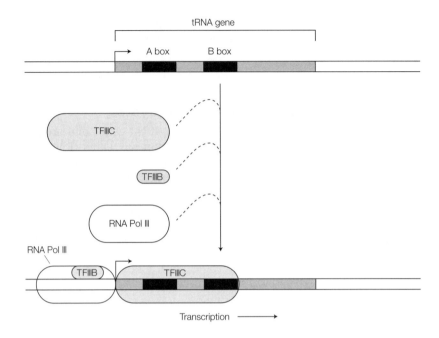

Figure 3. Initiation of transcription of a tRNA gene by RNA Pol III.

Pol III requires **transcription factor IIIC** (**TFIIIC**) as well as **TFIIIB**. TFIIIC binds to the A and B boxes whilst TFIIIB binds upstream of the A box. TFIIIB contains three subunits, one of which is **TBP**, the polypeptide required by all three eukaryotic RNA polymerases.

After synthesis, the pre-tRNA molecule folds up into the characteristic stem-loop structures (Figure 1) and non-tRNA sequence is cleaved from the 5′ and 3′ ends by ribonucleases. In prokaryotes, the CCA sequence at the 3′ end of the tRNA (which is the site of bonding to the amino acid) is encoded by the tRNA gene but this is not the case in eukaryotes. Instead, the CCA is added to the 3′ end by **tRNA nucleotidyl transferase** after the trimming reactions.

Another difference between prokaryotes and eukaryotes is that eukaryotic pre-tRNA molecules often contain a short **intron** in the loop of the anticodon arm (Figure 4). This intron must be removed in order to create a functional tRNA molecule. Its removal occurs by cleavage by a **tRNA splicing endonuclease** at each end of the intron and then ligation together of the tRNA ends by **tRNA ligase**. This RNA splicing pathway for intron removal is totally different from that used to remove introns from pre-mRNA molecules in eukaryotes (Section G7) and must have evolved independently.

Modification of tRNA

Transfer RNA molecules are notable for containing unusual nucleotides (Figure 5) such as **1-methylguanosine** (**m^{1}G**), **pseudouridine** (**Ψ**), **dihydrouridine** (**D**), **inosine** (**I**) and **4-thiouridine** (**S^{4}U**). These are created by modification of guanosine and uridine after tRNA synthesis. For example, inosine is generated by deamination of guanosine.

Figure 4. Processing of a typical eukaryotic pre-tRNA molecule.

Figure 5. Some modified nucleosides found in tRNA molecules.

H1 The genetic code

Key Notes

The genetic code is a triplet code

The genetic code is the rules that specify how the nucleotide sequence of an mRNA is translated into the amino acid sequence of a polypeptide. The nucleotide sequence is read as triplets called codons. The codons UAG, UGA and UAA do not specify amino acids and are called termination codons or stop codons. AUG codes for methionine and also acts as an initiation (start) codon.

Codons are recognized by anticodons in tRNA molecules

As the ribosome reads the mRNA, the correct amino acid for a codon is brought to the ribosome by a specific tRNA molecule, which binds to the codon through codon–anticodon base pairing. The amino acid is covalently bound to the 3′ end of the tRNA forming an aminoacyl tRNA molecule. Base pairing of the codon to the anticodon occurs in an antiparallel arrangement. The tRNA molecule therefore acts as an adaptor, mediating between the nucleotide sequence of the mRNA and the amino acid sequence of the polypeptide being synthesized.

The genetic code is degenerate

Most amino acids in proteins are specified by more than one codon (i.e. the genetic code is degenerate). Recognition of these codons often occurs using different tRNAs (with corresponding anticodons to these codons) that bind the same amino acid (isoaccepting tRNAs). In addition, codons that specify the same amino acid (synonyms) often differ only in the third base, the wobble position, where base pairing with the anticodon may be less stringent than for the first two positions of the codon, allowing one tRNA to recognize more than one codon.

Universality of the genetic code

The genetic code is not universal but is the same in most organisms. Exceptions are found in mitochondrial genomes where some codons specify different amino acids to that normally encoded by nuclear genes. In mitochondria, the UGA codon does not specify termination of translation but instead encodes for tryptophan. Similarly, in certain protozoa, UAA and UAG encode glutamic acid instead of acting as termination codons. The presence of a SElenoCysteine Insertion Sequence (SECIS element) in the mRNA can also direct that a particular UGA codon(s) leads to the incorporation of selenocysteine (Se-Cys) instead of acting as a stop codon.

Reading frames

The mRNA sequence can be read by the ribosome in three possible reading frames. Usually only one reading frame codes for a functional protein since the other two reading

	frames contain multiple termination codons. In some bacteriophage genomes, overlapping genes occur, which use different reading frames.
Open reading frames	An open reading frame (ORF) is a run of codons that starts with ATG and ends with a termination codon, TGA, TAA or TAG. Coding regions of genes contain relatively long ORFs unlike noncoding DNA where ORFs are comparatively short. The presence of a long open reading frame in a DNA sequence therefore may indicate the presence of a coding region. Computer analysis of the ORF can be used to deduce the sequence of the encoded protein.
Related topics	(G1) An introduction to RNA (G8) Transcription and processing of ribosomal RNA (G9) Transcription and processing of transfer RNA (H2) Translation in prokaryotes (H3) Translation in eukaryotes

The genetic code is a triplet code

During translation, the sequence of an mRNA molecule is read from its 5′ end by ribosomes, which then synthesize an appropriate polypeptide. Both in prokaryotes and in eukaryotes, the DNA sequence of a single gene is colinear with the amino acid sequence of the polypeptide it encodes. In other words, the nucleotide sequence of the coding DNA strand, 5′ to 3′, specifies in exactly the same order the amino acid sequence of the encoded polypeptide, N-terminal to C-terminal.

The relationship between the nucleotide sequence of the mRNA and the amino acid sequence of the polypeptide is called **the genetic code**. The sequence of the mRNA is read in groups of three nucleotides called **codons**, with each codon specifying a particular amino acid (Figure 1). However, three particular codons, UAG, UGA and UAA, do not encode an amino acid. Whenever one of these codons is encountered by a ribosome, it leads to termination of protein synthesis. Therefore, these three codons are called **termination codons** or **stop codons**. The codon AUG codes for methionine. Although methionine is found at internal positions in polypeptide chains, all eukaryotic polypeptides also start with methionine (Section H3) and all prokaryotic polypeptides start with a modified methionine (*N*-formylmethionine; Section H2). Therefore, the first AUG codon that is read by the ribosome in an mRNA is called the **initiation codon** or **start codon**.

Codons are recognized by anticodons in tRNA molecules

During translation, as the ribosome moves along the mRNA, it reads each codon in turn. The correct amino acid corresponding to the codon is brought to the ribosome by a **tRNA molecule**. The tRNA molecule with its covalently bound amino acid is called an **aminoacyl-tRNA**. As we have seen (Section G9), each tRNA molecule has a cloverleaf secondary structure. The amino acid is covalently bound to the A residue of the CCA sequence at the 3′end of the tRNA (Figure 2a).

Codon sequence

1st base	2nd base				3rd base
(5′end) ↓	U	C	A	G	(3′end) ↓
U	Phe	Ser	Tyr	Cys	U
	Phe	Ser	Tyr	Cys	C
	Leu	Ser	**Stop**	**Stop**	A
	Leu	Ser	**Stop**	Trp	G
C	Leu	Pro	His	Arg	U
	Leu	Pro	His	Arg	C
	Leu	Pro	Gln	Arg	A
	Leu	Pro	Gln	Arg	G
A	Ile	Thr	Asn	Ser	U
	Ile	Thr	Asn	Ser	C
	Ile	Thr	Lys	Arg	A
	Met	Thr	Lys	Arg	G
G	Val	Ala	Asp	Gly	U
	Val	Ala	Asp	Gly	C
	Val	Ala	Glu	Gly	A
	Val	Ala	Glu	Gly	G

Figure 1. The genetic code.

The correct nomenclature is, for example, tRNA^{Gly} for the tRNA that will accept glycine, whereas the corresponding aminoacyl-tRNA is Gly-tRNA^{Gly}, and is the aminoacyl-tRNA shown in Figure 2a.

Each codon in an mRNA molecule being read by a ribosome during translation binds the correct aminoacyl-tRNA because the codon is recognized by a triplet of bases, called an anticodon, in a specific tRNA molecule (Figure 2a). In this way, the tRNA functions as an adaptor molecule, translating the nucleotide sequence of the mRNA into a polypeptide sequence composed of the correct sequence of amino acids.

Each base in the codon base pairs with its complementary base in the anticodon. Note that in codon–anticodon interaction, the codon in the mRNA molecule in a 5′ to 3′ orientation is hydrogen bonded with the anticodon of the tRNA in a 3′ to 5′ orientation (Figure 2b). This antiparallel arrangement means that the first base of the codon pairs with the third base of the anticodon, the second base of each hydrogen bond together and the third base of the codon hydrogen bonds with the first base of the anticodon.

The genetic code is degenerate

Since RNA is composed of four types of nucleotides, there are $4^3=64$ possible codons, that is 64 possible triplets of nucleotides with different sequences. However, only 20 amino acids are commonly found in proteins (Section B1) so that, in most cases, a single amino acid is coded for by several different codons (Figure 1). The genetic code is therefore said to be **degenerate**. In fact, only methionine and tryptophan are represented by a single codon. Different codons that specify the same amino acid are called **synonyms**.

Figure 2. (a) Structure of an aminoacyl-tRNA, Gly-tRNA^{Gly}. (b) Codon–anticodon base pairing; the codon in the mRNA molecule (5′ to 3′) is hydrogen bonded by complementary base pairing with the anticodon of the tRNA in a 3′ to 5′ orientation.

As a result of the genetic code's degeneracy, a mutation that changes only a single nucleotide in DNA (**point mutation**), and hence changes only a single nucleotide in the corresponding mRNA, often has no effect on the amino acid sequence of the encoded polypeptide.

Of the 64 possible codons with a triplet code (see above), there are three termination or stop codons, and 61 other codons available for amino acid coding (so called '**sense**' codons). Bacteria have about 30–45 different tRNAs and eukaryotic cells have up to 50 tRNAs. Several of these tRNAs have different anticodons – and hence recognize different codons – but bind the same amino acid; these are known as **isoaccepting tRNAs**. For example, four isoaccepting tRNAs exist for arginine in *E. coli* and these are indicated using superscript numbers, e.g. tRNA^{Arg1}, tRNA^{Arg2}, etc.

However, there are not enough isoaccepting tRNAs to account for all of the sense codons. Another aspect is also important here. The pairing of the third base of a codon is less stringent than for the first two bases (i.e. there is some '**wobble base pairing**') so that in some cases a single tRNA may base pair with more than one codon. For example, tRNA^{Leu3} in *E. coli* has the anticodon 3′-GAG-5′ and recognizes both of the codons 5′-CUU-3′ and 5′-CUC-3′ (i.e. either U or C as the third base of the codon can base pair with G, the first base of the anticodon). The third position of the codon is therefore also called the **wobble position**.

Together, the existence of isoaccepting tRNAs and wobble pairing allows all of the 61 sense codons to be read correctly, e.g. *E. coli* has four tRNA^{Arg} isoacceptor tRNAs for reading the six arginine codons CGN (where N means any one of the four possible nucleotides) and AGA/G.

Universality of the genetic code

- For many years, it was thought that the genetic code is 'universal', namely that all living organisms used the same code. Now we know that the genetic code is almost the same in all organisms but there are some differences.

- Mitochondria contain DNA as double-stranded DNA circles which code for about 10–20 proteins. Surprisingly, in **mitochondrial mRNAs**, some codons have different meanings from their counterparts in mRNA in the cytosol. A few examples are given below (N denotes any of the four nucleotides A, G, C or U):

mammalian mitochondria	AUA = Met not Ile
	UGA = Trp not **Stop**
	AGA and AGG = **Stop** not Arg
yeast (*Saccharomyces cerevisiae*) mitochondria	AUA = Met not Ile
	UGA = Trp not **Stop**
	CUN = Thr not Leu

- Some unicellular organisms are also now known to use a variant genetic code. For example:

some ciliated protozoa	UAA and UAG = Glu not **Stop**

- Another variation occurs because several key enzymes in prokaryotes and eukaryotes contain selenocysteine (Se-Cys) in which a selenium atom replaces the sulfur atom in cysteine. These proteins are called selenoproteins (e.g. glutathione peroxidases, glycine reductases). There is no codon that codes just for Se-Cys. Instead, the mRNA contains a sequence called a SElenoCysteine Insertion Sequence (SECIS element) that causes a particular UGA codon(s) to incorporate Se-Cys instead of acting as a stop codon.

Reading frames

Since the sequence of an mRNA molecule is read in groups of three nucleotides (codons) from the 5′ end, it can be read in three possible **reading frames**, depending on which nucleotide is used as the first base of the first codon (Figure 3). Usually, only one reading frame (reading frame 3 in Figure 3) will produce a functional protein since the other two reading frames will include several termination (**Stop**) codons. The correct reading frame is set *in vivo* by recognition by the ribosome of the initiation codon, AUG, at the start of the coding sequence.

Usually one sequence of bases encodes only a single protein. However, in some bacteriophage DNAs, several genes overlap, with each gene being in a different reading frame. This organization of **overlapping genes** generally occurs when the genome size is smaller than can accommodate the genes necessary for phage structure and assembly using only one reading frame.

Open reading frames

In many cases these days, the protein encoded by a particular gene is deduced by cloning (Section I4) and then sequencing the corresponding DNA. The DNA sequence is then

5′   3′

Reading frame 1   U U A U G A G C G C U A A A U

Leu   **Stop**   Ala   Leu   Asn

Reading frame 2   U U A U G A G C G C U A A A U

Tyr   Glu   Arg   **Stop**

Reading frame 3   U U A U G A G C G C U A A A U

Met   Ser   Ala   Lys

Figure 3. Three potential reading frames for any given mRNA sequence depending on which nucleotide is 'read' first.

scanned using a computer program to identify runs of codons that start with ATG and end with TGA, TAA or TAG. These runs of codons are called **open reading frames** (**ORFs**) and identify potential coding regions.

Because genes carry out important cellular functions, the sequence of coding DNA (and of important regulatory sequences) is more strongly conserved in evolution that that of noncoding DNA. In particular, mutations that lead to the creation of termination codons within the coding region, and hence premature termination during translation, are selected against. This means that the coding regions of genes often contain comparatively long ORFs whereas in noncoding DNA, triplets corresponding to termination codons are not selected against and ORFs are comparatively short. Thus, when analyzing the ORFs displayed for a particular cloned DNA, it is usually true that a long ORF is likely to be coding DNA whereas short ORFs may not be. Nevertheless, one must be aware that some exons can be short and so some short ORFs may also be coding DNA. Computer analysis may be able to detect these by screening for the conserved sequences at exon–intron boundaries and the splice branchpoint sequence (Section G7). Finally, by referring to the genetic code, computer analysis can predict the protein sequence encoded by each ORF. This is called the **deduced protein sequence**.

H2 Translation in prokaryotes

Key Notes

Overview

During translation, the mRNA is read in a 5′ to 3′ direction and protein is made in an N-terminal to C-terminal direction. Translation relies upon aminoacyl-tRNAs that carry specific amino acids and recognize the corresponding codons in mRNA by anticodon–codon base pairing. Translation takes place in three phases; initiation, elongation and termination.

Synthesis of aminoacyl-tRNA

Each tRNA molecule has a cloverleaf secondary structure consisting of three stem-loops, one of which bears the anticodon at its end. The amino acid is covalently bound to the 2′-OH or 3′-OH group at the 3′ end by a Class I or Class II aminoacyl synthetase, respectively, to form aminoacyl-tRNA. The reaction, called amino acid activation, occurs in two steps and requires ATP to form an intermediate, aminoacyl-adenylate.

Initiation of protein synthesis

Each prokaryotic ribosome has three binding sites for tRNAs; an A site where the incoming aminoacyl-tRNA binds, a P site where the tRNA linked to the growing polypeptide chain is bound, and an E site which binds tRNA prior to its release from the ribosome. Translation in prokaryotes begins by the formation of a 30S initiation complex between the 30S ribosomal subunit, mRNA, initiation factors and fMet-tRNA$_f^{Met}$. The 30S subunit binds to the Shine–Dalgarno sequence, which lies 5′ to the AUG start codon and is complementary to the 16S rRNA of the small ribosomal subunit. The ribosome then moves in a 3′ direction along the mRNA until it encounters the AUG codon. The 50S ribosomal subunit now binds to the 30S initiation complex to form the 70S initiation complex. In this complex, the anticodon of the fMet-tRNA$_f^{Met}$ is base paired to the AUG initiation codon (start codon) in the P site.

Elongation

The elongation cycle consists of three steps: aminoacyl-tRNA binding, peptide bond formation, and translocation. In the first step, the aminoacyl-tRNA corresponding to the second codon binds to the A site on the ribosome as an aminoacyl-tRNA/EF-Tu/GTP complex. After correct base pairing between the anticodon of the tRNA and the codon of the mRNA, the GTP is hydrolyzed and EF-Tu/GDP is released. The EF-Tu is regenerated via the EF-Tu–EF-Ts exchange cycle. Peptide bond formation is catalyzed by peptidyl transferase between the C-terminus of the aminoacyl moiety in the P site and the amino group of the aminoacyl-tRNA in

	the A site. In the final step (translocation), EF-G/GTP binds to the ribosome, the deacylated tRNA moves from the P site to the E site, the dipeptidyl-tRNA in the A site moves to the P site, and the ribosome moves along the mRNA to place the next codon in the A site. The GTP is hydrolyzed to GDP and inorganic phosphate. When the next aminoacyl-tRNA binds to the A site in the next round of elongation, the deacylated tRNA is released from the E site.
Termination	The appearance of a UAA or UAG termination (stop) codon in the A site causes release factor RF-1 to bind whereas RF-2 recognizes UGA. RF-3 assists RF-1 and RF-2. The release factors trigger peptidyl transferase to transfer the polypeptide to a water molecule instead of to aminoacyl-tRNA. The polypeptide, mRNA, and free tRNA leave the ribosome and the ribosome dissociates into its subunits ready to begin a new round of translation.
Related topics	(G1) An introduction to RNA (G2) Transcription in prokaryotes (G3) Operons (G4) Transcription in eukaryotes: an overview (G5) Transcription of protein-coding genes in eukaryotes (G6) Regulation of transcription by RNA Pol II (G7) Processing of eukaryotic pre-mRNA (G8) Transcription and processing of ribosomal RNA (G9) Transcription and processing of transfer RNA (H1) The genetic code (H3) Translation in eukaryotes

Overview

Ribosomes exist normally as separate subunits that come together to form a ribosome when they bind to an mRNA near its 5′ end. The ribosome reads the nucleotide sequence from the 5′ to 3′ direction, synthesizing the corresponding protein from amino acids in an N-terminal (amino terminal) to C-terminal (carboxyl terminal) direction. The amino acids used are covalently bound to tRNA (transfer RNA) molecules to form aminoacyl-tRNAs. Each aminoacyl-tRNA bears a triplet of bases, called an **anticodon**. The ribosome reads each triplet codon of the mRNA in turn and an aminoacyl-tRNA molecule with an anticodon that is complementary to the codon binds to it via hydrogen bonding. A peptide bond is then formed between the incoming amino acid and the growing end of the polypeptide chain.

Overall, protein synthesis (or **translation**) takes place in three stages: **initiation, elongation** and **termination**. During initiation, the mRNA–ribosome complex is formed and the first codon (always AUG) binds the first aminoacyl-tRNA (called **initiator tRNA**). During the elongation phase, the other codons are read sequentially and the polypeptide grows by addition of amino acids to its C-terminal end. This process continues until a termination codon (stop codon), which does not have a corresponding aminoacyl-tRNA with which to base pair, is reached. At this point, protein synthesis ceases (termination phase) and the finished polypeptide is released from the ribosome. Usually at any one time, many ribosomes are translating an mRNA simultaneously, forming a structure called a **polyribosome** or **polysome**.

Synthesis of aminoacyl-tRNA

Synthesis of aminoacyl-tRNAs is crucially important for two reasons. First, each amino acid must be covalently linked to a tRNA molecule in order to take part in protein synthesis, which depends upon the 'adaptor' function of tRNA to ensure that the correct amino acids are incorporated. Second, the covalent bond that is formed between the amino acid and the tRNA is a high-energy bond that enables the amino acid to react with the end of the growing polypeptide chain to form a new peptide bond. For this reason, the synthesis of aminoacyl-tRNA is also referred to as **amino acid activation**. Amino acids that are not linked to tRNAs cannot be added to the growing polypeptide.

The attachment of an amino acid to a tRNA is catalyzed by an enzyme called **aminoacyl-tRNA synthetase**. A separate aminoacyl-tRNA synthetase exists for every amino acid, making 20 synthetases in total. The synthesis reaction occurs in two steps.

1. The first step is the reaction of an amino acid and ATP to form an **aminoacyl-adenylate** (also known as **aminoacyl-AMP**):

Aminoacyl-adenylate
(Aminoacyl-AMP)

2. In the second step, without leaving the enzyme, the aminoacyl group of aminoacyl-AMP is transferred to the terminal adenosine at the 3′ end of the tRNA molecule to form aminoacyl-tRNA:

$$\text{aminoacyl-AMP} + \text{tRNA} \rightleftharpoons \text{aminoacyl-tRNA} + \text{AMP}$$

This occurs by forming an ester bond between the –COOH group of the amino acid and the 2′-OH or 3′-OH of the ribose sugar of the adenosine. Whether the 2′-OH or 3′-OH is used depends on the type of aminoacyl-tRNA synthetase involved in the reaction; Class I aminoacyl-tRNA synthetases attach the amino acid to the 2′-OH group and Class II aminoacyl-tRNA synthetases use the 3′-OH group.

The sum of these two reactions is:

$$\text{amino acid} + \text{ATP} + \text{tRNA} \rightleftharpoons \text{aminoacyl-tRNA} + \text{AMP} + \text{PP}_i$$

The reaction is driven to completion (i.e. synthesis of aminoacyl-tRNA) by the subsequent hydrolysis of the pyrophosphate (PP_i) to inorganic phosphate (P_i):

$$PP_i + H_2O \rightarrow 2P_i$$

The overall reaction is therefore:

$$\text{amino acid} + \text{ATP} + \text{tRNA} + H_2O \rightarrow \text{aminoacyl-tRNA} + \text{AMP} + 2P_i$$

Figure 1. Schematic of a prokaryotic 70S ribosome showing the peptidyl-tRNA site (P site), aminoacyl-tRNA site (A site) and exit site (E site).

Initiation of protein synthesis

Each prokaryotic ribosome, shown schematically in Figure 1 (see Section G8 for details of ribosome structure), has *three* binding sites for tRNAs. The **aminoacyl-tRNA binding site** (or **A site**) is where, during elongation, the incoming aminoacyl-tRNA binds. The **peptidyl-tRNA binding site** (or **P site**) is where the tRNA linked to the growing polypeptide chain is bound. The **exit site** (or **E site**) is a binding site for tRNA following its role in translation and prior to its release from the ribosome.

All three sites (A, P and E) are formed by the rRNA molecules in the ribosome (Section G8) and bridge the large and small ribosomal subunits; anticodon–codon interaction is associated with the small subunit whereas the aminoacyl part of the tRNA is associated with the large subunit.

Note that the E site is unique to prokaryotic ribosomes. Eukaryotic ribosomes do not have an exit site; the tRNA simply dissociates from the ribosome after use.

The first codon translated in all mRNAs is AUG, which codes for methionine. This AUG is called the **start codon** or **initiation codon**. Naturally, other AUG codons also occur internally in an mRNA where they encode methionine residues internal to the protein. Two different tRNAs are used for these two types of AUG codon; $tRNA_f^{Met}$ is used for the initiation codon and is called the **initiator tRNA**, whereas $tRNA_m^{Met}$ is used for internal AUG codons. In prokaryotes, the first amino acid of a new protein is ***N*-formylmethionine** (abbreviated **fMet**). Hence the aminoacyl-tRNA used in initiation is **$fMet\text{-}tRNA_f^{Met}$** which is abbreviated as $fMet\text{-}tRNA_f$.

The use of *N*-formylmethionine, which has a formyl group (-COH) attached to its amino group, prevents the amino group being involved in peptide bond formation. So only the –COOH group of *N*-formylmethionine can form a peptide bond and this ensures that polypeptide synthesis occurs in the N to C direction (Section H1).

It is essential that the correct AUG is used as the initiation codon, since this sets the correct reading frame for translation (Section H1). A short sequence rich in purines (5′-AGGAGGU-3′), called the **Shine–Dalgarno sequence**, lies 5′ to the AUG initiation codon (Figure 2) and is complementary to part of the 16S rRNA in the small ribosomal subunit called the **anti-SD sequence**. The 30S ribosomal subunit binds to the Shine–Dalgarno sequence and then migrates in a 3′ direction along the mRNA until it encounters the AUG initiation codon. Therefore the Shine–Dalgarno sequence delivers the ribosomal subunit to the correct AUG for initiation of translation.

Figure 2. The Shine–Dalgarno sequence in prokaryotic mRNA.

The initiation of protein synthesis requires proteins called **initiation factors** (**IFs**). In prokaryotes, three initiation factors (IF-1, IF-2 and IF-3) are essential. Because of the complexity of the process, the exact order of binding of the components, and particularly of IF-1, is still unclear. One current model is shown in Figure 3 and is described below.

- Initiation begins with the binding of IF-1 and IF-3 to the small (30S) ribosomal sub-unit. Binding of IF-3 stops the 30S subunit binding to the 50S subunit in the absence of mRNA and fMet-tRNA$_f$^{Met}, which would result in a nonfunctional ribosome.

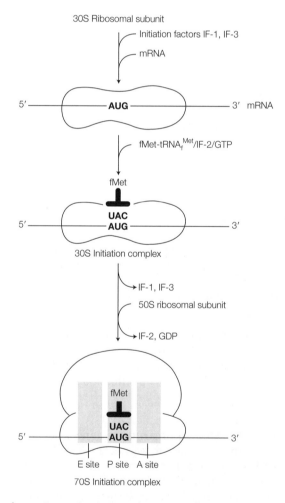

Figure 3. Initiation of protein synthesis in prokaryotic cells.

- The small subunit then binds to the mRNA via the Shine–Dalgarno sequence and moves 3′ along the mRNA until it locates the AUG initiation codon.

- The initiator tRNA charged with *N*-formylmethionine and in a complex with IF-2 and GTP (fMet-tRNA$_f^{Met}$/IF-2/GTP) now binds.

- This complex of mRNA, fMet-tRNA$_f^{Met}$, IF-1, IF-2, IF-3 and the 30S ribosomal subunit is called the **30S initiation complex**.

- IF-1 and IF-3 are released.

- The large (50S) ribosomal subunit now binds. The GTP bound to IF-2 is hydrolyzed to GDP and P$_i$, and triggers the release of IF-2. The resulting structure is a **70S initiation complex**.

One important point to note is that, unlike all other aminoacyl-tRNA molecules (which bind to the A site; see below), the binding of fMet-tRNA$_f^{Met}$ occurs directly into the P site.

Elongation

At the start of the first round of elongation (Figure 4), the initiation codon (AUG) is positioned in the P site with fMet-tRNA$_f^{Met}$ bound to it via codon–anticodon base pairing. The next codon in the mRNA is positioned in the A site. Elongation of the polypeptide chain occurs in three steps called the **elongation cycle**, namely **aminoacyl-tRNA binding**, **peptide bond formation** and **translocation**:

1. *Aminoacyl-tRNA binding*: in this first step, the corresponding aminoacyl-tRNA for the second codon binds to the A site via codon–anticodon interaction (Figure 4). But binding of the aminoacyl-tRNA is not a passive process; it requires **elongation factor EF-Tu** (which is a GTPase) and GTP. These bind to the aminoacyl-tRNA to form an aminoacyl-tRNA/EF-Tu/GTP complex, which then delivers the aminoacyl-tRNA to the A site. If the anticodon of the aminoacyl-tRNA base pairs correctly with the codon, the GTP is hydrolyzed to GDP, the aminoacyl-tRNA is transferred to the ribosome and EF-Tu is released as EF-Tu/GDP (Figure 4). However if the anticodon and codon cannot base pair correctly, GTP hydrolysis does not occur and the aminoacyl-tRNA is not transferred to the ribosome.

    Before the EF-Tu molecule can catalyze the binding of another charged tRNA to the ribosome, it must be regenerated by a process involving another elongation factor, **EF-Ts**. This regeneration is called the **EF-Tu–EF-Ts exchange cycle** (Figure 5). First, EF-Ts binds to EF-Tu and displaces the GDP. Then GTP binds to the EF-Tu and displaces EF-Ts. The EF-Tu/GTP is now ready to take part in another round of elongation.

    Note that EF-Tu will not bind to fMet-tRNA$_f^{Met}$, which prevents the initiator tRNA from binding to the A site and stops internal AUG codons from binding initiator tRNA. In contrast, IF-2 binds to fMet-tRNA$_f^{Met}$ but not to Met-tRNA$_m^{Met}$, (or any other aminoacyl-tRNA) which means that, of all the tRNAs available, it will only place the initiator tRNA in the P site in response to an initiation codon AUG.

2. *Peptide bond formation*: the second step, peptide bond formation, is catalyzed by **peptidyl transferase**. In this reaction the carboxyl end of the amino acid bound to the tRNA in the P site is uncoupled from the tRNA and becomes joined by a peptide bond to the amino group of the amino acid linked to the tRNA in the A site (Figure 4). A protein with peptidyl transferase activity has never been isolated. The reason is

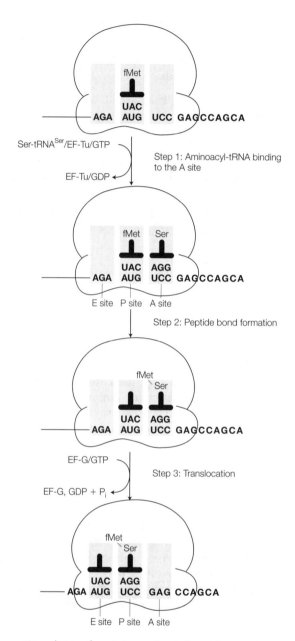

Figure 4. The elongation phase of protein synthesis in prokaryotes.

Figure 5. The EF-Tu–EF-Ts exchange cycle.

now clear; in *E. coli* at least, the peptidyl transferase activity is associated with part of the 23S rRNA in the large ribosomal subunit. In other words, peptidyl transferase is a **ribozyme**, a catalytic activity that resides in an RNA molecule (Section G8).

3. *Translocation*: in the third step, a complex of **elongation factor EF-G** (also called **translocase**) and GTP (i.e. EF-G/GTP) binds to the ribosome. Three concerted movements now occur, collectively called translocation:

- The deacylated tRNA moves from the P site to the E site

- The dipeptidyl-tRNA in the A site moves to the P site

- The ribosome moves along the mRNA (5′ to 3′) by three nucleotides to place the next codon in the A site

During the translocation events, GTP is hydrolyzed to GDP and inorganic phosphate, and EF-G is released ready to bind more GTP for another round of elongation.

After translocation, the A site is empty and ready to receive the next aminoacyl-tRNA. The A site and the E site cannot be occupied simultaneously. Therefore the deacylated tRNA must be released from the E site before the next aminoacyl-tRNA binds to the A site to start a new round of elongation.

Elongation continues, adding one amino acid to the C-terminal end of the growing polypeptide for each codon that is read, with the peptidyl-tRNA moving back and forth from the P site to the A site as it grows.

Termination

Eventually, one of three **termination codons** (also called **stop codons**) becomes positioned in the A site (Figure 6). These are UAG, UAA and UGA. Unlike other codons, prokaryotic cells do not contain aminoacyl-tRNAs complementary to stop codons. Instead, one of two **release factors** (**RF-1** and **RF-2**) binds instead. RF-1 recognizes UAA and UAG, whereas RF-2 recognizes UAA and UGA. A third release factor, **RF-3**, is also needed to assist RF-1 or RF-2 interaction with the ribosome. Thus either RF-1+RF-3 or RF-2+RF-3 bind depending on the exact termination codon in the A site. RF-1 (or RF-2) binds at or near the A site whereas RF-3/GTP binds elsewhere on the ribosome. The release factors cause the peptidyl transferase activity to transfer the polypeptide to a water molecule instead of to aminoacyl-tRNA, effectively cleaving the bond between the polypeptide and tRNA in the P site.

To understand this, it is important to realize that the ribosome normally excludes water from the reaction center since this could otherwise hydrolyze the peptidyl-tRNA ester bond and cause premature release of the polypeptide. The release factors appear to work by allowing a water molecule to access the peptidyl transferase reaction center so that hydrolysis now occurs. The free polypeptide now leaves the ribosome, followed by the mRNA and free tRNA, and the ribosome dissociates into 30S and 50S subunits ready to start translation again.

Figure 6. Termination of protein synthesis in prokaryotic cells.

H3 Translation in eukaryotes

Key Notes

Initiation by scanning

Eukaryotic ribosomes are larger (80S) and more complex than prokaryotic ribosomes (70S). Initiation is basically similar in prokaryotes and eukaryotes except that in eukaryotes at least 13 initiation factors have been identified (cf. three factors in prokaryotes), the initiating amino acid is methionine (cf. N-formylmethionine in prokaryotes), and most eukaryotic mRNAs do not contain internal ribosome binding sites so the AUG initiation codon is detected by ribosome scanning from the 5′ end instead. Initiation via scanning in eukaryotes involves the formation of a pre-initiation complex between the 40S ribosomal subunit, a ternary complex of Met-tRNA$_i^{Met}$/eukaryotic initiation factor (eIF)-2 /GTP and eIF-1, eIF-1A and eIF-3. The pre-initiation complex attaches to the 5′ end of the mRNA by interacting with eIF-4F that binds to the 5′ cap. This complex bound at the 5′ end is called the initiation complex. It now scans along the mRNA to locate the AUG initiation codon (often contained in a short sequence called the Kozak consensus). The 60S ribosomal subunit then binds to form the 80S initiation complex.

Initiation without scanning

Some mRNAs derived from eukaryotic protein-coding genes, as well as picornavirus RNA, contain internal ribosome entry sites (IRES) and initiation of protein synthesis can occur without scanning.

Elongation

Elongation in eukaryotes requires three eukaryotic initiation factors that have similar functions to the corresponding prokaryotic proteins.

Termination

A single eukaryotic release factor, eRF-1, recognizes all three termination codons but may be assisted by eRF-3.

Related topics

Initiation by scanning

The overall mechanism of protein synthesis in eukaryotes is basically the same as in pro-karyotes, with three phases defined as initiation, elongation and termination. However, there are some significant differences, particularly during initiation.

- Whereas a prokaryotic ribosome has a sedimentation coefficient (Section G8) of 70S and subunits of 30S and 50S, a eukaryotic ribosome has a sedimentation coefficient of 80S with subunits of 40S and 60S (Section G8). The composition of eukaryotic ribo-somal subunits is also more complex than prokaryotic subunits (Section G8) but the function of each subunit is essentially the same as in prokaryotes.

- In almost all cases, each eukaryotic mRNA has only one initiation site and hence encodes a single protein. In prokaryotes, many mRNAs are **polycistronic**, that is they encode several proteins. Each coding sequence in a prokaryotic mRNA has its own initiation and termination codons.

- Initiation of protein synthesis in eukaryotes involves many more initiation factors than in prokaryotes; at least 13 distinct **eukaryotic initiation factors** (**eIFs**) have been reported (Table 1) compared with the three initiation factors (IFs) in prokaryotes (Section H2).

- In eukaryotes, the initiating amino acid is **methionine**, not N-formylmethionine as in prokaryotes.

- As in prokaryotes, a special initiator tRNA is required for initiation and is distinct from the tRNA that recognizes and binds to codons for methionine at internal positions in the mRNA. When charged with methionine ready to begin initiation, this is known as **Met-tRNA$_i^{Met}$** (where the 'i' stands for initiation).

- The main difference between initiation of translation in prokaryotes and eukaryotes is that in bacteria a Shine–Dalgarno sequence (Section H2) lies 5′ to the AUG initiation codon and is the binding site for the 30S ribosomal subunit, marking this AUG as the one to use for initiation rather than any other AUG internal in the mRNA. The initiation

Table 1. Comparison of protein synthesis factors in prokaryotes and eukaryotes

Prokaryotic	Eukaryotic	Function
Initiation factors		
IF-1, IF-2, IF-3	At least 13 initiation factors; so far identified: eIF-1, e-IF-1A, eIF-2, eIF-2B, eIF-3, eIF-4A, eIF-4B, eIF-4E, eIF-4F, eIF-4G, eIF-4H, eIF-5, eIF-6	Individual factors have functions that differ between prokaryotes and eukaryotes (see the text)
Elongation factors		
EF-Tu	eEF-1α	Aminoacyl-tRNA delivery to ribosome
EF-Ts	eEF-1βγ	Recycling of EF-Tu or eEF-1α
EF-G	eEF-2	Translocation
Termination factors		
RF-1, RF-2, RF-3	eRF-1, eRF-3	Polypeptide chain release

complex is assembled directly over this initiation codon. In contrast, most eukaryotic mRNAs do not contain internal ribosome binding sites. Instead, a 40S ribosomal subunit attaches at the 5′ end of the mRNA and moves downstream (i.e. in a 5′ to 3′ direction) until it finds the AUG initiation codon. This process is called **scanning**.

The full details of initiation in eukaryotes are still not fully known but the process occurs broadly as follows.

1. The first step is the formation of a **pre-initiation complex** consisting of the 40S small ribosomal subunit, a so-called 'ternary complex' of Met-tRNA$_i^{Met}$, eIF-2 and GTP, and initiation factors eIF-1, eIF-1A and eIF-3.

2. The pre-initiation complex now binds to the 5′ end of the eukaryotic mRNA. This binding requires **eIF-4F** (a complex of eIF-4A, eIF-4E, and eIF-4G), also called the **cap binding complex**. The cap binding complex binds to the 5′ cap on the mRNA and so attaches the pre-initiation complex to the 5′ end of the mRNA. Once bound at the 5′ end, this is now called the **initiation complex**.

   The poly(A) tail present at the 3′ end of most eukaryotic mRNAs can also influence initiation. It appears to do so by interaction between eIF-4G and **poly(A) binding protein**, which is attached to the poly(A) tail, which of course implies that the mRNA bends back on itself to form a circular molecule to let this interaction occur. The role of this interaction is not yet clear but the existence of a poly(A) tail, and hence an intact mRNA at the 3′ end, may facilitate initiation.

3. The initiation complex now moves along the mRNA in a 5′ to 3′ direction until it locates the AUG initiation codon (i.e. scanning). The initiation codon is usually recognizable because it is often (but not always) contained in a short sequence called the **Kozak consensus** (5′-ACC**AUG**G-3′). The 5′ untranslated regions of eukaryotic mRNAs vary in length but can be several hundred nucleotides long and may contain secondary structures such as hairpin loops. The eIF-4A, and possibly also eIF-4B, is an ATP-dependent RNA helicase that unwinds any secondary structures, allowing the initiation complex to pass along the mRNA.

4. Once the complex is positioned over the initiation codon, the 60S large ribosomal subunit binds to form an **80S initiation complex**, a step that requires the hydrolysis of GTP and leads to the release of several initiation factors.

Initiation without scanning

A group of viruses called **picornaviruses** (including the common cold virus – rhinovirus) that infect eukaryotic cells have RNA genomes without caps. These RNA genomes are translated as mRNAs using an **internal ribosome entry site (IRES)**. The IRES sequences of different picornaviruses are more variable than the Shine–Dalgarno sequence in prokaryotes but function in a similar way, allowing initiation to occur internally in the mRNA.

Some mRNAs derived from eukaryotic protein-coding genes are now also known to contain IRESs, for example the mammalian immunoglobulin heavy-chain binding protein. In these cases, therefore, initiation can occur without scanning.

Elongation

The elongation stage of translation in eukaryotes requires three elongation factors, **eEF-1α**, **eEF-1βγ** and **eEF-2**, which have similar functions to their prokaryotic counterparts EF-Tu, EF-Ts and EF-G (Table 1).

During elongation in bacteria, the deacylated tRNA in the P site moves to the E site prior to leaving the ribosome (Section H2). In contrast, eukaryotic ribosomes appear not to have an E site and the deacylated tRNA is ejected directly from the ribosome.

Termination

Eukaryotes have just one release factor (eRF-1) that recognizes all three termination codons (UAA, UAG and UGA). A second release factor, eRF-3, may assist eRF-1 to terminate translation (Table 1).

H4 Protein targeting

Key Notes

Overview

Both in prokaryotes and eukaryotes, newly synthesized proteins must be delivered to a specific subcellular location or secreted from the cell for correct activity. This phenomenon is called protein targeting.

Secretory proteins

Secretory proteins have an N-terminal signal peptide, which targets the protein to be synthesized on the RER. The signal peptide is recognized by the SRP, which targets the mRNA–ribosome complex to the SRP receptor on the RER membrane. The SRP then dissociates and is recycled, allowing the growing polypeptide chain to be translocated across the membrane through the translocon, so-called co-translational translocation. Vesicles then bud off from the RER and carry the protein to the Golgi complex. Other vesicles then carry it to the plasma membrane. Fusion of these transport vesicles with the plasma membrane then releases the protein to the cell exterior.

Plasma membrane proteins

Plasma membrane proteins are also synthesized on the RER but become inserted into the RER membrane (and hence ultimately the plasma membrane). They may be single membrane-spanning proteins (Type I and Type II integral membrane proteins) or multiple membrane-spanning proteins (Type IV integral membrane protein). The orientation of the protein in the membrane is determined by topogenic sequences within the polypeptide chain. Type I proteins have a cleaved N-terminal signal sequence and a hydrophobic stop-transfer sequence, Type II have an uncleaved N-terminal signal sequence that doubles as the membrane-anchoring sequence, and Type IV have multiple signal sequences and stop-transfer sequences. Proteins destined to be anchored in the membrane by a GPI structure have both a cleaved N-terminal signal sequence and a C-terminal hydrophobic sequence that directs addition of the preformed GPI anchor.

Proteins of the endoplasmic reticulum

Proteins destined for the RER have an N-terminal signal peptide and are co-translationally translocated into the RER lumen or inserted into the RER membrane. C-terminal amino acid sequences (KDEL in soluble RER lumen proteins, KKXX or KXKXX in Type I integral membrane proteins) are recognized by specific receptor proteins and retain the proteins in the ER.

Lysosomal proteins

Lysosomal proteins are targeted to the lysosomes via the addition of a mannose 6-phosphate signal that is added

	in the *cis*-compartment of the Golgi and is recognized by a receptor protein in the *trans*-compartment of the Golgi. The protein is then transported by specialized vesicles that mature into a lysosome. The mannose 6-phosphate receptor recycles back to the Golgi for re-use.
Mitochondrial and chloroplast proteins	Most mitochondrial and chloroplast proteins are made on cytosolic ribosomes, released into the cytosol and then taken up into the organelle. Uptake into the mitochondrial matrix requires a matrix-targeting sequence and occurs at sites where the outer and inner mitochondrial membranes come into contact. The process is mediated by hsc70 and hsc60 proteins and requires both ATP hydrolysis and the electrochemical gradient across the inner mitochondrial membrane.
Nuclear proteins	Proteins destined for import into the nucleus typically require a nuclear localization signal, four to eight amino acids long and rich in positively charged amino acids, located internally in the protein. Uptake occurs via nuclear pores and requires ATP hydrolysis.
Related topics	(A1) Prokaryotic cells (E2) Membrane structure (A2) Eukaryotic cells (E4) Membrane transport: (B2) Protein structure and macromolecules function (H5) Protein glycosylation

Overview

Cells must ensure that each newly synthesized protein is sorted to its correct location where it can carry out the appropriate function. This process is called **protein targeting**. In a eukaryotic cell, the protein may be destined to stay in the cytosol, for example an enzyme involved in glycolysis (Section J3). Alternatively, it may need to be targeted to an organelle (such as a mitochondrion, lysosome, peroxisome, chloroplast or the nucleus) or be inserted into the plasma membrane or exported out of the cell (Section A2). In bacteria such as *E. coli*, the protein may stay in the cytosol, be inserted into the plasma membrane or the outer membrane, be sent to the space between these two membranes (the periplasmic space) or be exported from the cell (Section A1). In both prokaryotes and eukaryotes, if a protein is destined for the cytosol, it is made on ribosomes in the cytosol and released directly into the cytosol. If it is destined for other final locations, specific protein-targeting mechanisms are involved.

Secretory proteins

Proteins destined to be secreted from the eukaryotic cell are synthesized by ribosomes bound to the RER. As the protein is synthesized, it is translocated across the RER membrane into the lumen of the RER where it folds into its final conformation. The ER then buds off vesicles that carry the protein to the **Golgi apparatus** (Figure 1) (Section A2). The Golgi has a *cis* **face** (where vesicles enter) and a *trans* **face** (where vesicles leave). Thus the RER vesicles fuse with the *cis* **compartment** of the Golgi, releasing the protein into the Golgi lumen. The protein then moves through the Golgi complex to the

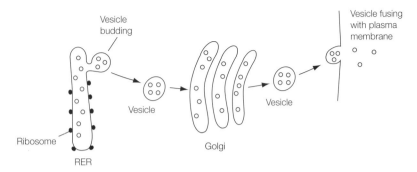

Figure 1. Synthesis and exocytosis of secretory proteins; see text for details. The ribosomes attached to the RER are shown as filled-in circles whereas the open circles in the lumen of the RER, vesicles and Golgi complex represent secretory protein molecules.

trans **compartment**, being modified *en route* by the addition of carbohydrate residues (glycosylation; Section H5). Finally, vesicles bud from the *trans* compartment and carry the glycosylated secretory proteins to the plasma membrane where the vesicles fuse, releasing their contents to the cell exterior. This fusion and extracellular release of proteins is called **exocytosis** (see Section E4 for more details).

Translocation across the ER membrane

A typical secretory protein differs from a cytosolic protein by having a sequence about 13–35 amino acids long at its N-terminal end called a **signal sequence** or **signal peptide**. The signal peptides of different secretory proteins differ in amino acid sequence but there are some common features: the center of the sequence usually consists of 10–15 hydrophobic amino acids flanked by several relatively hydrophilic residues including one or more positively charged (Arg, Lys) residues near the N-terminus. The signal peptide directs the secretory protein to the RER membrane and so targets the protein to cross into the RER lumen and ultimately be secreted. A simplified version of the mechanism is shown in Figure 2.

The mRNA for the secretory protein binds to a free cytoplasmic ribosome and protein synthesis begins. The first part of the protein made is the N-terminal signal peptide. An SRP, which is a large complex of a 7S RNA and six proteins, binds to the signal peptide and stops further protein synthesis. This stops the secretory protein from being released prematurely into the cytosol. The ribosome–mRNA–SRP complex now binds to an **SRP receptor**, a protein on the surface of the RER to form the ribosome–translocon complex. In a concerted series of reactions, the SRP is released from the signal peptide, and translation now continues once more, with the nascent polypeptide passing through an aqueous pore in the membrane created by the **translocon**. The translocon is a protein complex, the transmembrane pore of which is made from the **Sec61** protein in mammals and **SecY** in prokaryotes. As it passes through the translocon, the signal peptide is cleaved off by **signal peptidase** on the lumenal face of the RER (Figure 2) and degraded, with the rest of the protein being translocated across the RER membrane into the lumen. Since transport across the RER membrane occurs during protein synthesis, the process is called **co-translational translocation**. The protein is then transported through the Golgi to the cell exterior as described above. The released SRP is recycled ready for binding to another signal peptide (the **SRP cycle**) (Figure 2).

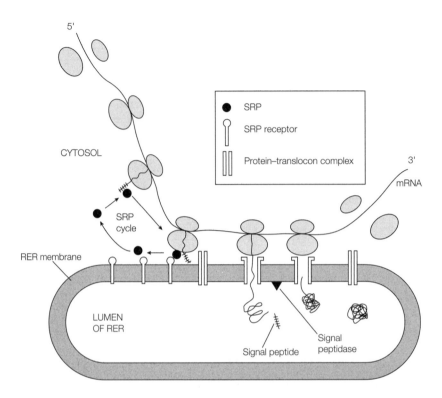

Figure 2. A simplified version of the signal hypothesis (see the text for details).

Plasma membrane proteins

Integral plasma membrane proteins are also synthesized by ribosomes on the RER, but become inserted in the RER membrane rather than transported into the lumen. During transport to the Golgi and then to the cell surface, these proteins stay anchored in the membrane, the final vesicles that fuse with the plasma membrane then becoming new plasma membrane (Figure 3). Note that, after insertion in the RER membrane, one part of the protein faces in towards the RER lumen but eventually this faces outward on the cell surface. It is this part of the protein that receives the carbohydrate during glycosylation in the RER and Golgi complex so that the carbohydrate is exposed on the cell surface (Section H5).

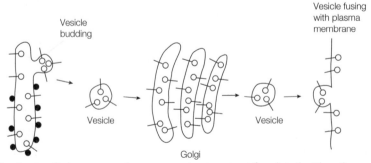

Figure 3. Synthesis of plasma membrane proteins; see text for details. The ribosomes attached to the RER are shown as filled-in circles whereas the newly synthesized plasma membrane proteins are shown as open circles on stalks.

Translocation of the plasma membrane protein across the RER membrane occurs during synthesis by the same mechanism as that for secretory proteins. However, by definition, the protein is destined to remain anchored in the membrane and not enter the RER lumen entirely. There are several ways in which this is achieved, depending on the type of membrane protein. Some integral membrane proteins are **single membrane-spanning proteins**, that is the polypeptide chain crosses the membrane only once, whereas in other cases the protein is a **multiple membrane-spanning protein** (Section E2). The orientation of the protein in the membrane and the number of times it spans the lipid bilayer depend on specific **topogenic sequences** within the polypeptide chain. These topogenic sequences are regions of predominantly hydrophobic amino acids, and fall into three types: **N-terminal signal sequences**, **internal signal sequences** and **stop-transfer sequences**.

In the single membrane-spanning **Type I integral membrane proteins** (Figure 4a), in addition to the N-terminal signal sequence, which is cleaved from the protein by signal peptidase as in secretory proteins, there is a second hydrophobic sequence located internally in the protein. Thus the protein starts to cross the RER membrane during synthesis just like a secretory protein, but then transfer is stopped before the entire protein is translocated and the protein stays inserted in the membrane through the interaction of the hydrophobic stop-transfer sequence with the translocon and then the hydrophobic interior of the bilayer. In the single membrane-spanning **Type II membrane proteins** (Figure 4b), there is just an N-terminal signal sequence as found in secretory proteins. However, in this case the signal sequence is not cleaved from the membrane protein by signal peptidase and doubles as the membrane anchor. Multiple membrane-spanning proteins (referred to as **Type IV integral membrane proteins)** (Figure 4c), which cross the membrane several times, have multiple alternating internal signal sequences and stop-transfer sequences to organize this arrangement during translocation. The final orientation of the N-terminus and the C-terminus depends on whether the N-terminal signal sequence is cleaved and whether the final topogenic sequence is an internal signal sequence or a stop-transfer sequence, respectively.

Those proteins that are anchored in the membrane through a covalently attached GPI structure at the C-terminus (Figure 4d; Section E2) possess both an N-terminal signal sequence to direct them to the RER membrane and a second hydrophobic sequence at the very C-terminus. The N-terminal signal sequence is cleaved off by signal peptidase, while the C-terminal sequence directs the addition of a preformed GPI structure to an internal amino acid residue near the C-terminus once the protein has been translocated across the RER membrane. The GPI structure is built up by sequential addition of sugars (glucosamine and mannose) and ethanolamine phosphate to phosphatidylinositol (Section E2) in the RER membrane. A **transamidase** enzyme then cleaves off the C-terminal signal sequence and concomitantly adds on the complete GPI anchor.

Proteins of the endoplasmic reticulum

The RER contains many proteins, called **chaperones** (Section B2), which have the role of assisting nascent proteins to fold correctly into their native conformation. RER-resident proteins are made on the RER and either pass into the lumen (as do secretory proteins) or are anchored in the membrane (as for Type I integral membrane proteins). However, these proteins contain a **retention signal** at the C-terminus that is recognized by specific receptor proteins, which retain these proteins in the RER, preventing them from moving along the secretory pathway to the Golgi. In the case of soluble proteins in the lumen of the RER, the retention signal is Lys-Asp-Glu-Leu (or **KDEL** using the one-letter amino

Figure 4. Insertion of integral membrane proteins into the RER membrane during synthesis: (a) Type I integral membrane protein with a cleavable N-terminal signal sequence and a stop-transfer sequence; (b) Type II integral membrane protein with an uncleaved N-terminal signal sequence; (c) Type IV integral membrane protein with multiple signal and stop-transfer sequences; (d) GPI-anchored membrane protein with a cleavable N-terminal signal sequence and a C-terminal GPI anchor addition sequence.

acid code) at the C-terminus. In the case of Type I integral membrane proteins in the RER membrane, the retention signal is KKXX or KXKXX in the cytosolic C-terminus.

Lysosomal proteins

Lysosomal enzymes and lysosomal membrane proteins are synthesized on the RER and transported to the *cis* compartment of the Golgi complex. *En route* they are N-glycosylated and then in the *cis*-Golgi a phosphate residue is added to one or more mannose residues to form **mannose 6-phosphate**. The mannose 6-phosphate is the signal that targets the lysosomal protein to its correct destination. It is recognized by **mannose 6-phosphate receptor** proteins in the *trans* compartment of the Golgi, which bind to the lysosomal protein and package it in transport vesicles that bud from the Golgi apparatus (Figure 5). The transport vesicles then fuse with sorting vesicles, the contents of which are acidic. The low pH causes dissociation of the lysosomal protein from its receptor and a phosphatase removes the phosphate from the mannose 6-phosphate, preventing it from re-binding to the receptor. Vesicles bud from the sorting vesicle to return the receptor to the Golgi for re-use (**receptor recycling**) and the lysosomal protein is now delivered to the lysosome by vesicle fusion with it (Figure 5). In **I-cell disease**, this selective targeting of the proteins to the lysosome goes awry because of a deficiency of the enzyme required for phosphorylation of the mannose residues in the *cis*-Golgi.

Not all lysosomal proteins take the normal route of protein targeting; some end up being exported by the cell and must be retrieved. This **scavenger pathway** works as follows. The lysosomal glycoprotein binds to mannose 6-phosphate receptors in the plasma membrane and is internalized again by endocytosis (Figure 5). This process, called **receptor-mediated endocytosis**, creates an endocytic vesicle (or **endosome**) that then delivers the lysosomal protein to the lysosome by fusion (Section E4).

Mitochondrial and chloroplast proteins

Mitochondria and chloroplasts contain their own DNA, ribosomes, mRNA, etc., and carry out protein synthesis, but very few mitochondrial or chloroplast proteins are made in this way. Rather, the large majority of mitochondrial and chloroplast proteins are encoded by the nuclear genome, are synthesized in the cytosol on ribosomes, released after synthesis and then translocated across the organelle membrane. Thus, this process is referred to as **post-translational translocation**. The protein may need to be targeted to any one of several locations; for mitochondria, this could be the **outer mitochondrial membrane**, the **inner membrane**, the **intermembrane space** or the mitochondrial **matrix**. Chloroplasts have the same subcompartments plus two other potential destinations, the **thylakoid membrane** and the **thylakoid space** (Sections A2 and L3).

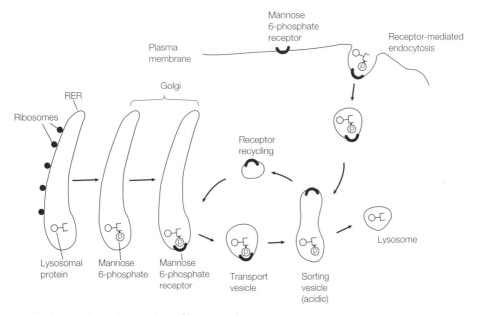

Figure 5. Synthesis and targeting of lysosomal proteins.

Proteins are targeted to the mitochondrial matrix by an N-terminal **matrix-targeting sequence**. This sequence is typically 25–50 amino acids long and rich in hydrophobic amino acids, the hydroxylated amino acids Ser and Thr, and the positively charged amino acids Arg and Lys. These matrix-targeting sequences probably assume an α-helical conformation with the positively charged amino acids on one side of the helix and the hydrophobic amino acids on the other, such that these sequences are amphipathic. After synthesis by cytosolic ribosomes, the protein is released into the cytosol but is kept in an unfolded state by **chaperone proteins** from the **hsc70** family of proteins, which bind to it during synthesis (Section B2). This is necessary since folded proteins cannot be imported into mitochondria. The hsc70 then transfers the unfolded protein to an **import receptor** in the outer mitochondrial membrane that is believed to slide along the membrane until it reaches a site where the inner membrane and outer membrane are in contact (a **contact site**). At this point, the protein passes into the matrix via a protein translocon formed from the components of both membranes (Figure 6). Proteins in the outer mitochondrial membrane involved in targeting and import are designated **Tom** proteins for *t*ranslocon of the *o*uter *m*embrane, those in the inner membrane are designated **Tim** proteins). As it passes through the pore, the cytoplasmic hsc70 is released, the signal peptide is cleaved off by a signal peptidase, and the protein is bound in the matrix by **mitochondrial hsc70**. The hsc70 is then replaced by **mitochondrial hsc60**, which assists the protein to fold correctly into its final active state. Import of proteins into the mitochondrion requires energy from the electrochemical gradient across the inner membrane (Section L2) as well as ATP hydrolysis. Protein import into the mitochondrial inner membrane and intermembrane space needs two signals; the protein is first imported into the matrix as described above and then a second signal sequence directs the protein back into the inner membrane or across it into the intermembrane space.

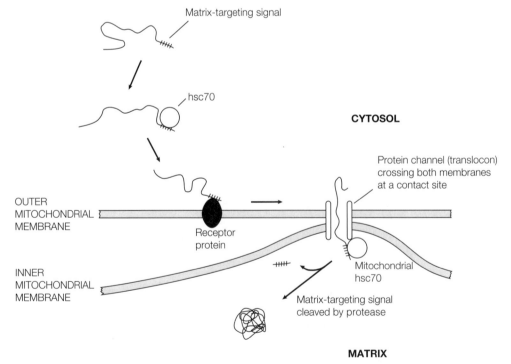

Figure 6. Uptake of proteins into the mitochondrial matrix; see text for details.

Protein import into chloroplasts follows similar mechanisms to those in mitochondria but the signals used are different since mitochondria and chloroplasts are present together in some plant cells and yet proteins become targeted to the correct destination.

Nuclear proteins

The nucleus has an inner and an outer membrane (Section A2) and is perforated by 3000–4000 **nuclear pores**. Each pore consists of a **nuclear pore complex** of more than 100 different proteins organized in a hexagonal array. Although small molecules can pass through the pore by free diffusion, large proteins entering the nucleus require **a nuclear localization signal**. This is four to eight amino acids long and is rich in the positively charged amino acids Lys and Arg as well as usually containing Pro, and is located internally within the polypeptide chain. The protein is taken through the pore in an ATP-requiring step and enters the nucleus without cleavage of the localization signal.

H5 Protein glycosylation

Key Notes

Protein glycosylation: overview	Many proteins synthesized by the ribosomes of the RER contain short chains of carbohydrates (oligosaccharides) and are called glycoproteins. The oligosaccharides are of two main types; O-linked (to the OH side chain of Ser or Thr) and N-linked (to the NH_2 side chain of Asn). To be modified by N-linked glycans the Asn residues must be in the consensus sequence Asn-X-Ser or Asn-X-Thr.
Synthesis of O-linked oligosaccharides	O-linked oligosaccharides are synthesized by the sequential addition of monosaccharides to the protein as it passes through the Golgi complex. These oligosaccharides usually consist of only four to five sugar residues.
Synthesis of N-linked oligosaccharides	N-linked oligosaccharides are initially synthesized on a dolichol phosphate carrier that is anchored to the RER membrane. The completed precursor structure with the composition $(Glc)_3(Man)_9(GlcNAc)_2$ is then transferred to the protein by the enzyme oligosaccharide transferase. Before leaving the RER the three terminal glucose residues are removed. The resulting high-mannose type oligosaccharide may be trimmed down to a pentasaccharide core with the composition $(Man)_3(GlcNAc)_2$ and additional monosaccharides added in the Golgi to produce a complex type of oligosaccharide.
Related topics	(H4) Protein targeting (J1) Monosaccharides and disaccharides

---

Protein glycosylation: overview

Most proteins made by ribosomes on the RER (Section H4) are glycoproteins, that is they contain short chains of carbohydrates (oligosaccharides) covalently linked to them during passage through the RER and Golgi complex. Two main types of oligosaccharide linkage exist:

- **O-linked oligosaccharides** are commonly attached to the protein via O-glycosidic bonds to OH groups of Ser or Thr side chains (Figure 1a). The residues that are modified are often in a region of the polypeptide chain that is rich in Ser and/or Thr residues. In plant glycoproteins, the OH groups of hydroxyproline (Hyp) residues can be glycosylated, while the hydroxylysine (Hyl) residues in the mammalian protein collagen can be O-glycosylated (Section B4).

- **N-linked oligosaccharides** are linked to the protein via N-glycosidic bonds to the NH_2 groups of Asn side chains (Figure 1b) where the Asn occurs in the sequence Asn-X-Ser or Asn-X-Thr where X is any amino acid except Pro. However, not all occurrences of

Figure 1. Structures of oligosaccharide linkages: (a) O-linked glycosidic bond between GalNAc and Ser (Thr) residue; (b) N-linked glycosidic bond between GlcNAc and an Asn residue.

Asn in such sequences are glycosylated; the sites that are glycosylated depend on the cell type synthesizing the protein. If the Asn is N-glycosylated the Ser or Thr residue in this motif is not O-glycosylated.

Synthesis of O-linked oligosaccharides

The synthesis of O-linked oligosaccharides occurs by the sequential addition of monosaccharide units to the newly synthesized protein as it passes through the Golgi complex. First, N-acetylgalactosamine (GalNAc) is transferred to the relevant Ser or Thr residue of the protein by GalNAc transferase, an enzyme that uses UDP-GalNAc as the precursor (Figure 2). Other monosaccharides [galactose, N-acetylglucosamine (GlcNAc), sialic acid, fucose] (Section J1) are then added using the corresponding sugar nucleotides as precursors. The exact type and number (usually only four to five) of monosaccharides added depends on the protein that is being modified.

Figure 2. Synthesis of O-linked oligosaccharide. The example shown is an O-linked oligosaccharide in human immunoglobulin A (IgA).

Synthesis of N-linked oligosaccharides

In contrast to O-linked oligosaccharides, which are built up sequentially on the protein, N-linked oligosaccharides are synthesized as a large, branched precursor structure that is then added en bloc to the acceptor Asn residue. The oligosaccharide is made on a lipid carrier called **dolichol phosphate**. This consists of 22 isoprene (C5) units (Section K5) with a terminal phosphate group and it is located in the RER membrane.

Synthesis of the oligosaccharide starts by the dolichol phosphate accepting monosaccharides from the cytosolic face of the RER membrane (Figure 3) but when the $(Man)_5(GlcNAc)_2$–dolichol phosphate intermediate has formed, this flips orientation and now accepts further monosaccharides from the lumenal side of the RER membrane (Figure 3). All of these subsequent transfers are from dolichol phosphate-linked monosaccharides that are made on the cytoplasmic side of the RER membrane then likewise flipped across to act as donors. The final oligosaccharide, with the composition $(Glc)_3(Man)_9(GlcNAc)_2$ is linked to the dolichol by a high-energy pyrophosphate bond (Figure 3). This provides the energy for transfer of the oligosaccharide to the protein, a reaction catalyzed by a membrane-bound **oligosaccharide transferase** enzyme and which occurs in the lumen of the RER essentially as soon as the growing polypeptide chain appears through the translocon (Figure 4).

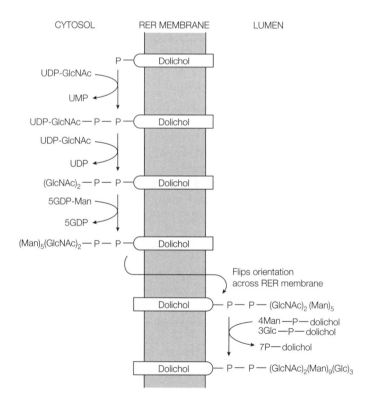

Figure 3. Synthesis of N-linked oligosaccharides on a dolichol phosphate carrier in the RER membrane.

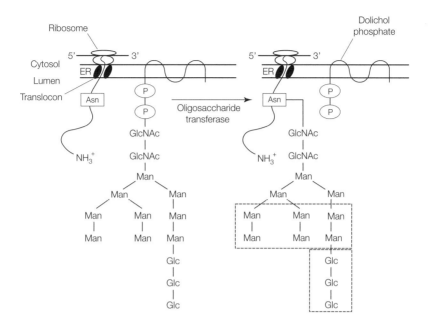

Figure 4. Transfer of the N-linked oligosaccharide from the dolichol phosphate carrier to an acceptor Asn residue in a growing polypeptide chain by oligosaccharide transferase. The terminal three glucose residues and six mannose residues that are removed during trimming of the oligosaccharide are indicated.

Whilst the protein is still in the RER, the three glucose residues are removed (Figures 4 and 5). Interestingly, glucose residues are added back to the protein if it is unfolded or wrongly folded. Thus, only when the protein is correctly folded are all glucose residues finally removed and the protein may continue along the secretory pathway. The folded glycoprotein with its **high-mannose type oligosaccharides** (Figure 3 in Section J2) is now transported to the Golgi complex via vesicles (Section H4). As it moves through the Golgi complex, the oligosaccharide is often '**trimmed**' or '**processed**' with the terminal six mannose residues being removed to leave the **pentasaccharide core** (Figures 4 and 5). Further mannose residues and other monosaccharides may then be added to the oligosaccharide in the Golgi to generate the **complex type of oligosaccharide** (Figure 5; Figure 3 in Section J2).

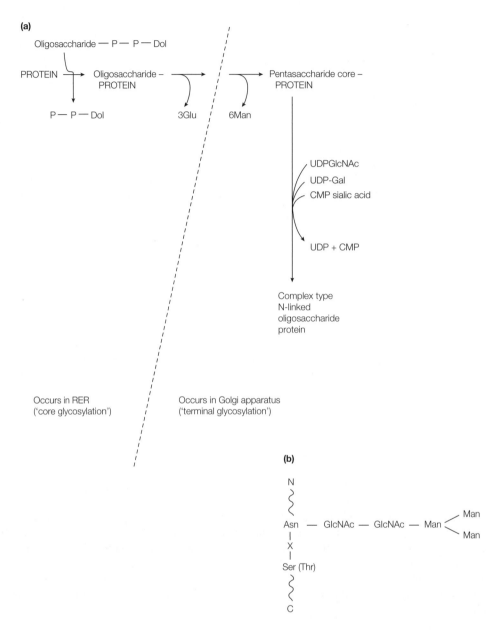

Figure 5. (a) Transfer of the N-linked oligosaccharide to a protein and its further processing in the RER and Golgi (see text for details); (b) structure of the pentasaccharide core of N-linked oligosaccharides.

I1 The power of recombinant DNA approaches

Key Notes	
Genomics	Genomics refers to studies of an organism's genome. The analysis of gene function (functional genomics) relies upon a range of techniques including reverse genetics and RNA interference (RNAi).
Transcriptomics and proteomics	Transcriptomics is the study of the transcriptome (the expressed RNA) whereas proteomics is the study of the proteome (the proteins that are synthesized).
Metabolomics	Metabolomics is the study of the small molecule components of a cell, i.e. the metabolome.
Transgenic organisms	Transgenic organisms are those that have been modified by the insertion of a cloned gene(s).
Related topics	(C3) Protein sequencing and peptide synthesis (I2) Restriction enzymes (I3) Nucleic acid hybridization (I4) DNA cloning (I5) DNA sequencing (I6) Polymerase chain reaction

Genomics

We are living in an unprecedented era of biological discovery and the application of biological knowledge. The first genome sequenced, in 1977, was a small bacterial virus (bacteriophage) called φX174 – with just 11 genes and about 5000 base pairs of DNA, i.e. A paired with T, G paired with C. In contrast, the human genome consists of about 3.2 billion base pairs! The **Human Genome Project** – to sequence the human genome – officially began in 1990. Advances in automated DNA sequencing allowed the announcement of the first draft (about 90%) of the human genome on 26 June 2000. By 2003, the sequence was essentially complete.

In June 2010, just 10 years after the first draft of the human genome was announced, the Wellcome Trust in the UK launched a project to sequence 10 000 human genomes in the next 3 years! This so-called UK10K project will seek to identify rare genetic variants that are important in human genetic disease by focusing on the gene-containing regions of these genomes (so-called '**exomes**'). The pace of technological change is breathtaking; it took 10 years to sequence just one genome and will now take just 3 years to sequence 10 000 genomes.

This vast, and ever-increasing, wealth of DNA sequence data is a major asset for **genomics**, the study of an organism's genome. One of the key goals is to understand the functions of the large numbers of new genes predicted by genome sequencing, a field known as **functional genomics**. One approach to revealing the function of a gene is called **reverse genetics**, whereby a mutation can be created in a cloned gene and the modified gene can be introduced into a host cell to monitor the effects of the mutation. The approach is called *reverse* genetics since starting with a gene and then creating a mutant is the reverse of traditional genetics whereby mutants were used to identify genes. Another useful approach is **RNA interference (RNAi)**, whereby a double-stranded RNA (dsRNA) corresponding to the sense and antisense strands of the gene under investigation is introduced into cells. The dsRNA is degraded *in vivo* but the resulting fragments base pair with mRNA from the target gene and cause that to be degraded also, in effect terminating expression of the gene. Monitoring the cellular effects of this loss of function provides evidence of the role of the gene *in vivo*.

Transcriptomics and proteomics

By analogy with the term 'genome', the **transcriptome** is all of the RNA sequence transcribed from a cell's genome and the **proteome** is all of the expressed proteins of that cell. Whereas all of the cells of an organism such as a human contain essentially the same genome, the transcriptome and proteome of different cell types varies depending on the genes that are expressed. For example, the transcriptome and proteome of a liver cell are different from that of a neuron.

Transcriptomics refers to studies of the transcriptome and includes, for example, the use of DNA microarrays (Section I3) to determine expression profiles. **Proteomics** is the study of the proteome. The methodology used to study a proteome is called **protein profiling** or **expression proteomics.**

Protein profiling relies heavily on using two-dimensional gel electrophoresis (Section C2) to separate the proteins expressed by a cell or tissue, followed by mass spectrometry to produce **peptide mass fingerprints** of each protein (Section C3). The expressed proteins can then be identified by comparing these fingerprints with databases of fingerprints, including those predicted from DNA sequence data. Given the large scale of such techniques, many of these procedures are necessarily automated.

Metabolomics

Together, genomics, transcriptomics and proteomics are powerful approaches to increasing our understanding not only of how cells function normally but also what the key changes are in disease. So these techniques will find increasing use in the diagnosis of specific diseases and in the identification and assessment of new chemotherapeutic agents.

Metabolomics is perhaps the newest of these fields of study and relates to the study of the small molecule components of the cell. The complete collection of metabolites in a cell or tissue under specific conditions is called the **metabolome**. By analyzing the metabolites of a cell, one can generate a **metabolic profile** that indicates the cell's metabolic activity; information that should prove useful for a range of applications, including clinical diagnostics and drug discovery.

The study of metabolomes relies on methods to separate the various classes of small molecules (such as gas liquid chromatography, high performance liquid chromatography and capillary electrophoresis) and powerful methods for metabolite identification and quantification (for example, mass spectrometry, infrared spectroscopy, nuclear magnetic resonance).

Once this information is combined with information about reaction rates for the various metabolic pathway steps, it is possible to construct models of the **metabolic flux**, i.e.

the flow of metabolites through the various pathways. Ideally, one would like to be able to monitor the metabolic flux of healthy tissues and compare this with various disease states. This then may lead to the design of drugs to reverse or overcome undesirable metabolic activity associated with these disease states. Right now, these approaches to study eukaryotic cells are in their infancy with much of the work centered on simple organisms such as yeast. But if techniques for large-scale metabolomic analyses can be devised along the lines of those employed for genomics, proteomics and transcriptomics, then we can expect rapid future progress.

Transgenic organisms

As the functions of individual genes become known, the power of this new biology can be used to modify organisms in predictable and desirable ways. Organisms that have been modified by the insertion of a cloned gene are called **transgenic organisms**; **transgenic plants** and **transgenic animals** are both possible. Such approaches are not only of academic importance but are increasingly finding commercial applications. For example, modified plants with improved pest resistance or virus resistance have obvious attractions for agriculture. However, there are important ethical considerations that must be applied to the use of this new technology both for plants and, even more controversially, for animals (including humans).

This wealth of knowledge about the genome and its expression, and the application of that knowledge, relies upon a wide range of recombinant DNA tools and techniques. Many of the present-day experimental approaches are extremely sophisticated and are beyond the scope of this introductory text. However, the following sections provide an understanding of some of the core methodology: the ability to cut DNA at specific sites using restriction endonucleases (**restriction enzymes**; Section I2), procedures that allow the detection of specific DNA (and RNA) sequences with great accuracy (**nucleic acid hybridization**; Section I3), methods for preparing specific DNA sequences in large amounts in pure form (**DNA cloning**; Section I4) and rapid **DNA sequencing** (Section I5). More recently, the development of the **polymerase chain reaction (PCR)** (Section I6) has revolutionized the field of molecular biology and of key applications such as forensics. Finally, this section looks at our ability to modify DNA sequences at will by **site-directed mutagenesis** (Section I7), allowing us to design new proteins.

Since this book is focused on the subject of biochemistry, there are many exciting aspects of recombinant DNA technology that it cannot include because of lack of space. For a more extensive description, the reader is recommended to read the companion book, *Instant Notes in Molecular Biology*.

I2 Restriction enzymes

Key Notes

Restriction enzyme digestion	Restriction enzymes recognize specific recognition sequences and cut the DNA to leave cohesive ends or blunt ends. Restriction enzymes have a three-letter name based on the genus and species name of the bacterium from which they were isolated, together with a Roman numeral designed to indicate the identity of the enzyme in cases when the bacterium contains several different restriction enzymes. Of the three types of restriction enzymes (Types I, II and III), Type II enzymes are of most use in recombinant DNA work because they always cut at the same site relative to the recognition sequence. The ends of the restricted DNA molecules can be joined together by ligation to create new recombinant DNA molecules.
Gel electrophoresis	DNA fragments in a restriction digest can be separated by size by electrophoresis in polyacrylamide or agarose gel. Polyacrylamide gel is used to separate smaller DNA molecules whilst agarose gel has larger pore sizes and so can separate larger DNA fragments. The exact size range of DNA fragments that can be separated by a given gel depends on the gel concentration used, since gel pore size decreases as the gel concentration increases.
Restriction maps	A map showing the position of cut sites for a variety of restriction enzymes is called the restriction map for that DNA molecule. Restriction maps are useful in recombinant DNA experiments for planning where and how to cut specific DNA molecules and sometimes also to monitor the progress of an experiment.
Restriction fragment length polymorphisms	A restriction fragment length polymorphism (RFLP) is a common difference between the DNA of individuals in a population (i.e. a polymorphism) that affects the sizes of fragments produced by a specific restriction enzyme. If the RFLP lies near a gene, mutations in which can cause a human genetic disease, it can be used as a marker for that gene. In the past, RFLPs have proved valuable both for screening patients for the gene defect and also in studies directed at cloning the gene. However, RFLPs are becoming less commonly used in such work as the genes themselves are identified. The polymerase chain reaction (PCR) is increasingly the method of choice for screening.
Related topics	(F1) An introduction to DNA (I4) DNA cloning (I3) Nucleic acid hybridization

Restriction enzyme digestion

Restriction enzymes recognize specific nucleotide sequences (**recognition sequences**) in double-stranded DNA, which are usually four, five or six nucleotides long, and then cut both strands of the DNA at specific locations.

There are three types of restriction enzyme: Types I, II and III. Types I and III enzymes do not cut the DNA at a specific location relative to the recognition sequence and so are less useful for recombinant DNA work in the laboratory. However, Type II enzymes always cut at the same location within the recognition sequence or close to it, which makes them powerful experimental tools.

Restriction enzymes are isolated from bacteria, where they play a role in protecting the host cell against virus infection. Over 2500 Type II restriction enzymes have now been isolated and have been named according to the bacterial species from which they were isolated. The first three letters of the enzyme name are the first letter of the genus name and the first two letters of the species name. Since each bacterium may contain several different restriction enzymes, a Roman numeral is also used to identify each enzyme. *Eco*RI, for example, was the first enzyme isolated from _Escherichia coli_.

There are basically three ways in which the DNA can be cut by a Type II enzyme; a staggered cut to leave a **5′ overhang** (i.e. a short single-stranded region of DNA is left that has a 5′ end and overhangs the end of the double-stranded DNA), a staggered cut to leave a **3′ overhang**, or a cut in the same place on both strands to leave a **blunt end** (Figure 1). For enzymes that cut in the staggered manner, the single-stranded tails are called **cohesive ends** or **sticky ends** because they allow any two DNA fragments produced by the same restriction enzyme to form complementary base pairs (Figure 2). The cut ends can then be joined together (**ligated**) by an enzyme called **DNA ligase**. The new DNA molecule that has been made by joining the DNA fragments is called a **recombinant DNA molecule** (Figure 2). Blunt-ended DNA molecules can also be joined together by DNA ligase but the reaction is less favorable.

One very useful feature is that some restriction enzymes have different recognition sequences but generate the same cohesive ends, which allow DNA fragments generated by these two different enzymes to be ligated together. For example, *Sau*3AI has a four base pair recognition sequence of 5′-GATC-3′, whereas *Bam*HI recognizes the six base pair recognition sequence 5′-GGATCC-3′ but each generates a 5′-GATC-3′ sticky end (Figure 3) that can then be ligated together to form a recombinant DNA molecule.

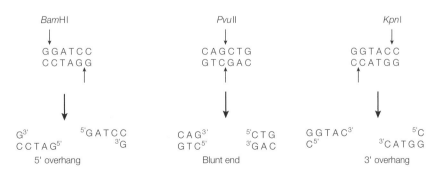

Figure 1. The three types of cleavage by commonly used restriction enzymes.

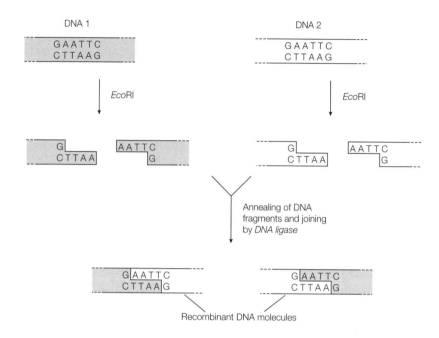

Figure 2. Using a restriction enzyme, *Eco*RI, to create recombinant DNA.

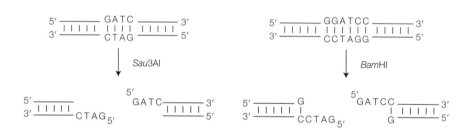

Figure 3. An example of two restriction enzymes with different recognition sequences that produce the same cohesive ends after cleavage; *Sau*3AI (recognition site 5′GATC3′) and *Bam*HI (recognition site 5′GGATCC3′).

Gel electrophoresis

When a DNA molecule is cut by a restriction enzyme, the DNA fragments (called **restriction fragments**) from that **restriction digest** can be separated by **gel electrophoresis** (Figure 4). Electrophoresis on a polyacrylamide gel will separate small DNA fragments of less than about 500 bp in size, but agarose gels (which have larger pores) are needed to separate larger DNA fragments. In each case, the pore size of the gel decreases with increasing gel concentration so the gel concentration that is used will determine the range of sizes of DNA fragments that can be separated in any given gel.

Figure 4. Agarose gel electrophoresis of DNA fragments. DNA fragments of known size were electrophoresed in lane 1 (the sizes in bp are given on the left). A restriction digest of the sample DNA was electrophoresed in lane 2. By comparison with the migration positions of fragments in lane 1, it can be seen that the two sample DNA fragments have sizes of approximately 9000 and 2000 bp. The sizes could be determined more accurately by plotting the data from lane 1 as a standard curve of log DNA size vs. migration distance and then using this to estimate the size of the sample DNA fragments from their measured migration distances.

During electrophoresis of DNA that has been digested by a restriction enzyme, the various restriction fragments separate into a series of bands in the gel. Since small fragments travel further in the gel than larger fragments, the size of each fragment can be determined by measuring its migration distance relative to standard DNA fragments of known size (Figure 4).

The DNA can be located in the gel after electrophoresis by staining with ethidium bromide that binds to the DNA and fluoresces under UV light. Alternatively, if the DNA is labeled with a radioisotope such as ^{32}P, the bands can be detected after electrophoresis by laying the gel against an X-ray film whereby the radioactivity causes silver grains to be formed in the film emulsion, giving black images corresponding to the radioactive bands (**autoradiography**).

Restriction maps

Any double-stranded DNA will be cut by a variety of restriction enzymes that have different recognition sequences. By separating the restriction fragments and measuring their sizes by gel electrophoresis, it is possible to deduce where on the DNA molecule each restriction enzyme cuts. A **restriction map** of the DNA molecule can be drawn showing the location of these cut sites (**restriction sites**) (Figure 5).

Restriction maps are important experimentally during recombinant DNA work, for example to identify specific DNA molecules by their distinctive map, to plan where particular DNA molecules should best be cut and to monitor the progress of the experiment.

Figure 5. A typical restriction map of a DNA molecule. The cleavage sites of different restriction enzymes, indicated by letters, are shown. For example, E denotes sites where *Eco*RI cuts.

Restriction fragment length polymorphisms

Analysis of human genomic DNA has revealed that there are many differences in DNA sequence between individuals that have no obvious effect, often because the changes lie in introns or between genes. Some of these changes are very common in individuals in a population and are called **polymorphisms**. Some polymorphisms affect the size of fragments generated by a particular restriction enzyme, for example by changing a nucleotide in the recognition sequence and so eliminating a cut site. Instead of two restriction fragments being generated from this region, a single large restriction fragment is now formed (Figure 6). Alternatively, the polymorphism may result from the insertion or deletion of sequences between two cut sites, so increasing or decreasing the size of that restriction fragment produced. A polymorphism that affects restriction fragment sizes is called a **restriction fragment length polymorphism (RFLP)**. Provided a DNA probe (Section I3) exists for a sequence of DNA within the affected region, so that this sequence can be detected by hybridization, RFLPs can be detected by Southern blotting (Section I3).

The value of RFLPs has been in the ability to use these as markers for particular human genetic diseases. Consider a polymorphism that happens to be very near to the site of a change in a key gene that results in a human genetic disease (Figure 6). Because these

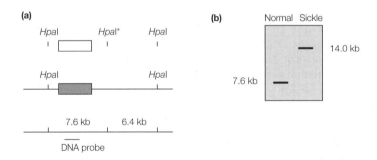

Figure 6. Analysis of human genetic disease using RFLPs. The analysis concerns two individuals from a family, one of whom has normal β-globin and one of whom has an abnormal β-globin gene that leads to sickle-cell anemia. (a) The sickle β-globin is associated with a nucleotide change that results in the loss of the *Hpa*I site marked with an asterisk. The presence or absence of this *Hpa*I* site is detected by hybridization and Southern blotting (Section I3) using a DNA probe for the 7.6 kb fragment; (b) normal DNA with three *Hpa*I sites yields a 7.6 kb fragment detected by the DNA probe but sickle DNA yields a 14.0 kb fragment due to loss of the *Hpa*I* site.

two changes, the polymorphism and the genetic defect, lie close together on the same chromosome, they will tend to be co-inherited. Identifying such a **closely linked** RFLP has two major advantages. First, experiments can be directed to cloning DNA near the RFLP in the hope of identifying the gene itself, which can then be sequenced and studied. Second, even in the absence of the gene, the RFLP acts as a screening marker for the disease; individuals who have the RFLP have a high probability of having the associated gene defect. Of course other RFLPs that are located a very long way from the gene, or even on a different chromosome, will essentially be **unlinked** (i.e. because of the high probability of cross-over events during meiosis to produce germ-line cells, the gene and RFLP will have only a 50:50 chance of being co-inherited). Thus a large amount of very painstaking work has to be carried out to identify a useful RFLP for a particular human genetic disease. Large numbers of individuals in family groups, some of whom suffer from the disease, need to be screened for a range of likely RFLPs to attempt to locate an RFLP that is routinely co-inherited with the gene defect.

As the genes themselves are identified and sequenced, so the need for RFLP markers declines since specific DNA probes (Section I3) for the most common types of gene defect can be employed. In addition, the use of the polymerase chain reaction (PCR; Section I6) in screening for human genetic disease is increasingly the method of choice rather than RFLP analysis since it is much faster to perform and requires far less clinical material for analysis.

I3 Nucleic acid hybridization

Key Notes

The hybridization reaction	Double-stranded DNA denatures into single strands as the temperature rises but renatures into a double-stranded structure as the temperature falls. Any two single-stranded nucleic acid molecules can form double-stranded structures (hybridize) provided they have sufficient complementary nucleotide sequence to make the resulting hybrid stable under the reaction conditions.
Monitoring specific nucleic acid sequences	The concentration of a specific nucleic acid sequence in a sample can be measured by hybridization with a suitable labeled DNA probe. After hybridization, nuclease is used to destroy unhybridized probe and the probe remaining is a measure of the concentration of the target sequence. The hybridization conditions can be altered to ensure that only identical sequences (high stringency conditions) or identical plus related sequences (low stringency conditions) will hybridize with the probe and hence be detected.
Southern blotting	Southern blotting involves electrophoresis of DNA molecules in an agarose gel and then blotting the separated DNA bands on to a nitrocellulose filter. The filter is then incubated with a labeled DNA probe to detect those separated DNA bands that contain sequences complementary to the probe.
Northern blotting	Northern blotting is analogous to Southern blotting except that the sample nucleic acid that is separated by gel electrophoresis is RNA rather than DNA.
***In situ* hybridization**	For *in situ* hybridization, a tissue sample is incubated with a labeled nucleic acid probe, excess probe is washed away and the location of hybridized probe is examined. The technique enables the spatial localization of gene expression to be determined as well as the location of individual genes on chromosomes.
DNA microarrays (DNA Chips)	A DNA microarray is a large number of DNA fragments or oligonucleotides arrayed in known positions on a glass slide or silicon. After hybridization with fluorescently labeled target complementary DNA (cDNA), examination using automated scanning-laser microscopy indicates which DNA sequences are expressed.
Related topics	(F1) An introduction to DNA (I4) DNA cloning (I2) Restriction enzymes

The hybridization reaction

As double-stranded DNA is heated, a temperature is reached at which the two reaction strands separate. This process is called **denaturation**. The temperature at which half of the DNA molecules have denatured is called the **melting temperature** or T_m for that DNA. If the temperature is now lowered and falls below the T_m, the two complementary strands will form hydrogen bonds with each other once more to re-form a double-stranded molecule. This process is called **renaturation** (or **reannealing**).

In fact, double-stranded structures can form between any two single-stranded nucleic acid molecules (DNA–DNA, DNA–RNA, RNA–RNA) provided they have sufficient complementary nucleotide sequence to make the double-stranded molecule stable under the conditions used. The general name given to this process is **hybridization** and the double-stranded nucleic acid product is called a **hybrid**.

Monitoring specific nucleic acid sequences

The rate of formation of double-stranded hybrids depends on the concentration of the two single-stranded species. This can be used to measure the concentration either of specific DNA or RNA sequences in a complex mixture. The first task is to prepare a single-stranded **DNA probe** (i.e. a DNA fragment that is complementary to the nucleic acid being assayed). This can be one strand of a DNA restriction fragment, cloned DNA or a synthetic oligonucleotide. It must be labeled in order to be able to detect the formation of hybrids between it and the target nucleic acid. Whereas most labeling used to be via the incorporation of a radioisotope, nonradioactive chemical labels may be used instead. For example, a DNA probe can be labeled with digoxygenin, a steroid, by using digoxygenin-labeled dUTP during DNA synthesis. Hybrids containing the digoxygenin-labeled DNA probe can then be detected using antidigoxygenin antibody linked to a fluorescent dye.

Irrespective of the method of probe labeling, the DNA probe is incubated with the nucleic acid sample (the **'target'** DNA or RNA) and then nuclease is added to degrade any unhybridized single-stranded probe. The amount of labeled probe remaining indicates the concentration of the target nucleic acid in the sample.

The hybridization conditions (e.g. temperature, salt concentration) can be varied in order to govern the type of hybrids formed. The conditions may be arranged so that only perfectly matched hybrids are stable and hence assayed (conditions known as **high stringency**). Alternatively, the conditions may be such that even poorly complementary hybrids are stable and will be detected (**low stringency**). Thus by varying the reaction conditions it is possible to detect and quantify only those target sequences that are identical to the DNA probe or, alternatively, to detect and quantify related sequences. Hybridization of nucleic acid probes with genomic DNA, for example, can be used to measure the copy number of particular DNA sequences in the genome. Hybridization of a DNA probe with cellular RNA as the target will indicate the concentration of the corresponding RNA transcript and hence give information about the level of gene expression.

Southern blotting

Gel electrophoresis is widely used to separate and size DNA molecules during recombinant DNA experiments. After gel electrophoresis, there is often a need to detect one or more DNA fragments containing a specific nucleotide sequence. This is easily carried out by Southern blotting – a technique so called after the name of its inventor, Edwin Southern.

In the original capillary blotting method, after electrophoresis of the restriction fragments through an agarose gel, the gel is soaked in alkali to denature the DNA to single strands and the pH is then neutralized. The gel is placed in contact with a nitrocellulose or nylon membrane filter sheet arranged so that buffer flows through the gel and carries the DNA fragments to the membrane (Figure 1). The membrane binds the single-stranded DNA and so the band pattern in the gel is now transferred to it. The membrane filter is peeled from the gel, baked at high temperature to fix the DNA to it, and then incubated with a radiolabeled DNA probe. After hybridization, the probe will have bound only to DNA fragments with complementary sequences. These can be visualized by washing away excess probe and then placing the filter against an X-ray film for autoradiography. The images produced on the autoradiogram indicate those bands that contain the probe sequence (Figure 1).

Figure 1. Southern blotting. The procedure shown is the original method of Southern using capillary action to blot the DNA bands from the gel to the nitrocellulose membrane. Electrolytic transfer is now often used instead.

Southern blotting is still very widely used although in many laboratories the DNA may be transferred from the gel to the nitrocellulose or nylon membrane sheet by electrophoretic transfer rather than capillary blotting.

Northern blotting

Northern blotting follows much the same procedure as Southern blotting except that the sample analyzed by gel electrophoresis and then bound to the filter is RNA not DNA. Therefore Northern blotting detects RNA molecules that are complementary to the DNA probe. If cellular RNA is electrophoresed, for example, a DNA probe for a specific mRNA could be used to detect whether that mRNA was present in the sample. The migration distance of the RNA in the gel would also allow estimation of its size.

Note that there was no inventor called 'Northern'; the name was coined simply because the original technique for DNA was called Southern and is a geographical pun!

In situ hybridization

It is also possible to incubate radioactive or fluorescent nucleic acid probes with sections of tissues or even chromosomes, wash away excess probe and then detect where the probe has hybridized. This technique (***in situ* hybridization**) has proved to be very powerful in determining which cells in a complex tissue such as the mammalian brain express a particular gene and for locating specific genes on individual chromosomes.

DNA microarrays (DNA chips)

The techniques described above are limited in that only a relatively small number of samples can be analyzed at any one time. In contrast, **DNA microarrays** (DNA chips, gene chips) can analyze the expression of tens of thousands of genes simultaneously. A DNA microarray is a large number of DNA sequences that are spotted onto a glass slide or silicon in a pre-determined grid pattern; given the large numbers of DNAs involved, this is done robotically. In other cases, instead of spotting DNA fragments, the DNA microarray may be produced by synthesizing thousands of oligonucleotides on the glass slide *in situ*; as many as 300 000 oligonucleotides per square centimeter (high-density arrays). The microarray, containing DNA fragments or oligonucleotides, can then be used to explore the expression of each of these DNA sequences in the tissue or sample of interest by hybridization. In the terminology of hybridization (see above), the DNAs or oligonucleotides on the microarray are the 'probe' and the RNA in the tissue or sample is the 'target'.

To understand how DNA microarrays are used, consider the following typical application (Figure 2). Imagine that the goal is to determine which genes are regulated by a newly discovered hormone. Cells are exposed to the hormone (test sample) or left untreated (control sample), RNA is isolated from each sample and is used to synthesize complementary DNA (cDNA) using reverse transcriptase. When synthesizing cDNA from the test sample RNA, one of the nucleotide precursors is labeled with, for example, a red fluorescent dye, so that the resulting cDNA is also tagged with this label. The control sample cDNA is similarly labeled but with, for example, a green fluorescent dye. The two cDNAs are mixed together and allowed to hybridize to the DNA microarray. Any cDNA that does not hybridize is washed away and the DNA microarray is examined using an automated scanning-laser microscope. Laser excitation of the microarray with light of the appropriate wavelength to excite the relevant fluorophore and measurement of the intensity of the resulting fluorescence for each DNA or oligonucleotide spot allows the extent of hybridization with test (red) and control (green) cDNA to be determined. Since the exact location of every DNA or oligonucleotide in the microarray is known, these data immediately indicate which genes are activated by the hormone (red spots), which genes are expressed only in the absence of the hormone (green spots) and which genes are unaffected and are expressed both in the absence and presence of the hormone (yellow spots=red+green).

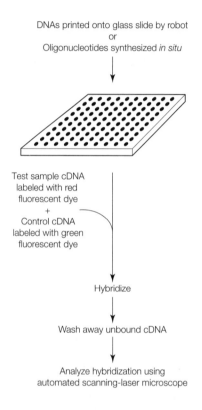

Figure 2. A typical use of a DNA microarray to determine the expression of thousands of genes simultaneously (see the text for details).

DNA microarrays are now widely used to examine changes in gene expression in both plants and animals. For example, in humans, they can be used to determine how particular diseases affect the pattern of gene expression (the **expression profile**) in various tissues, or the identity (from the expression profile) of the infecting organism. Thus, in clinical medicine alone, DNA microarrays have huge potential for diagnosis.

I4 DNA cloning

Key Notes

The principle of DNA cloning

Most DNA fragments cannot self-replicate in a cell and must therefore be joined (ligated) to a vector (virus or plasmid DNA) that can replicate autonomously. Each vector typically will join with a single fragment of such DNA. If a complex mixture of DNA fragments is used, a population of recombinant DNA molecules is produced. This is then introduced into the host cells, each of which will typically contain only a single type of recombinant DNA. Identification of the cells that contain the DNA fragment of interest allows the purification of large amounts of that single recombinant DNA and hence the foreign DNA fragment.

The basics of DNA cloning

To clone into a plasmid vector, both the plasmid and the DNA to be cloned are cut with the same restriction enzyme and mixed together. The cohesive ends of each DNA reanneal and are ligated together. The resulting recombinant DNA molecules are introduced into bacterial host cells. If the vector contains an antibiotic resistance gene(s) and the host cells are sensitive to this antibiotic(s), plating on nutrient agar containing the relevant antibiotic will allow only those cells that have been transfected and contain plasmid DNA to grow.

DNA libraries

Genomic DNA libraries are made from the genomic DNA of an organism. A complete genomic DNA library contains all of the nuclear DNA sequences of that organism. A cDNA library is made using complementary DNA (cDNA) synthesized from mRNA by reverse transcriptase. It contains only those sequences that are expressed as mRNA in the tissue or organism of origin.

Screening DNA libraries

Genomic libraries and cDNA libraries can be screened by hybridization using a labeled DNA probe complementary to part of the desired gene. The probe may be an isolated DNA fragment (e.g. restriction fragment) or a synthetic oligonucleotide designed to encode part of the gene as deduced from a knowledge of the amino acid sequence of part of the encoded protein. In addition, expression cDNA libraries may be screened using a labeled antibody to the protein encoded by the desired gene or by using any other ligand that binds to that protein.

Related topics

(F1) An introduction to DNA
(I2) Restriction enzymes

(I3) Nucleic acid hybridization

The principle of DNA cloning

Consider an experimental goal that is to make large amounts of a particular DNA fragment in pure form from a mixture of DNA fragments from a mouse. Although the DNA fragments can be introduced into bacterial cells, most or all will lack the ability for self-replication and will quickly be lost. However, two types of DNA molecule are known which can replicate autonomously in bacterial cells; bacteriophages and plasmids. Plasmids are small circular double-stranded DNA molecules that exist free inside bacterial cells, often carry particular genes that confer drug resistance, and are self-replicating. If a recombinant DNA molecule is made by joining a mouse DNA fragment to plasmid (or bacteriophage) DNA, then the mouse DNA is replicated when the plasmid or phage DNA is replicated. In this role, the plasmid or phage DNA is known as a **vector**.

A population of recombinant DNA molecules can be made, each recombinant molecule containing one of the mouse DNA fragments in the original mixture. This can then be introduced into a population of bacteria such that each bacterial cell contains, in general, a different type of recombinant DNA molecule. If we can identify the bacterial cell that contains the recombinant DNA bearing the mouse DNA fragment we want, the cell can be multiplied in culture and large amounts of the recombinant DNA isolated. The mouse DNA of interest can then be recovered from this in pure form; it is then said to have been **cloned**.

The vector that was used to achieve this cloning is called a **cloning vector**. Vectors are not limited to bacterial cells; animal and plant viruses can also act as vectors.

The basics of DNA cloning

There are a wide variety of different procedures for cloning DNA into either plasmid or viral vectors and the reader is referred to the companion textbook *Instant Notes in Molecular Biology* for more detail. However, the basic scheme of events is often the same. To clone into a plasmid vector, the circular plasmid DNA is cut with a restriction enzyme (Section I2) that has only a single recognition site in the plasmid. This creates a linear plasmid molecule with cohesive ends (Figure 1). The simplest cloning strategy is now to cut the DNA that is to be cloned (in the above example, mouse DNA) with the same restriction enzyme. Alternatively, different restriction enzymes can be used, provided they create the same cohesive ends (Section I2). The restricted mouse DNA and linear plasmid DNA are now mixed. The cohesive ends of the mouse DNA anneal with the ends of the plasmid DNA and are joined covalently by DNA ligase. The resulting **recombinant plasmid DNA** is introduced into bacterial host cells that have been treated to become permeable to DNA. This uptake of DNA by the bacterial cells is called **transfection**; the bacterial cells are said to have been **transfected** by the recombinant plasmid. The bacterial cells are now allowed to grow and divide, during which time the recombinant plasmids will replicate many times within the cells.

One useful procedure is to select as cloning vector a plasmid that carries one or more antibiotic resistance genes plus a host that is sensitive to those antibiotics (Figure 1). Then, after transfection, the cells are grown in the presence of the antibiotic(s). Only cells containing plasmid DNA will be resistant to the antibiotic(s) and can grow. If the cells are spread on an agar plate, each cell will multiply to form a bacterial colony where all the cells of that colony contain the same recombinant plasmid DNA bearing the same foreign DNA fragment. Thus all that is now needed is to identify the particular bacterial colony that contains the mouse DNA sequence of interest.

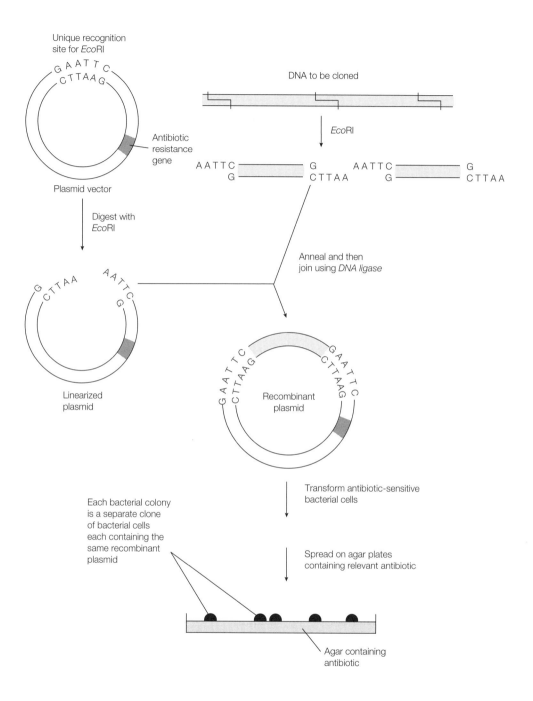

Figure 1. A simple method of DNA cloning using a plasmid vector. In the text, the DNA to be cloned is assumed to be mouse DNA as an example.

DNA libraries

A DNA library is a collection of cloned DNA fragments in a cloning vector that can be searched for a DNA of interest. If the goal is to isolate particular gene sequences, two types of library are useful.

- **Genomic DNA libraries**. A genomic DNA library is made from the genomic DNA of an organism. For example, a mouse genomic library could be made by digesting mouse nuclear DNA with a restriction nuclease to produce a large number of different DNA fragments but all with identical cohesive ends. The DNA fragments would then be ligated into linearized plasmid vector molecules or into a suitable virus vector. This library would contain all of the nuclear DNA sequences of the mouse and could be searched for any particular mouse gene of interest. Each clone in the library is called a **genomic DNA clone**. Not every genomic DNA clone would contain a complete gene since in many cases the restriction enzyme will have cut at least once within the gene. Thus some clones will contain only a part of a gene.

- **cDNA libraries**. A cDNA library is made by using the reverse transcriptase of a retrovirus to synthesize **complementary DNA (cDNA)** copies of the total mRNA from a cell (or perhaps a subfraction of it). The single-stranded cDNA is converted into double-stranded DNA and inserted into the vector. Each clone in the library is called a **cDNA clone**. Unlike a complete genomic library that contains all of the nuclear DNA sequences of an organism, a cDNA library contains only sequences that are expressed as mRNA. Different tissues of an animal, that express some genes in common but also many different genes, will thus yield different cDNA libraries.

Screening DNA libraries

Genomic libraries are screened by hybridization (Section I3) with a DNA probe that is complementary to part of the nucleotide sequence of the desired gene. The probe may be a DNA restriction fragment or perhaps part of a cDNA clone. Another approach is possible if some of the protein sequence for the desired gene is known. Using the genetic code, one can then deduce the DNA sequence of this part of the gene and synthesize an oligonucleotide with this sequence to act as the DNA probe.

When using a plasmid vector, a simple procedure for screening would be to take agar plates bearing bacterial colonies that make up the genomic library and overlay each plate with a nitrocellulose membrane (Figure 2). This is peeled off and is a **replica** of the plate in that some of the bacterial cells in each colony will have adhered to it and in the same pattern as the colonies on the plate. This filter is often called a '**colony lift**'. It is treated with alkali to lyse the bacterial cells and denature the DNA and then hybridized with a radiolabeled DNA probe. After washing away unreacted probe, autoradiography of the filter shows which colonies have hybridized with the probe and thus contain the desired sequences. These are then recovered from the agar plate.

When a bacteriophage is used as the cloning vector, the gene library is screened as an array of plaques in a bacterial lawn. A hybridization screening method is used similar to that described for plasmid screening (Figure 2); in this case the replica filter is called a '**plaque lift**'.

For cDNA libraries, screening can similarly be carried out by hybridization. In addition, it is possible to make the cDNA library using a vector that will actually transcribe the inserted cDNA and then translate the resulting mRNA to form protein corresponding to the cloned gene. A library made with such an **expression vector** is an **expression cDNA**

library. It can be screened using a labeled antibody that recognizes the specific protein and hence identifies those bacteria that contain the desired gene and are synthesizing the protein. Not just antibody but any ligand that binds to the target protein can be used as a probe. For example, labeled hormone may be used to identify clones synthesizing hormone receptor proteins.

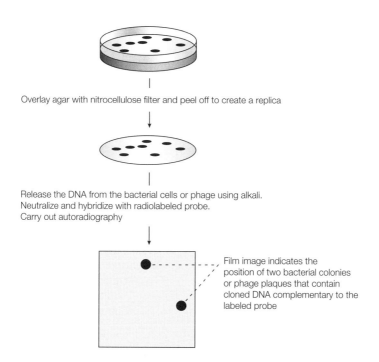

Figure 2. Screening a gene library by hybridization.

15 DNA sequencing

Key Notes

Several methods for DNA sequencing

DNA can be sequenced by several methods but the method usually used is the chain termination procedure. The (single-stranded) DNA to be sequenced serves as the template for the synthesis of a complementary strand when supplied with a specific primer and a suitable DNA polymerase.

Choice of DNA polymerase

The DNA polymerase used for chain termination sequencing must have high processivity and negligible (ideally zero) exonuclease activity. Klenow polymerase was the enzyme used originally but has since been superceded by other enzymes, such as Sequenase.

Chain termination method

Four incubation mixtures are set up, each containing the DNA template, a specific DNA primer, *E. coli* DNA polymerase I and all four dNTPs. In addition, each mixture contains a different dideoxynucleoside triphosphate analog, ddATP, ddCTP, ddGTP or ddTTP. Incorporation of a dideoxy analog prevents further elongation and so produces a chain termination extension product. The products are electrophoresed on a polyacrylamide gel and the DNA sequence is read from the band pattern produced.

Automated DNA sequencing

Automated DNA sequencing uses the chain termination method but with an oligonucleotide primer labeled with a fluorescent dye. Each of the four reactions receives a primer labeled with a different dye. After incubation, the reaction mixtures are pooled and electrophoresed on one lane of a polyacrylamide gel. The order in which the different fluorescently labeled termination products elute from the gel gives the DNA sequence. More advanced systems use multiple capillary sets in which sample preparation, loading and data analysis are automated for maximum throughput.

Related topics

| (F1) An introduction to DNA | (I3) Nucleic acid |
| (F3) DNA replication in bacteria | hybridization |

Several methods for DNA sequencing

Several methods have been devised to sequence DNA including the **chemical degradation method** (also called the *Maxam–Gilbert method* after its inventors), the **chain termination method** (also known as the *Sanger dideoxy method* after its inventor) and most recently **pyrosequencing**.

The chain termination method is now the method usually used because of its speed and simplicity. The chemical degradation method is useful for some stretches of DNA that form intrastrand base pairing, which can prove difficult to sequence by the chain termination method. Pyrosequencing does not need electrophoresis or separation of fragments and so is faster than the other two methods but it can generate only a few tens of base pairs of sequence per experiment. It is useful in situations where a large number of short sequences is required quickly. These additional methods are described elsewhere; here we will focus only on the main sequencing approach, the chain termination method.

The DNA to be sequenced using the chain termination method is prepared as a single-stranded molecule so that it can act as a template for DNA synthesis (Section F3) in the sequencing reaction. A suitable DNA polymerase (see below) is used to copy this DNA template but needs a primer to start synthesis (Section F3). The primer used can be either a DNA restriction fragment complementary to the single-stranded template or it can be a short sequence of complementary DNA that has been synthesized chemically (a **synthetic oligonucleotide**).

Choice of DNA polymerase

The DNA polymerase that is used in chain termination sequencing needs to synthesize DNA complementary to the DNA template by extending the primer. The enzyme must have high **processivity** (i.e. synthesize long stretches of DNA without stopping). However, most DNA polymerases also have exonuclease activity, which means they can degrade DNA as well as synthesize it. The ideal DNA polymerase for DNA sequencing should therefore have negligible or zero levels of 5′ to 3′ exonuclease activity and negligible or zero levels of 3′ to 5′ exonuclease activity.

The first enzyme developed for DNA sequencing was **Klenow polymerase**, which is *E. coli* DNA polymerase I from which the 5′ to 3′ exonuclease activity had been removed by genetic engineering, i.e. using recombinant DNA techniques. Unfortunately, Klenow polymerase has relatively low processivity and so has largely been replaced by other enzymes, especially a bacteriophage T7 DNA polymerase, which has high processivity but which has been engineered to lack exonuclease activity (available commercially as 'Sequenase™').

Chain termination method

An incubation mixture is set up containing the single-stranded DNA template, the primer, DNA polymerase I and all four deoxyribonucleoside triphosphates (dATP, dGTP, dCTP, dTTP), one of which is radioactively labeled, plus a single 2′3′ dideoxyribonucleoside triphosphate analog, say ddGTP. In this incubation, the DNA polymerase begins copying template molecules by extending the bound primer. As the new DNA strand is synthesized, every time that dGTP should be incorporated there is a chance that ddGTP will be incorporated instead. If this happens, no further chain elongation can occur because dideoxy analogs lack the 3′-OH group needed to make the next 3′5′ phosphodiester bond. Thus this particular chain stops at this point. In this first incubation mixture, a large population of templates is being copied and each new strand will stop randomly at positions where a G must be added to the newly synthesized strand. Thus, for every G in the complementary sequence there will be some new DNA strands that have terminated at that point (Figure 1).

In fact, four incubation mixtures are set up, each containing the same components except that each contains a different dideoxy analog; one of ddATP, ddCTP, ddTTP or ddGTP (Figure 1). This produces four sets of chain-terminated fragments corresponding to the positions of A, C, T and G in the sequence. After the incubation, all four reaction mixtures are electrophoresed in parallel lanes of a polyacrylamide gel and then subjected

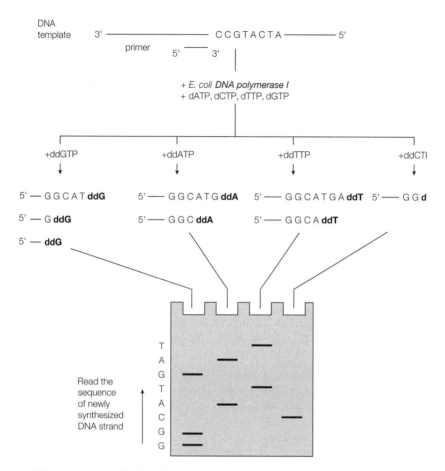

Figure 1. DNA sequencing by the chain termination (Sanger) method.

to autoradiography. The DNA sequence is determined simply by reading the band pattern on the autoradiogram from the bottom of the gel towards the top (i.e. reading the DNA sequence as it is synthesized from the primer; Figure 1). The sequence read off the gel is the sequence of the synthesized DNA strand and hence is the complementary sequence to the original DNA template strand.

Automated DNA sequencing

Automated DNA sequencing is now commonplace, based on the chain termination method but using a fluorescent dye attached to an oligonucleotide primer instead of using radioactive labeling. A different fluorescent dye is attached to the primer for each of the four sequencing reactions but, after incubation, all four mixtures are combined and electrophoresed on one gel lane. Laser detection systems then distinguish the identity of each termination product as it elutes from the gel. The sequence in which the different fluorescent products elute from the gel gives the DNA sequence.

Modern automated DNA sequencing systems designed to generate almost a million bases of sequence per day use sequencing gels contained in multiple capillary tubes (rather than in a slab gel format). Preparation and loading of the samples onto the capillary gels is carried out by robots, providing 24-hour operation, and data analysis is also automated.

16 Polymerase chain reaction

Key Notes

Principles of PCR

The polymerase chain reaction (PCR) allows an extremely large number of copies to be synthesized of any given DNA sequence provided two oligonucleotide primers are available that hybridize to the flanking sequences on the complementary DNA strands. The reaction requires the target DNA, the two primers, all four deoxyribonucleoside triphosphates and a thermostable DNA polymerase such as *Taq* DNA polymerase. A PCR cycle consists of three steps; denaturation, primer annealing and elongation. This cycle is repeated for a set number of times depending on the degree of amplification required.

Applications of PCR

PCR has made a huge impact in molecular biology, with applications in many areas including medical diagnosis, forensic medicine, and the creation of specific mutations. PCR screening of human genetic diseases is based upon the analysis of microsatellites (also called STRs), which are repeats of short nucleotide sequences in the human genome. Variations in the numbers of repeats between individual genomes allows STRs to be used as screening markers.

Real-time PCR

Real-time PCR allows the quantification of DNA (or RNA) target molecules in biological samples by measuring the rate of amplification during the PCR reaction. One approach is to include a dye in the PCR reaction mixture that binds to double-stranded DNA and fluoresces in the bound state. A more specific approach is to use primers that have a fluorescent reporter covalently attached at one end of the primer and a quenching agent at the other end. As primer is used in the PCR reaction, the 5' to 3' exonuclease activity of *Taq* polymerase removes one end of the primer and allows the fluorescent reporter module to fluoresce since it is no longer in close proximity to the quenching molecule. The increase in fluorescence therefore monitors the rate of the amplification reaction. Real-time PCR can be used to measure the amount of target molecule in the test sample either relative to other samples or in absolute terms.

Related topics

(F1) An introduction to DNA
(F3) DNA replication in bacteria
(I2) Restriction enzymes

(I3) Nucleic acid hybridization
(I4) DNA cloning
(I5) DNA sequencing

Principles of PCR

The polymerase chain reaction (PCR) is an extremely simple yet immensely powerful technique. It allows enormous amplification of any specific sequence of DNA provided short sequences either side of it are known. The technique is shown in Figure 1. A PCR reaction contains the target double-stranded DNA, two primers that hybridize to flanking sequences on opposing strands of the target, all four deoxyribonucleoside triphosphates and a DNA polymerase. Because, as we shall see, the reaction periodically becomes heated to high temperature, PCR depends upon using a heat-stable DNA polymerase. Many such heat-stable enzymes from thermophilic bacteria (bacteria that live in high-temperature surroundings) are now available commercially. The first one used was *Taq* polymerase from the thermophilic bacterium *Thermus aquaticus.*

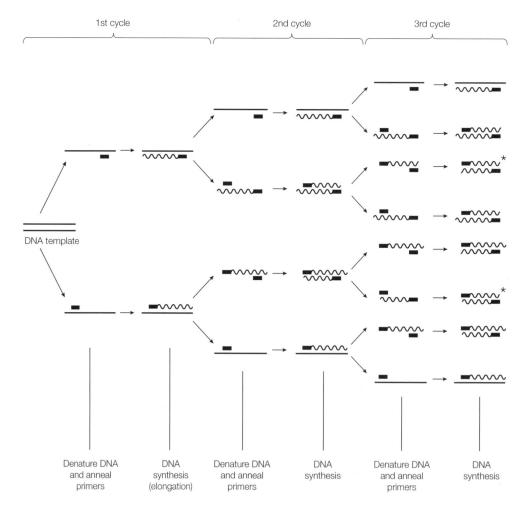

Figure 1. The polymerase chain reaction (PCR). Asterisks indicate the first PCR products that arise (in the third cycle), which consist of DNA sequence only between the two primer sites.

PCR consists of three steps:

1. **Denaturation**. The reaction mixture is heated to 94°C for a short time period (about 15–30 sec.) to denature the target DNA into single strands that can act as templates for DNA synthesis.

2. **Primer annealing**. The mixture is rapidly cooled to a defined temperature (50–60°C), which allows the two primers to bind to the sequences on each of the two strands flanking the target DNA. This **annealing temperature** is calculated carefully to ensure that the primers bind only to the desired DNA sequences. One primer binds to each strand (Figure 1). The two parental strands do not re-anneal with each other because the primers are in large excess over parental DNA.

3. **Elongation**. The temperature of the mixture is raised to 72°C (the optimum temperature for *Taq* polymerase) and kept at this temperature for a pre-set period to allow *Taq* polymerase to elongate each primer by copying the single-stranded templates. At the end of this incubation, both single-stranded template strands have been made partially double stranded. The new strand of each double-stranded DNA extends for a variable distance downstream.

The three steps of the PCR cycle are repeated. Thus in the second cycle, the four strands denature, bind primers and are extended. No other reactants need to be added. The three steps are repeated once more for a third cycle (Figure 1) and so on for a set number of additional cycles. By the third cycle, some of the PCR products (indicated by asterisks in Figure 1) represent DNA sequence only between the two primer sites and the sequence does not extend beyond these sites. As more and more reaction cycles are carried out, this type of double-stranded DNA molecule becomes the majority species present. After 20 cycles, the original DNA has been amplified a million-fold and this rises to a billion-fold (1000 million) after 30 cycles. At this point, the vast majority of the products are identical in that the DNA amplified is only that between the two primer sites.

Automated **thermocyclers** are now routinely used to cycle the reaction without manual interference so that a billion-fold amplification of the DNA sequence between the two primer sites (30 cycles) can take less than 1 hour!

Applications of PCR

PCR already has very widespread applications, and new uses are being devised on a regular basis. Some (and certainly not all) of the applications are as follows:

● PCR can amplify a single DNA molecule from a complex mixture, largely avoiding the need to use DNA cloning to prepare that molecule. However, it does require some sequence data flanking the gene in order to amplify it and so cannot be used to specifically amplify genes for which this information is not known.

● Variants of the technique can amplify a specific single RNA molecule from a complex mixture. Since *Taq* polymerase cannot copy RNA, the first step is to use reverse transcriptase to make a DNA copy of the RNA, which is then amplified as usual by the *Taq* polymerase. This variant is therefore called **reverse transcriptase PCR** or **RT-PCR**.

● PCR is exquisitely sensitive and can amplify vanishingly small amounts of DNA or RNA. Using appropriate primers, very small amounts of specific bacteria and viruses can be detected in tissues, making PCR invaluable for medical diagnosis. Importantly, given the sensitivity of PCR analysis, this diagnosis can occur at a time very early in the disease state. For example, the early diagnosis of viral-induced cancer offers the potential of a far higher success rate for treatment.

- PCR is also invaluable for screening for human genetic diseases and is rapidly replacing the use of RFLPs (Section I2). The PCR screen is based on the analysis of **microsatellites** (also called **short tandem repeats** or **STRs**), which are repeats of short nucleotide sequences; STRs used as DNA markers for PCR are typically 6 bp or less in size and are repeated from 10 to more than 30 times. About 5×10^5 STRs with repeats of 6 bp or less occur scattered throughout the human genome and so most regions of the genome have suitable STRs for use as markers in PCR. The STRs can be used in the same way that RFLPs were used in the past (Section I2); the different sizes of STR give different sizes of amplified DNA fragments that can then be used as screening markers. The STRs are usually analyzed using a fluorescent label attached to one or both of the PCR primers followed by capillary electrophoresis using polyacrylamide gel. The fluorescent PCR products migrate through the gel past a fluorescence detector; the time taken to reach the detector compared to the time taken for known size markers then allows a linked computer to calculate the size of each PCR product. The method is very fast, reliable and uses very small amounts of clinical material.

- Because of its extreme sensitivity, PCR is also now fundamentally important to forensic medicine. Variation in the number of repeats in specific STRs from person to person, coupled with the large number of STRs available for use as markers, allows individual, identification. For example, it is possible to use PCR to amplify the DNA from a single human hair, or from biological material left on a cigarette or on a postage stamp, or from a microscopic drop of blood. Any of these items left at the scene of a crime can generate enough DNA using PCR for detailed characterization and identification of the relevant individuals.

- By using suitable primers, it is possible to use PCR to create point mutations in the target DNA, which greatly facilitates the analysis of gene expression and function (Section I7).

Real-time PCR

Traditional PCR is not useful for quantifying the amount of original RNA or DNA template since the amount of PCR product synthesized is determined more by the amount of primers and other reagents in the reaction mixture than the amount of template. However, quantitation is possible using **real-time PCR**.

Note that the abbreviation RT-PCR is usually reserved for reverse transcriptase PCR (see above) and not real-time PCR but this is not always universally adhered to and may be confusing. Real-time PCR is so called because the amplified DNA is detected as the reaction progresses in real time, as opposed to traditional PCR where the PCR product is analyzed at the end of the amplification.

There are two main methods for carrying out real-time PCR:

- The first approach is to include a dye in the PCR reaction mixture that binds only to double-stranded DNA and which fluoresces strongly when bound, e.g. SYBR Green I. As the PCR proceeds, the dye binds to the double-stranded amplified DNA and fluoresces. So the rate of reaction can be monitored by measuring the increase in fluorescence over time (using a real-time PCR thermocycler) and this rate is proportional to the amount of original template. However, because the dye binds to any double-stranded DNA irrespective of its sequence, it is a nonspecific quantitation method.

- The second method is for the primers used in the PCR reaction mixture to be labeled at one end with a fluorescent reporter molecule and at the other end with a fluorescence quencher molecule. Because the quenching molecule is close to the fluorescence reporter, it prevents the intact primer from fluorescing. However, as the primer is used in the PCR reaction during amplification, the 5′ to 3′ exonuclease activity of *Taq* polymerase removes one end of the primer. So the fluorescent reporter is no longer near the quencher and now fluoresces. Multiple probes with different colored fluorescence labels for the detection of several different DNA sequences in the same reaction (**multiplex PCR**) can also be used.

Real-time PCR can be used to quantify nucleic acids relatively or absolutely. Relative quantification measures the fold-difference (e.g. 2×, 3×) in the target molecule for PCR between two or more samples. Absolute quantification measures the exact amount of the target molecule in the sample by comparing the rate of amplification with that of known standards.

17 Site-directed mutagenesis

Key Notes

Why use site-directed mutagenesis?	Site-directed mutagenesis uses recombinant DNA techniques to create defined mutations in the target gene.
Oligonucleotide-directed mutagenesis	An oligonucleotide is synthesized that contains the desired point mutation – a single nucleotide change. This is then hybridized to a recombinant M13 phage single-stranded DNA template containing the target gene sequence. The oligonucleotide is used as a primer by DNA polymerase to synthesize a complete copy of the DNA template but now containing the required point mutation. After transformation into *E. coli* and replication, 50% of the recombination DNA molecules contain the original gene sequence but the other 50% of the copies contain the mutated gene sequence. As well as point mutations, oligonucleotide-directed mutagenesis can also be used to create small deletions or small insertions in the target sequence.
Cassette mutagenesis	In cassette mutagenesis, an appropriate region of the target gene is removed from a recombinant DNA molecule using restriction enzyme digestion and is replaced by a double-stranded oligonucleotide that has been synthesized to include the desired mutation(s).
PCR site-directed mutagenesis	Two PCR reactions are carried out, each using an oligonucleotide primer bearing the desired mutation and a primer that hybridizes to a sequence flanking the gene sequence of interest. The two products of these PCR reactions each correspond to about 50% of the original target DNA sequence. These are then denatured, mixed together and undergo another PCR reaction, which leads to the synthesis of full length molecules containing the mutation.
***De novo* synthesis**	Gene sequences bearing defined mutations can be synthesized from overlapping synthetic oligonucleotides, using DNA polymerase to fill the gaps.
Protein engineering	Protein engineering uses recombinant DNA techniques to create novel proteins. The rational design approach uses knowledge of the structure-function relationships of the target protein to decide which amino acids should be mutated using recombinant DNA technology to obtain the desired phenotypic change. Directed evolution works by creating large numbers of mutations throughout the target gene sequence and then selecting those that display improvements in the desired characteristic(s). The process

is then repeated, hopefully culminating in the selection of a protein that fulfills the desired outcomes. Novel proteins can also be produced by fusing regions of genes that encode specific protein domains.

Related topics   (H1) The genetic code   (I4) DNA cloning
(I2) Restriction enzymes   (I5) DNA sequencing
(I3) Nucleic acid   (I6) Polymerase chain
hybridization   reaction

Why use site-directed mutagenesis?

Classically a great deal has been learned about gene function by analyzing mutations. Typically, the organism under study is exposed to a mutagen to generate random mutations and those mutants that show a particular phenotype are selected for further study. A range of methods, including DNA sequencing, are used to identify which gene(s) have been mutated and the nature of the changes. This gives valuable information on both gene and protein function. However, this classic approach is extremely laborious and involves screening large numbers of random mutations in the hope of finding useful mutations in the gene of interest. In contrast, recombinant DNA techniques allow defined changes to be made to the gene sequence and generate the desired mutations with certainty. These techniques are collectively called **site-directed mutagenesis** or *in vitro* **mutagenesis**.

There are many methods in use for site-directed mutagenesis, some of which are complex, and so not all of these can be covered here. However, the following description explains the main types of procedure used.

Oligonucleotide-directed mutagenesis

Assume that the goal is to replace a particular cysteine residue in a target protein with glycine, and that the cysteine residue in question is encoded in the gene by the codon TGC. We know from the genetic code (Section H1) that one of the glycine codons is GGC so the required mutation is to replace the T in the specific TGC codon in the gene with G to convert the codon to GGC.

The first step is for the target gene to be cloned into M13 phage as the vector, which has the useful feature of producing a single-stranded DNA template (Figure 1). Next an oligonucleotide is synthesized that contains the required mutation and this is hybridized to the recombinant M13 phage single-stranded DNA. As shown in Figure 1, and in more detail in Figure 2, all of the nucleotides of the oligonucleotide will base pair to their complementary nucleotides in the gene except for the mutation because this is a mismatch. The oligonucleotide is now used as a primer for DNA polymerase, which copies the entire recombinant M13 DNA template to produce a double-stranded DNA molecule, one strand of which contains the desired mutation (Figure 1). This is introduced into *E. coli* and produces multiple copies, half of which contain the original target gene sequence and half of which bear the mutated gene. The resulting recombinant phage can be plated onto agar to produce plaques and the mutant phage plaques can be identified by hybridization using the oligonucleotide bearing the mutation as a radioactive probe (Section I4).

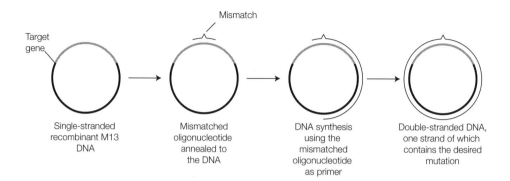

Figure 1. Oligonucleotide-directed mutagenesis. The target gene contained in a recombinant M13 phage is hybridized to an oligonucleotide containing the single nucleotide change required to generate the desired mutation. DNA synthesis using the oligonucleotide as a primer creates a double-stranded DNA molecule, one strand of which is wild-type and the other bears the point mutation.

There are a range of variations of this basic protocol including the use of double-stranded DNA plasmids instead of M13 phage but in each protocol variant, the principle of using a mismatched oligonucleotide to generate the desired mutated gene remain the same.

Oligonucleotide-directed mutagenesis used as described here generates **point mutations**, that is, the substitution of one nucleotide for another. However, it can also be used to generate small **insertions** or **deletions** such as the insertion of three additional nucleotides encoding an additional amino acid or the deletion of three nucleotides removing a specific amino acid from the protein encoded by the target gene.

Cassette mutagenesis

In cassette mutagenesis, a recombinant plasmid containing the target gene is cut with restriction enzymes to excise a fragment bearing the gene sequence that is to be mutated (Figure 3). A double-stranded oligonucleotide that has been synthesized to include the desired mutation(s) is then prepared with cohesive ends that are complementary to the ends of the cut recombinant plasmid. This is hybridized to the cut plasmid and ligated together using T4 DNA ligase producing a recombinant plasmid containing the desired mutation in the gene (Figure 3).

Figure 2. An oligonucleotide containing a single nucleotide change from the original DNA sequence base pairs with the target gene except at the position of the changed nucleotide since this represents a mismatch.

Figure 3. Cassette mutagenesis.

As well as the introduction of point mutations, cassette mutagenesis can also introduce deletions or insertions and these can be larger than is possible via oligonucleotide-directed mutagenesis.

PCR site-directed mutagenesis

A range of methods for PCR site-directed mutagenesis are available. The approach shown in Figure 4 uses two primers that are at the boundaries of the target gene sequence of interest (primers 1 and 4) and two partially overlapping primers, which each contain the desired mutation (primers 2 and 3).

Two PCR reactions are carried out, one using primers 1 and 2 and the other using primers 3 and 4. In the example shown in Figure 4 where the desired mutation is approximately at the midpoint of the target gene sequence, each of the two PCR reactions amplifies approximately half of the target sequence. The two PCR products are then denatured and mixed together; they anneal because of the overlapping sequence of incorporated primers 2 and 3. The annealed DNA then undergoes 3′ extension and amplification in a final PCR reaction using primers 1 and 4, producing full-length amplified DNA containing the desired mutation.

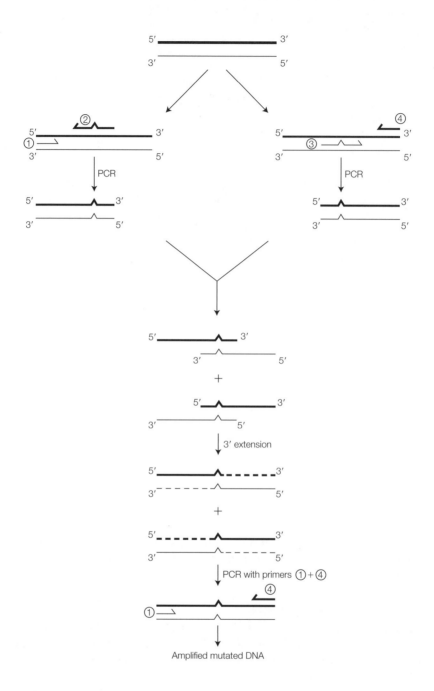

Figure 4. PCR site-directed mutagenesis. The various steps are described in the text. Adapted from Strachan, T. and Read, A. (2011) *Human Molecular Genetics*, 4th Edn; Garland Science.

De novo synthesis

Yet another approach to introducing mutation is to synthesize the gene *de novo* by preparing a series of overlapping oligonucleotides, each about 150 nucleotides long or more, hybridizing these together and filling the gaps using DNA polymerase. Finally, the synthetic gene bearing the desired mutations is inserted into a vector for amplification and then testing in a suitable host organism.

Protein engineering

The power of recombinant DNA technology to modify the sequence of a gene at will opens up the potential to create novel proteins that are better suited to specific research, clinical or industrial use. For example, one could increase the stability of a specific enzyme, or modify its substrate specificity. This new field is called **protein engineering**. Two broad approaches are used for protein engineering, **rational design** and **directed evolution**.

Rational design draws upon knowledge of the native protein to identify what changes might lead to the desired outcome and then those changes are engineered using site-directed mutagenesis. The problem with this approach is that we still do not know enough about the structure-function relationships of most proteins to be able to predict what mutations we need to carry out to achieve this goal. However, as our knowledge of protein structure increases in years to come, this approach promises great potential.

In directed evolution, large numbers of variants of the gene of interest are created by generating mutations using recombinant DNA techniques. These are expressed and then a selection process is used to pick out those mutated genes which yield proteins that have more of the desired qualities. These selected mutated genes are then subjected to further rounds of mutation and selection, hopefully yielding variant proteins even better equipped with the desired properties. The process is called directed evolution because it has parallels with the natural evolution of beneficial mutations. The big disadvantage of directed evolution is the huge effort required for producing and screening large numbers of protein variants so this is a high-throughput approach. However, the advantage is that it requires no knowledge of what aspects of protein structure will lead to the desired functionality.

Finally, novel proteins can be created by fusing together pieces of genes that encode specific protein domains from different proteins. For example, immunotoxins are novel proteins that have been created by fusing a bacterial or plant toxin gene segment with an antibody or growth factor gene. The goal is for the antibody or growth factor to bind to specific cells in the body and then be internalized by them so that the enzymatic fragment of the toxin can then kill the cell. The technique is looking very promising with one such fusion protein, between human interleukin-2 and truncated diphtheria toxin, already in use to treat a specific cancer called cutaneous T-cell lymphoma.

J1 Monosaccharides and disaccharides

Key Notes

Aldoses and ketoses

A monosaccharide has the general formula $(CH_2O)_n$ and contains either an aldehyde group (an aldose) or a ketone group (a ketose). The free aldehyde or ketone group can reduce cupric ions (Cu^{2+}) to cuprous ions (Cu^+) and hence such a monosaccharide is called a reducing sugar.

Stereoisomers

The D and L stereoisomers of sugars refer to the configuration of the asymmetric carbon atom furthest from the aldehyde or ketone group. The sugar is said to be a D isomer if the configuration of the atoms bonded to this carbon atom is the same as for the asymmetric carbon in D-glyceraldehyde.

Ring structures

Tetroses and larger sugars can cyclize by reaction of the aldehyde or ketone group with a hydroxyl group on another carbon atom of the sugar. Glucose mainly cyclizes to form a six-membered pyranose ring whilst other sugars form five-membered furanose rings. Two forms (anomers) of D-glucopyranose exist, depending on whether the hydroxyl group attached to the anomeric carbon atom (C-1) lies below the plane of the ring (the α form) or above the plane of the ring (β form). In solution, the α and β forms interconvert via the open-chain form (mutarotation). The pyranose ring can exist in either boat or chair conformations but the chair form predominates since the side groups, which are usually OH groups, are less sterically hindered in this conformation.

Disaccharides

A disaccharide is formed when two monosaccharides become joined by a glycosidic bond. The bond may be an α- or β-bond depending on the configuration of the anomeric carbon atom involved in the bond. Usually the anomeric carbon atom of only one of the two monosaccharides is involved in the bond so that the disaccharide still has one free aldehyde or ketone group and is still reducing. However, in sucrose both anomeric carbon atoms are bonded together so that sucrose is a nonreducing disaccharide.

Sugar derivatives

The hydroxyl groups of sugars can be replaced by other groups to form a wide range of biologically important molecules including phosphorylated sugars, amino sugars and nucleotides.

Nomenclature

The names of simple sugars and sugar derivatives can all be abbreviated. This also allows an abbreviated description of the component sugars present in disaccharides, for example.

Related topics	(H5) Protein glycosylation	(J4) Gluconeogenesis
	(J2) Polysaccharides and oligosaccharides	(J5) Pentose phosphate pathway
	(J3) Glycolysis	

Aldoses and ketoses

A carbohydrate is composed of carbon (*carbo-*), and hydrogen and oxygen (*-hydrate*). The simplest carbohydrates are the **monosaccharides** that have the general formula $(CH_2O)_n$ where n is 3 or more. A monosaccharide or simple **sugar**, consists of a carbon chain with a number of hydroxyl (OH) groups and either one **aldehyde group** ($-C{\scriptstyle\diagup}^{H}_{\diagdown O}$ often written as –CHO) or one **ketone group** ($\diagdown C = O$). A sugar that bears an aldehyde aldehyde group is called an **aldose** whereas a sugar with a ketone group is a **ketose**. The smallest carbohydrates, for which $n=3$, are called **trioses**.

The above terms can be combined. Thus glyceraldehyde (Figure 1) is a triose that has an aldehyde group and so is an aldose. Thus it can also be called an **aldotriose**. Similarly, dihydroxyacetone (Figure 1) is a **ketotriose**.

Figure 1. Structures of glyceraldehyde and dihydroxyacetone.

Sugars that contain a free aldehyde or ketone group in the open-chain configuration can reduce cupric ions (Cu^{2+}) to cuprous ions (Cu^+) and hence are called **reducing sugars**. This is the basis of the Fehling's and Benedict's tests for reducing sugars. The **reducing end** of such a sugar chain is thus the end that bears the aldehyde or ketone group.

Note that glyceraldehyde and dihydroxyacetone have the same chemical composition, $C_3H_6O_3$, but differ in structure (i.e. they are **structural isomers**).

Stereoisomers

Glyceraldehyde has a single asymmetric carbon atom (the central one) and so two **stereoisomers** (also called **optical isomers**) are possible, that is two forms of glyceraldehyde, denoted as D- and L-**glyceraldehyde**, which are mirror images of each other (Figure 2). Stereoisomers also exist for amino acids (Section B1).

Figure 2. D- and L-glyceraldehyde are mirror images of each other (stereoisomers or optical isomers).

Sugars with four, five, six or seven carbons are called **tetroses, pentoses, hexoses** and **heptoses**, respectively. In these cases, the sugars may have more than one asymmetric carbon atom. The convention for numbering carbon atoms and naming configurations is as follows:

- The carbon atoms are numbered from the end of the carbon chain starting with the aldehyde (which is carbon 1, C-1) or ketone group (which is usually C-2)

- The symbols D and L refer to the configuration of the asymmetric carbon atom furthest from the aldehyde or ketone group

Thus, for example, glucose, an aldohexose, exists as D and L forms (Figure 3a). The furthest asymmetric carbon from the aldehyde group is C-5. D-Glucose (Figure 3a) is called D because the configuration of the atoms bonded to C-5 is the same as for D-glyceraldehyde (Figure 2). Similarly D-fructose (a ketohexose; Figure 3b) is designated D because the configuration at C-5 matches that for D-glyceraldehyde. D sugars that differ in configuration at only a single asymmetric carbon atom are called **epimers**. Thus D-glucose and D-galactose are epimers, differing only in their configuration at C-4 (Figure 4).

Figure 3. (a) D- and L-glucose; (b) D- and L-fructose.

Figure 4. The epimers D-glucose and D-galactose.

Ring structures

The aldehyde or ketone group can react with a hydroxyl group to form a covalent bond. Formally, the reaction between an aldehyde and the hydroxyl group of a sugar (an alcohol) creates a **hemiacetal** (Equation 1) whereas a ketone reacts with a hydroxyl group (alcohol) to form a **hemiketal** (Equation 2).

For tetroses and larger sugars, the reaction can take place within the same molecule so that the straight chain form of the sugar cyclizes. For example, if the C-5 hydroxyl group of glucose reacts with the aldehyde group, a six-membered ring is formed whereas if the C-4 hydroxyl reacts with the aldehyde group, a five-membered ring is formed. Figure 5 shows the cyclization of D-glucose to form a six-carbon ring. Because of their similarity to the ring compound called pyran (Figure 5b), six-membered ring structures are called **pyranoses**.

Figure 5. (a) Cyclization of the open-chain form of D-glucose; (b) the structure of pyran.

(a)

α-D-Fructose   β-D-Fructose   Furan **(b)**

Figure 6. (a) Cyclization of the open-chain form of D-fructose; (b) the structure of furan.

Figure 6 shows the cyclization of the ketohexose, fructose, to form a five-membered ring. Because of their similarity to the ring compound called furan (Figure 6b), five-membered ring structures are called **furanoses**. In general, aldoses and ketoses with five or more carbons can form either pyranose or furanose rings so that in solution a mixture of these exist. Which is the more stable ring form, and hence predominant, depends on the chemical structure of the monosaccharide, including the nature of the substituent groups. Usually aldohexoses such as glucose exist mainly in the pyranose ring form.

The ring structures shown in Figures 5 and 6 are called **Haworth projections**, in which the plane of the ring can be imagined as approximately perpendicular to the plane of the paper with the thick lines of the ring in the diagram pointing towards the reader. Note that during cyclization of glucose (an aldose), a new asymmetric center is formed, at C-1, the carbon that carried the carbonyl moiety. So **two isomers** of D-glucose exist (Figure 5a), α-D-glucose (in which the OH group at C-1 lies below the plane of the ring) and β-D-glucose (in which the OH group at C-1 lies above the plane of the ring). The pyranose ring structure of β-D-glucose may be written as β-D-glucopyranose. The carbonyl carbon (C-1 in this case) is called the **anomeric carbon atom** and so the α and β forms are called **anomers**. In aqueous solution, the α- and β-forms rapidly interconvert via the open-chain structure, to give an equilibrium mixture (Figure 5a). This process is called **mutarotation**.

In the case of the ketose, fructose, the anomeric carbon atom (that carried the carbonyl moiety) is C-2 and hence two isomers (anomers) exist that differ in their configuration about that carbon atom (Figure 6a), i.e. the α/β designation refers to the configuration about C-2 not C-1.

The pyranose ring of a six-carbon sugar can exist in either a **boat** or a **chair** configuration (Figure 7). The substituents attached to the ring carbons that extend parallel to the symmetry axis are said to be axial (a) whilst those that extend outward from this axis are said to be equatorial (e); (Figure 7). In the boat form, there is considerable steric hindrance

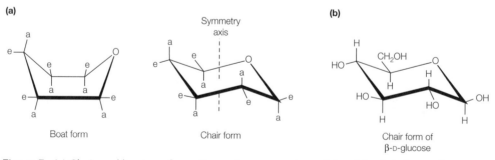

Boat form   Chair form   Chair form of β-D-glucose

Figure 7. (a) Chair and boat conformations of pyranose rings; (b) stable chair form of β-D-glucose.

between the various groups attached to the carbon atoms of the ring and hence this form is less favorable energetically. Hence the chair form predominates, as shown for β-D-glucose in Figure 7, where all the axial positions are occupied by hydrogen atoms.

Disaccharides

The aldehyde or ketone group on the anomeric carbon atom of one monosaccharide can react with the hydroxyl group of a second monosaccharide to form a **disaccharide**. The covalent bond formed is called a **glycosidic bond**. Lactose (Figure 8a) is a disaccharide formed between the anomeric carbon (C-1) of D-galactose and C-4 of D-glucose. Since the anomeric carbon of the galactose molecule is involved in the bond and is in the β configuration, this is called a **β(1→4) bond**, which can be abbreviated as β1–4. Maltose (Figure 8b) is a disaccharide formed between the C-1 and C-4 positions of two glucose units. However, here the configuration of the anomeric carbon atom involved is the α form and hence the bond is called an **α(1→4) bond** or abbreviated as α1–4. For lactose and maltose, one of the anomeric carbons has been used to form the bond, leaving the second anomeric carbon free. Thus both lactose and maltose have a **reducing end**. In contrast, sucrose (Figure 8c) is a disaccharide formed by bond formation between the anomeric C-1 of glucose and the anomeric C-2 of fructose so that sucrose lacks a free reducing group.

Figure 8. Structure of common disaccharides.

Sugar derivatives

The hydroxyl groups of simple monosaccharides can be replaced with other groups to form a range of sugar derivatives. For example, **phosphorylated sugars** such as glucose 6-phosphate (Figure 9) are important metabolites in glycolysis (Section J3). In **amino sugars**, one or more hydroxyl groups are each replaced by an amino group (which is often acetylated), for example acetyl β-D-*N*-acetylglucosamine (Figure 9). This and other sugar derivatives are common components of many glycoproteins. Nucleotides (Section F1), such as ATP, each consist of a sugar in which the anomeric carbon atom has formed a covalent bond with a nitrogenous base (Figure 9). Since the bond is between the anomeric carbon of the sugar and a nitrogen atom of the base, it is called an **N-glycosidic bond**. Other modifications include oxidation of one of the carbons to form a carboxylate group, so generating a **uronic acid.** For example, oxidation of carbon-6 of D-glucose in this way yields D-glucuronic acid (Figure 9). Uronic acids are important components of many polysaccharides. Aldoses and ketoses can also be reduced to yield polyhydroxy alcohols called **alditols** (Figure 9) such as sorbitol (D-glucitol), D-mannitol and D-xylitol, which are sweet-tasting and are used to flavor sugarless gum and mints, and glycerol, which is an important lipid component. Alditols are linear molecules that cannot cyclize.

Nomenclature

Simple sugars have three-letter abbreviations [e.g. Glc (glucose), Gal (galactose), Man (mannose), Fuc (fucose)]. Sugar derivatives can also be abbreviated, such as GlcNAc (*N*-acetylglucosamine), GalNAc (*N*-acetylgalactosamine). This also allows an abbreviated form of description for sugars that are bonded together and the nature of the covalent bonds. Thus, for example, lactose (Figure 8a) can be represented as Galβ1 →4Glc.

Figure 9. Examples of sugar derivatives.

J2 Polysaccharides and oligosaccharides

Key Notes

Polysaccharides	Long chains of monosaccharides joined together are collectively called polysaccharides. The major storage polysaccharides are glycogen (in animals), starch (in plants) and dextran (in yeast and bacteria). Cellulose is a structural polysaccharide found in plant cell walls.
Glycogen	Glycogen is a branched-chain polysaccharide containing glucose residues linked by $\alpha 1$–4 bonds with $\alpha 1$–6 branchpoints. The branched nature of glycogen makes it more accessible to glycogen phosphorylase during degradation, since this enzyme degrades the molecule by sequential removal of glucose residues from the nonreducing ends.
Starch	Starch is a mixture of unbranched amylose (glucose residues joined by $\alpha 1$–4 bonds) and branched amylopectin (glucose residues joined $\alpha 1$–4 but with some $\alpha 1$–6 branchpoints).
Dextran	Dextran consists of glucose residues linked mainly by $\alpha 1$–6 bonds but with occasional branchpoints that may be formed by $\alpha 1$–2, $\alpha 1$–3 or $\alpha 1$–4 bonds.
Cellulose	Cellulose is a straight-chain polymer of glucose units linked by $\beta 1$–4 bonds. The polysaccharide chains are aligned to form fibrils that have great tensile strength. Cellulases, enzymes that degrade cellulose, are absent in mammals but are produced by some bacteria, fungi and protozoa.
Oligosaccharides	Short chains of monosaccharides linked by glycosidic bonds are called oligosaccharides. Oligosaccharides found in glycoproteins are linked to a serine or threonine residue (O-linked oligosaccharide) or to an asparagine residue (N-linked oligosaccharide). All N-linked oligosaccharides have a common pentasaccharide core. High-mannose N-linked oligosaccharides have additional mannose residues linked to the core whilst the complex type N-linked oligosaccharides have branches comprising combinations of GlcNAc, Gal, sialic acid and L-fucose.
Related topics	(H5) Protein glycosylation (J6) Glycogen metabolism (J1) Monosaccharides and disaccharides

Polysaccharides

Polysaccharides are large chains of sugar units joined together. Depending on the polysaccharide, the chains may be linear or branched. In animals, excess glucose is stored as a large branched polysaccharide called **glycogen**, whereas in most plants the storage form of glucose is the polysaccharide called **starch**. Bacteria and yeasts store glucose as yet another type of polysaccharide called **dextran**. In each case these are nutritional reserves; when required, they are broken down and the monosaccharide products are metabolized to yield energy (Section J3). In contrast, **cellulose** is a structural polysaccharide used to make plant cell walls.

Glycogen

A glycogen molecule consists entirely of glucose units, most of which are linked in long chains by α1–4 bonds. However, every 10 units or so, the chain is branched by the formation of an α1–6 glycosidic bond (Figure 1). Each straight-chain segment of glycogen forms an open helix conformation, which increases its accessibility to the enzymes of glycogen metabolism. Each chain terminates in a nonreducing end, that is an end with a free 4′-OH group. Since the enzyme that degrades glycogen (**glycogen phosphorylase**; Section J6) catalyzes the sequential removal of glycosyl units from the nonreducing end of a glycogen chain, the numerous branches, each with a nonreducing end, greatly increase the accessibility of the polysaccharide to degradation. This allows rapid mobilization of stored glycogen in times of need. The α1–6 branches themselves are removed by a **debranching enzyme** (Section J6).

Starch

Starch exists in plants as insoluble starch granules in chloroplasts. Each starch granule contains a mixture of two polysaccharide forms, **amylose** and **amylopectin**. Amylose is an unbranched polymer of glucose residues joined in α1–4 linkages. Amylopectin is the branched form; most of the constituent glucose residues are joined in α1–4 linkages but additional α1–6 bonds occur every 25–30 residues, creating the branchpoints.

Figure 1. The α1–4 linkages in the straight chain and α1–6 branchpoint linkages in glycogen.

Dextran

Dextran is a glucose polymer where the glucose residues are mainly linked by α1–6 bonds. However, a few branches also occur. These are typically formed by α1–2, α1–3 or α1–4 bonds depending on the bacterial or yeast species that is the source of the dextran.

Cellulose

Cellulose is an unbranched polysaccharide of glucose units linked by β1–4 bonds (Figure 2). Unlike glycogen where the α1–4 linkages lead to a helical conformation of the polysaccharide, the β linkage between glucose residues in cellulose creates very long straight chains that are arranged together in fibrils. In plant cell walls the cellulose fibrils are embedded in (and cross-linked to) a matrix of other polysaccharides. In wood, this matrix also contains **lignin**, a complex polymer of phenolic residues (Section A2). This composite has a very high tensile strength. Mammals, including humans, lack enzymes capable of digesting the β1–4 linkages of cellulose and so cannot digest plant cell walls. However, some bacteria produce **cellulases**; enzymes that degrade cellulose. Ruminant animals such as cattle have cellulase-producing bacteria in their digestive tracts and so can digest cellulose. In addition, some fungi and protozoa produce and secrete cellulase.

Cellulose

Figure 2. The repeating unit of cellulose showing the β1–4 linkage.

Oligosaccharides

Oligosaccharides are short chains of monosaccharides linked together by glycosidic bonds. In the case of oligosaccharides linked to proteins (glycoproteins) or lipids (glycolipids), the oligosaccharide is not a repeating unit but consists of a range of different monosaccharides joined by a variety of types of bonds. In glycoproteins, two main types of oligosaccharide linkages exist:

- **O-linked oligosaccharides** attached to the protein via **O-glycosidic bonds**, to the OH groups of serine or threonine side-chains (Section H5, Figure 1);

- **N-linked oligosaccharides** attached to the protein via **N-glycosidic bonds**, to the NH_2 groups of asparagine side-chains (Section H5, Figure 1). All N-linked oligosaccharides have a common pentasaccharide core of two GlcNAc and three Man residues but the nature of the side-chains differs (Figure 3). In the **high-mannose type** of N-linked oligosaccharide, typically two to six additional Man residues are joined to the pentasaccharide core (e.g. Figure 3a). The **complex type** of N-linked oligosaccharide contain two to five outer branches attached to the Man of the polysaccharide core; these branches contain different combinations of GlcNAc, Gal, sialic acid (*N*-acetylneuraminic acid), mannose and L-fucose. Figure 3b shows a complex oligosaccharide with two outer branches.

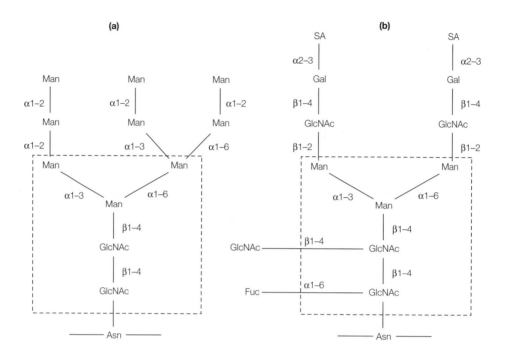

Figure 3. Examples of (a) high-mannose type and (b) complex type oligosaccharides. In each case, the sugars that comprise the common pentasaccharide core are boxed. SA, sialic acid.

J3 Glycolysis

Key Notes

Overview

Glycolysis is a set of reactions that takes place in the cytoplasm of prokaryotes and eukaryotes. The roles of glycolysis are to produce energy (both directly and by supplying substrate for the citric acid cycle and oxidative phosphorylation) and to produce intermediates for biosynthetic pathways.

The pathway

Glucose is phosphorylated to glucose 6-phosphate (by hexokinase), which is converted to fructose 6-phosphate (by phosphoglucose isomerase) and then to fructose 1,6-bisphosphate (by phosphofructokinase, PFK). The fructose 1,6-bisphosphate is split into glyceraldehyde 3-phosphate and dihydroxyacetone phosphate (by aldolase) and these two trioses are interconverted by triose phosphate isomerase. Glyceraldehyde 3-phosphate is converted to 1,3-bisphosphoglycerate (by glyceraldehyde 3-phosphate dehydrogenase), which reacts with ADP to give 3-phosphoglycerate and ATP (catalyzed by phosphoglycerate kinase). The 3-phosphoglycerate is converted to 2-phosphoglycerate (by phosphoglycerate mutase) and then to phosphoenolpyruvate (PEP) by enolase. Finally, PEP and ADP react to form pyruvate and ATP (catalyzed by pyruvate kinase).

Fates of pyruvate

Under aerobic conditions, pyruvate can be converted by pyruvate dehydrogenase to acetyl coenzyme A (acetyl CoA), which can then enter the citric acid cycle. Under anaerobic conditions, pyruvate is converted to lactate by lactate dehydrogenase (LDH). The NAD^+ regenerated by this reaction allows glycolysis to continue, despite the lack of oxygen. When oxygen becomes available, the lactate is converted back to pyruvate. In anaerobic conditions, yeast and other organisms carry out alcoholic fermentation that converts pyruvate to acetaldehyde and then to ethanol, regenerating NAD^+ that allows glycolysis to continue.

Energy yield

Two ATPs are used in glycolysis and four ATPs are synthesized for each molecule of glucose so that the net yield is two ATPs per glucose. Under aerobic conditions, the two NADH molecules arising from glycolysis also yield energy via oxidative phosphorylation. The formation of lactate under aerobic conditions or alcoholic fermentation each yield two molecules of ATP per molecule of glucose.

Metabolism of fructose

Fructose can be metabolized by two routes. In adipose tissue and muscle, hexokinase can phosphorylate fructose to fructose 6-phosphate that then enters glycolysis. In liver, most of the enzyme present is glucokinase not hexokinase and this does not phosphorylate fructose. In this tissue, fructose is metabolized instead by the fructose 1-phosphate pathway.

Metabolism of galactose

Galactose enters glycolysis via the galactose–glucose interconversion pathway, a four-step reaction sequence. Lack of the second enzyme in this pathway, galactose 1-phosphate uridylyl transferase, leads to the disease galactosemia through the accumulation of toxic products, including galactitol formed by the reduction of galactose. Lactose intolerance (hypolactasia) is an inability to metabolize lactose due to a lack of the enzyme lactase.

Regulation of glycolysis

The main control step is that catalyzed by PFK but hexokinase and pyruvate kinase are additional control sites. PFK is allosterically inhibited by ATP, but this inhibition is relieved by AMP. Citrate also inhibits PFK. The build-up of fructose 6-phosphate stimulates the formation of fructose 2,6-bisphosphate that in turn stimulates PFK. The enzyme that synthesizes fructose 2,6-bisphosphate (phosphofructokinase 2; PFK2) and the enzyme that hydrolyzes it back to fructose 6-phosphate (fructose bisphosphatase 2; FBPase2) are also regulated hormonally by glucagon that causes glycolysis to slow down when the blood glucose level falls. PFK is also inhibited by H^+ ions, thus preventing excessive formation of lactate under anaerobic conditions. Hexokinase is inhibited by glucose 6-phosphate, which builds up after PFK is inhibited. Pyruvate kinase is activated by fructose 1,6-bisphosphate but allosterically inhibited by ATP and alanine. Like PFK, it is also regulated hormonally by glucagon.

Related topics

(J1) Monosaccharides and disaccharides
(J2) Polysaccharides and oligosaccharides
(J4) Gluconeogenesis

(J5) Pentose phosphate pathway
(J6) Glycogen metabolism
(L1) Citric acid cycle

Overview

Glycolysis is a series of reactions (Figure 1) that takes place in the cytoplasm of all prokaryotes and eukaryotes. Glycolysis converts one molecule of glucose into two molecules of pyruvate [which are then converted to acetyl coenzyme A (acetyl CoA) ready for entry into the citric acid cycle]. Two ATP molecules are needed for early reactions in the glycolytic pathway but four ATPs are generated later, giving a net yield of two ATPs per molecule of glucose degraded.

Overall, glycolysis has a dual role. The first is to generate ATP. Although only two ATPs per glucose are made directly from the reactions of the glycolytic pathway, it also feeds substrates into the citric acid cycle and oxidative phosphorylation, where most ATP is made. The second role is to produce intermediates that act as precursors for a number of biosynthetic pathways. Thus acetyl CoA, for example, is the precursor for fatty acid synthesis (Section K3).

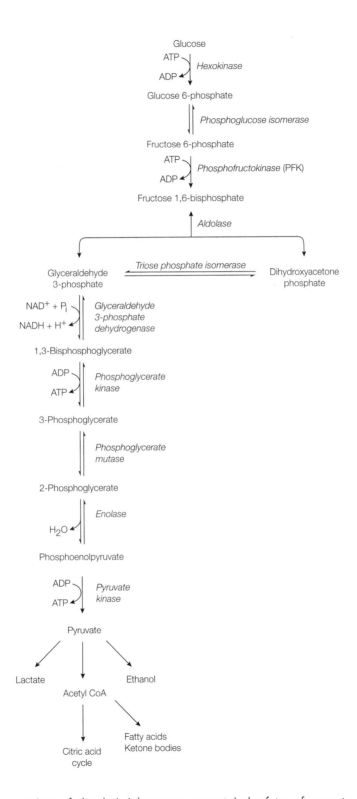

Figure 1. The reactions of glycolysis (glucose to pyruvate) plus fates of pyruvate.

The pathway

The individual steps in glycolysis are described below.

1. Glucose is phosphorylated by ATP to form glucose 6-phosphate and ADP. The reaction is catalyzed by the enzyme **hexokinase**.

2. Glucose 6-phosphate is converted to fructose 6-phosphate by **phosphoglucose isomerase**.

This isomerization involves the conversion of an aldose to a ketose, a conversion that is better seen by viewing the open-chain representations of these molecules.

3. Fructose 6-phosphate is phosphorylated by ATP to form fructose 1,6-bisphosphate and ADP. The enzyme catalyzing this step is **phosphofructokinase** (PFK).

4. **Aldolase** splits fructose 1,6-bisphosphate (a six-carbon molecule) into two three-carbon molecules, glyceraldehyde 3-phosphate and dihydroxyacetone phosphate.

Fructose 1,6-bisphosphate → **Dihydroxyacetone phosphate** + **Glyceraldehyde 3-phosphate** (Aldolase)

5. Glyceraldehyde 3-phosphate is the only molecule that can be used for the rest of glycolysis. However, the dihydroxyacetone phosphate formed in the previous step can rapidly be converted to glyceraldehyde 3-phosphate by **triose phosphate isomerase**. This is an equilibrium reaction; as the glyceraldehyde 3-phosphate is used by the rest of glycolysis, more dihydroxyacetone phosphate is converted to glyceraldehyde 3-phosphate as replacement. Thus effectively, for each molecule of fructose 1,6-bisphosphate that is cleaved in step 4, two molecules of glyceraldehyde 3-phosphate continue down the pathway.

Dihydroxyacetone phosphate (a ketose) → **Glyceraldehyde 3-phosphate** (an aldose) (Triose phosphate isomerase)

6. Glyceraldehyde 3-phosphate is converted to 1,3-bisphosphoglycerate. The reaction is catalyzed by **glyceraldehyde 3-phosphate dehydrogenase** and uses inorganic phosphate and NAD^+. The other product is NADH. The energy for creating this new high-energy phosphate bond comes from the oxidation of the aldehyde group of the glyceraldehyde 3-phosphate.

Glyceraldehyde 3-phosphate → **1,3-Bisphosphoglycerate (1,3-BPG)** (Glyceraldehyde 3-phosphate dehydrogenase)

7. The newly created high-energy phosphate bond of 1,3-bisphosphoglycerate is now used to synthesize ATP. **Phosphoglycerate kinase** catalyzes the transfer of the phosphoryl group from the 1,3-bisphosphoglycerate to ADP, generating ATP and 3-phosphoglycerate.

1,3-Bisphosphoglycerate   **3-Phosphoglycerate**

8. 3-Phosphoglycerate is converted to 2-phosphoglycerate by **phosphoglycerate mutase**. Thus the reaction is a movement of the phosphate group to a different carbon atom within the same molecule.

3-Phosphoglycerate   **2-Phosphoglycerate**

9. **Enolase** catalyzes the dehydration of 2-phosphoglycerate to form phosphoenolpyruvate (PEP). This reaction converts the low-energy phosphate ester bond of 2-phosphoglycerate into the high-energy phosphate bond of PEP.

2-Phosphoglycerate   **Phosphoenolpyruvate**

10. In the last reaction, **pyruvate kinase** catalyzes the physiologically irreversible transfer of the phosphoryl group from PEP to ADP to form ATP and pyruvate.

Phosphoenolpyruvate   **Pyruvate**

Substrate-level phosphorylation

There are two distinct methods by which cells synthesize ATP. In **oxidative phos-phorylation**, involving the electron transport chain, the generation of ATP is linked to the oxidation of NADH and $FADH_2$ to NAD^+ and FAD, respectively (Section L2), and occurs via the generation of a proton gradient across the inner mitochondrial membrane. In contrast, the two ATP synthetic reactions in glycolysis (catalyzed by phosphoglycerate kinase and pyruvate kinase) involve the direct transfer of a phosphate from a sugar–phosphate intermediate to ADP; these reactions are examples of **substrate-level phosphorylation**. A third example of substrate-level phosphorylation is the synthesis of GTP by succinate dehydrogenase in the citric acid cycle (Section L1). The GTP can be used to phosphorylate ADP to form ATP.

Fates of pyruvate

(1) Entry into the citric acid cycle. Glycolysis releases relatively little of the energy present in a glucose molecule (see **Energy yield** below); much more is released by the metabolism of pyruvate via the citric acid cycle and oxidative phosphorylation. Following this route (under aerobic conditions), pyruvate is converted to acetyl CoA by the enzyme **pyruvate dehydrogenase** and the acetyl CoA then enters the citric acid cycle. The pyruvate dehydrogenase reaction is an **oxidative decarboxylation** (Section L1 for details):

Pyruvate dehydrogenase

$$\text{pyruvate} + NAD^+ + CoA \longrightarrow \text{acetyl CoA} + CO_2 + NADH$$

(2) Conversion to fatty acid or ketone bodies. When the cellular energy level is high (i.e. ATP is in excess), the rate of the citric acid cycle (Section L1) decreases and acetyl CoA begins to accumulate. Under these conditions, the acetyl CoA formed from pyruvate can be used for fatty acid synthesis or the synthesis of ketone bodies (Sections K2 and K3).

(3) Conversion to lactate. The NAD^+ used during glycolysis (in the formation of 1,3-bisphosphoglycerate by glyceraldehyde 3-phosphate dehydrogenase; Figure 1) must be regenerated if glycolysis is to continue. Under aerobic conditions, NAD^+ is regenerated by the reoxidation of NADH via the electron transport chain (Section L2). When oxygen is limiting, as in muscle during vigorous contraction, the reoxidation of NADH to NAD^+ by the electron transport chain becomes insufficient to maintain glycolysis. Under these conditions, NAD^+ is regenerated instead by conversion of the pyruvate to lactate by **lactate dehydrogenase**:

Lactate dehydrogenase

$$\text{pyruvate} + NADH + H^+ \rightleftharpoons \text{lactate} + NAD^+$$

When sufficient oxygen becomes available once more, NAD^+ levels rise through operation of the electron transport chain. The lactate dehydrogenase reaction then reverses to regenerate pyruvate that is converted by pyruvate dehydrogenase to acetyl CoA, which can enter the citric acid cycle (see above). Thus the operation of lactate dehydrogenase in mammals is a mechanism for the reoxidation of NADH to NAD^+ hence allowing glycolysis to continue, and ATP to be made, under anaerobic conditions. The process is even more sophisticated in the case of vigorously contracting skeletal muscle. Here the lactate produced is transported in the bloodstream to the liver where it is converted back to glucose and can return once again via the bloodstream to the skeletal muscle to be metabolized to yield energy.

This is the Cori cycle and is described in Section J4. Finally, in some microorganisms, lactate is the normal product from pyruvate.

(4) Conversion to ethanol. In yeast and some other microorganisms under anaerobic conditions, the NAD^+ required for the continuation of glycolysis is regenerated by a process called **alcoholic fermentation**. The pyruvate is converted to acetaldehyde (by **pyruvate decarboxylase**) and then to ethanol (by **alcohol dehydrogenase**), the latter reaction reoxidizing the NADH to NAD^+:

$$\text{pyruvate} + H^+ \xrightarrow{\textit{Pyruvate decarboxylase}} \text{acetaldehyde} + CO_2$$

$$\text{acetaldehyde} + \text{NADH} + H^+ \xrightleftharpoons{\textit{Alcohol dehydrogenase}} \text{ethanol} + NAD^+$$

Energy yield

1. Glycolysis in aerobic conditions

Early in glycolysis, two ATPs are required for the conversion of glucose to glucose 6-phosphate by hexokinase and for the conversion of fructose 6-phosphate to fructose 1,6-bisphosphate by PFK. However, fructose 1,6-bisphosphate then gives rise to two three-carbon units, each of which generates two ATPs in subsequent steps (catalyzed by phosphoglycerate kinase and pyruvate kinase) giving a net yield of two ATPs per original glucose molecule (Figure 1).

The overall reaction is:

$$\text{Glucose} + 2P_i + 2\text{ADP} + 2NAD^+ \rightarrow 2 \text{ pyruvate} + 2\text{ATP} + 2\text{NADH} + 2H^+ + 2H_2O$$

Note that, under aerobic conditions, the two NADH molecules that are synthesized are reoxidized via the electron transport chain generating ATP. Given the cytoplasmic location of these NADH molecules, each is reoxidized via the glycerol 3-phosphate shuttle (Section L2) and produces approximately 1.5 ATPs during oxidative phosphorylation or via the malate–aspartate shuttle (Section L2) and produces approximately 2.5 ATPs during oxidative phosphorylation.

2. Formation of lactate under anaerobic conditions

The overall reaction is:

$$\text{Glucose} + 2P_i + 2\text{ADP} \rightarrow 2 \text{ lactate} + 2\text{ATP} + 2H_2O$$

Note that NAD^+ and NADH are not shown in the above reaction because the NADH generated from NAD^+ in glycolysis is reoxidized back to NAD^+ in the conversion of pyruvate to lactate. So there is no net change in NAD^+ or NADH levels during glycolysis and then conversion of the pyruvate to lactate. In this case, just two ATPs are produced by the conversion of glucose to lactate.

3. Alcoholic fermentation

The overall reaction is:

$$\text{Glucose} + 2P_i + 2\text{ADP} + 2H^+ \rightarrow 2 \text{ ethanol} + 2\text{ATP} + 2CO_2 + 2H_2O$$

NAD$^+$ and NADH are not shown in the above reaction because the NADH generated from NAD$^+$ in glycolysis is reoxidized back to NAD$^+$ in the conversion of pyruvate to ethanol so there is no net change in NAD$^+$ or NADH levels, and the net energy yield is two molecules of ATP per molecule of glucose.

Metabolism of fructose

Fructose is an abundant sugar in the human diet; sucrose (table sugar) is a disaccharide, which when hydrolyzed yields fructose and glucose (Section J1) and fructose is also a major sugar in fruits and honey. There are two pathways for the metabolism of fructose; one occurs in muscle and adipose tissue, the other in liver.

1. In muscle and adipose tissue, fructose can be phosphorylated by hexokinase (which is capable of phosphorylating both glucose and fructose) to form fructose 6-phosphate, which then enters glycolysis.

2. In liver, the cells contain mainly glucokinase instead of hexokinase and this enzyme phosphorylates only glucose. Thus in liver, fructose is metabolized instead by the **fructose 1-phosphate pathway** (Figure 2).

● Fructose is converted to fructose 1-phosphate by **fructokinase**.

● Fructose 1-phosphate is then split into glyceraldehyde and dihydroxyacetone phosphate by **fructose 1-phosphate aldolase**. The dihydroxyacetone feeds into glycolysis at the triose phosphate isomerase step (Figure 1).

● The glyceraldehyde is phosphorylated by **triose kinase** to glyceraldehyde 3-phosphate and so also enters glycolysis.

Figure 2. The fructose 1-phosphate pathway.

Metabolism of galactose

The hydrolysis of the disaccharide lactose (in milk) yields galactose and glucose. Thus galactose is also a major dietary sugar for humans. Galactose and glucose are epimers that differ in their configuration at C-4 (Section J1, Figure 4) so the entry of galactose into glycolysis requires an epimerization reaction. This occurs via a four-step pathway called the **galactose–glucose interconversion pathway** (Figure 3):

1. Galactose is phosphorylated by **galactokinase** to give galactose 1-phosphate.

2. **Galactose 1-phosphate uridylyl transferase** catalyzes the transfer of a uridyl group from uridine diphosphate (UDP)-glucose to galactose 1-phosphate to form UDP-galactose and glucose 1-phosphate.

3. The UDP-galactose is converted back to UDP-glucose by **UDP-galactose 4-epimerase**. Thus, overall, UDP-glucose is not consumed in the reaction pathway.

4. Finally the glucose 1-phosphate is converted to glucose 6-phosphate by **phosphoglucomutase**. The glucose 6-phosphate then enters glycolysis.

Lactose intolerance or **hypolactasia** is a genetic disease that occurs because of an inability to metabolize lactose and so patients suffering from this disease cannot tolerate milk in their diet. Usually hypolactasia is caused by a lack of the enzyme **lactase** that splits lactose into glucose and galactose. Any lactose in consumed milk is metabolized by bacteria in the colon, converting it to lactate and generating methane and hydrogen gas.

Figure 3. The galactose–glucose interconversion pathway.

The excessive gas causes an uncomfortable feeling of fullness as it distends the gut and also leads to undesirable flatulence. The lactate and any undigested lactose also cause water to enter the intestine by osmosis, which leads to diarrhea. The disease is usually treated simply by avoiding food that contains substantial amounts of lactose.

Galactosemia is another genetic disease caused by an inability to convert galactose to glucose. Toxic substances accumulate such as **galactitol**, formed by the reduction of galactose, and lead to dire consequences for the individual. Children who have the disease fail to thrive, may vomit or have diarrhea after drinking milk, and often have an enlarged liver and jaundice. The formation of cataracts in the eyes, mental retardation and an early death from liver damage are also possible. Most cases of galactosemia are due to a deficiency of the galactose 1-phosphate uridylyl transferase enzyme and hence these individuals cannot metabolize galactose. The disease is treated by prescribing a galactose-free (and lactose-free) diet, which prevents cataracts and liver damage but not mental retardation, which may be irreversible. Since such patients have normal levels of UDP-galactose 4-epimerase, they can still synthesize UDP-galactose from UDP-glucose and so can still synthesize, for example, oligosaccharides in glycoproteins that involve Gal residues.

Regulation of glycolysis

Phosphofructokinase

The most important control step of glycolysis is the irreversible reaction catalyzed by phosphofructokinase (PFK). The enzyme is regulated in several ways.

1. ATP/AMP ratio. When ATP levels fall, it can be regenerated from ADP by **adenylate kinase** by the following reaction:

$$2ADP \rightleftharpoons ATP + AMP$$

   So a rise in AMP concentration signals a low energy state.

   PFK is allosterically inhibited by ATP but this inhibition is reversed by AMP. This allows glycolysis to be highly responsive to the energy needs of the cell, speeding up when ATP is in short supply (and the AMP level is high) so that more ATP can be made, and slowing down when sufficient ATP is already available.

2. Citrate. PFK is also inhibited by citrate, the first product of the citric acid cycle proper (Section L1). A high level of citrate signals that there is a plentiful supply of citric acid cycle intermediates already and hence no additional breakdown of glucose via glycolysis is needed.

3. Fructose 2,6-bisphosphate. Fructose 2,6-bisphosphate (F-2,6-BP) is synthesized (Figure 4) from fructose 6-phosphate by an enzyme called **phosphofructokinase 2** (**PFK2**), a different enzyme from PFK. F-2,6-BP is hydrolyzed back to fructose 6-phosphate (Figure 4) by **fructose bisphosphatase 2** (**FBPase2**). Amazingly, both PFK2 and FBPase2 are activities catalyzed by the same polypeptide; i.e. this is a **bi-functional enzyme**. Fructose 6-phosphate stimulates the synthesis of F-2,6-BP and inhibits its hydrolysis (Figure 4). F-2,6-BP in turn strongly activates PFK and hence stimulates glycolysis. The overall effect is that when fructose 6-phosphate levels are high, PFK (and hence glycolysis) is stimulated. PFK2 and FBPase2 are also controlled by covalent modification (Section D5). When blood glucose levels fall, the hormone glucagon is released into the bloodstream and triggers a cAMP cascade (Section J7) that leads to phosphorylation of the PFK2/FBPase2 polypeptide at a single serine

residue. This activates FBPase2 and inhibits PFK2, lowering the level of F-2,6-BP and hence decreasing the rate of glycolysis. The reverse is true as glucose levels rise; the phosphate group is removed from the PFK2/FBPase2 polypeptide by a phosphatase, thus inhibiting FBPase2 and activating PFK2, raising the level of F-2,6-BP and hence increasing the rate of glycolysis.

F-2,6-BP is also important in preventing glycolysis (glucose degradation) and gluconeogenesis (glucose synthesis) operating simultaneously. This is called **reciprocal regulation** and is described in Section J4.

4. H^+ ions. PFK is inhibited by H^+ ions and hence the rate of glycolysis decreases when the pH falls significantly. This prevents the excessive formation of lactate (i.e. lactic acid) under anaerobic conditions (see above) and hence prevents the medical condition known as **acidosis** (a deleterious drop in blood pH).

Hexokinase

Hexokinase, which catalyzes the first irreversible step of glycolysis, is inhibited by glucose 6-phosphate. Thus when PFK is inhibited, fructose 6-phosphate builds up and so does glucose 6-phosphate since these two metabolites are in equilibrium via phosphoglucose isomerase (Figure 1). The hexokinase inhibition then reinforces the inhibition at the PFK step. At first sight, this seems unusual since it is usually the first irreversible step of a pathway (the committed step) that is the main control step. On this basis, it may appear that hexokinase should be the main control enzyme, not PFK. However, glucose 6-phosphate, the product of the hexokinase reaction, can also feed into glycogen synthesis (Section J6) or the pentose phosphate pathway (Section J5). Thus the first irreversible step that is unique to glycolysis is that catalyzed by PFK and hence this is the main control step.

Pyruvate kinase

Pyruvate kinase catalyzes the third irreversible step in glycolysis. It is activated by fructose 1,6-bisphosphate. ATP and the amino acid alanine allosterically inhibit the enzyme so that glycolysis slows when supplies of ATP and biosynthetic precursors (indicated by the levels of Ala) are already sufficiently high. In addition, in a control similar to that for PFK (see above), when the blood glucose concentration is low, glucagon is released and stimulates phosphorylation of the enzyme via a cAMP cascade (Section J7). This covalent modification inhibits the enzyme so that glycolysis slows down in times of low blood glucose levels.

Figure 4. Synthesis and degradation of fructose 2,6-bisphosphate.

J4 Gluconeogenesis

Key Notes

Overview

Gluconeogenesis synthesizes glucose from noncarbohydrate precursors and is important for the maintenance of blood glucose levels during starvation or during vigorous exercise. The brain and erythrocytes depend almost entirely on blood glucose as an energy source. Gluconeogenesis occurs mainly in the liver and to a lesser extent in the kidney. Most enzymes of gluconeogenesis are cytosolic, but pyruvate carboxylase and glucose 6-phosphatase are located in the mitochondrial matrix and bound to the smooth endoplasmic reticulum, respectively.

The pathway

Pyruvate is converted to oxaloacetate (by pyruvate carboxylase). The oxaloacetate is decarboxylated and phosphorylated to phosphoenolpyruvate (PEP) by phosphoenolpyruvate carboxykinase (PEP carboxykinase). PEP is converted to fructose 1,6-bisphosphate by a direct reversal of several reactions in glycolysis. Next, fructose 1,6-bisphosphate is dephosphorylated to fructose 6-phosphate (by fructose 1,6-bisphosphatase) and this is then converted to glucose 6-phosphate (by phosphoglucose isomerase). Finally, glucose 6-phosphate is dephosphorylated (by glucose 6-phosphatase) to yield glucose.

Energy used

The synthesis of one molecule of glucose from two molecules of pyruvate requires six molecules of ATP.

Transport of oxaloacetate

Oxaloacetate, the product of the first step in gluconeogenesis, must leave the mitochondrion and enter the cytosol where the subsequent enzyme steps take place. Since the inner mitochondrial membrane is impermeable to oxaloacetate, it is converted to malate by mitochondrial malate dehydrogenase. This leaves the mitochondrion and is converted back to oxaloacetate by cytosolic malate dehydrogenase.

Pyruvate carboxylase activation

Oxaloacetate, the product of the pyruvate carboxylase reaction, functions both as an important citric acid cycle intermediate in the oxidation of acetyl CoA and as a precursor for gluconeogenesis. The activity of pyruvate carboxylase depends on the presence of acetyl CoA so that more oxaloacetate is made when acetyl CoA levels rise.

Reciprocal regulation of glycolysis and gluconeogenesis

If glycolysis and gluconeogenesis operated simultaneously, the net effect would be the hydrolysis of two ATP and two GTP molecules. This is prevented by reciprocal regulation at the enzyme steps that are distinct in each pathway. AMP activates phosphofructokinase (PFK) (glycolysis) but inhibits fructose 1,6-bisphosphatase (gluconeogenesis). ATP and citrate inhibit PFK but citrate stimulates fructose 1,6-bisphosphatase. Glycolysis and gluconeogenesis are also responsive to starvation conditions via the concentration of fructose 2,6-bisphosphate (F-2,6-BP). During starvation, glucagon is released into the bloodstream and inhibits the synthesis of F-2,6-BP. In the fed state, insulin is released into the bloodstream and causes the accumulation of F-2,6-BP. Since F-2,6-BP activates PFK and inhibits fructose 1,6-bisphosphatase, glycolysis is stimulated and gluconeogenesis is inhibited in the fed animal and vice versa, during starvation. In liver, ATP and alanine inhibit pyruvate kinase (glycolysis) whilst ADP inhibits pyruvate carboxylase and PEP carboxykinase (gluconeogenesis). Thus glycolysis is inhibited in times when ATP and biosynthetic intermediates are in excess whilst gluconeogenesis is inhibited in times when the ATP level is low (and ADP is high). Pyruvate kinase is also stimulated by fructose 1,6-bisphosphate so its rate increases when glycolysis is active. During starvation, glycogen secretion into the bloodstream activates a cAMP cascade that leads to the phosphorylation and inhibition of pyruvate kinase (glycolysis).

The Cori cycle

During vigorous exercise, pyruvate produced by glycolysis in muscle is converted to lactate by lactate dehydrogenase. The lactate diffuses into the bloodstream and is carried to the liver. Here it is converted to glucose by gluconeogenesis. The glucose is released into the bloodstream and becomes available for uptake by muscle (as well as other tissues, including brain). This cycle of reactions is called the Cori cycle.

Related topics

(H5) Protein glycosylation
(J1) Monosaccharides and disaccharides
(J2) Polysaccharides and oligosaccharides
(J3) Glycolysis

(J5) Pentose phosphate pathway
(J6) Glycogen metabolism
(L1) Citric acid cycle

Overview

Gluconeogenesis synthesizes glucose from non-carbohydrate precursors, including lactate and pyruvate, citric acid cycle intermediates, the carbon skeletons of most amino acids and glycerol. This is extremely important since the brain and erythrocytes rely almost exclusively on glucose as their energy source under normal conditions. The liver's store of glycogen is sufficient to supply the brain with glucose for only about half a day during fasting. Therefore gluconeogenesis is especially important in periods of starvation or vigorous exercise. During starvation, the formation of glucose via gluconeogenesis particularly uses amino acids from protein breakdown and glycerol from fat breakdown.

During exercise, the blood glucose levels required for brain and skeletal muscle function are maintained by gluconeogenesis in the liver using lactate produced by the muscle.

The main site of gluconeogenesis is the liver, although it also occurs to a far lesser extent in the kidneys. Very little gluconeogenesis occurs in brain or muscle. Within liver cells, the first enzyme of gluconeogenesis, pyruvate carboxylase, is located in the mitochondrial matrix. The last enzyme, glucose 6-phosphatase, is bound to the smooth endoplasmic reticulum. The other enzymes of the pathway are located in the cytosol.

The pathway

In glycolysis (Section J3), glucose is metabolized to pyruvate. In gluconeogenesis, pyruvate is metabolized to glucose. In principle, gluconeogenesis is therefore a reversal of glycolysis. Indeed, some of the reactions of glycolysis are reversible and so the two pathways have these steps in common. However, three steps in glycolysis are essentially irreversible; those catalyzed by the enzymes hexokinase, phosphofructokinase (PFK) and pyruvate kinase (Section J3). Indeed it is the large negative free-energy change in these reactions that normally drives glycolysis forward towards pyruvate formation. Therefore, in gluconeogenesis, these three steps have to be reversed by using other reactions as shown in Figure 1; gluconeogenesis is *not* a simple reversal of glycolysis.

Precursors for gluconeogenesis

Glycerol can act as a substrate for glucose synthesis by conversion to dihydroxyacetone phosphate, an intermediate in gluconeogenesis (Figure 1). In order for **lactate**, **pyruvate**, **citric acid cycle intermediates** and the carbon skeletons of most **amino acids** to act as precursors for gluconeogenesis, these compounds must first be converted to oxaloacetate. Some of the carbon skeletons of the amino acids give rise to oxaloacetate directly. Others feed into the citric acid cycle as intermediates (Sections L1 and M2) and the cycle then converts these molecules to oxaloacetate. Lactate is converted to pyruvate by the lactate dehydrogenase reaction (Section J3) and some amino acids also give rise to pyruvate (Section M2). Therefore, for these precursors, the first step in the gluconeogenic pathway is the conversion of pyruvate to oxaloacetate.

The steps in gluconeogenesis (Figure 1) are as follows:

1. Pyruvate is converted to oxaloacetate by carboxylation using the enzyme **pyruvate carboxylase** that is located in the mitochondrial matrix.

Pyruvate   Oxaloacetate

This enzyme uses **biotin** as an **activated carrier of CO_2**, the reaction occurring in two stages:

$$\text{E-biotin} + \text{ATP} + \text{HCO}_3^- \longrightarrow \text{E-biotin-CO}_2 + \text{ADP} + \text{P}_i$$

$$\text{E-biotin-CO}_2 + \text{pyruvate} \longrightarrow \text{E-biotin} + \text{oxaloacetate}$$

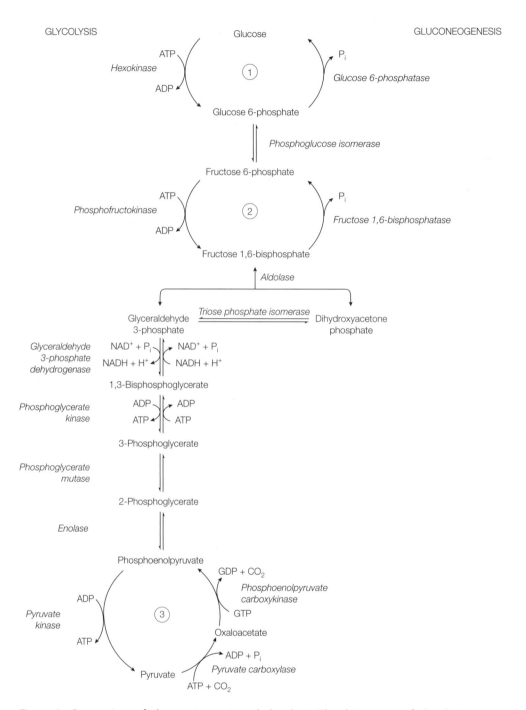

Figure 1. Comparison of gluconeogenesis and glycolysis. The three steps of glycolysis that are irreversible are numbered: (1) hexokinase in glycolysis is reversed by glucose 6-phosphatase in gluconeogenesis; (2) PFK in glycolysis is reversed by fructose 1,6-bisphosphatase in gluconeogenesis; (3) pyruvate kinase in glycolysis is reversed by two sequential reactions in gluconeogenesis catalyzed by pyruvate carboxylase and PEP carboxykinase.

2. The oxaloacetate is now acted on by **phosphoenolpyruvate carboxykinase (PEP carboxykinase)**, which simultaneously decarboxylates and phosphorylates it to form PEP, releasing CO_2 and using GTP in the process:

Oxaloacetate   **Phosphoenolpyruvate**

So reversal of the glycolytic step from PEP to pyruvate requires two reactions in gluconeogenesis, pyruvate to oxaloacetate by pyruvate carboxylase and oxaloacetate to PEP by PEP carboxykinase. Given that the conversion of PEP to pyruvate in glycolysis synthesizes ATP, it is not surprising that the overall reversal of this step needs the input of a substantial amount of energy, one ATP for the pyruvate carboxylase step and one GTP for the PEP carboxykinase step.

3. PEP is converted to fructose 1,6-bisphosphate in a series of steps that are a direct reversal of those in glycolysis (Section J3), using the enzymes enolase, phosphoglycerate mutase, phosphoglycerate kinase, glyceraldehyde 3-phosphate dehydrogenase, triose phosphate isomerase and aldolase (Figure 1). This sequence of reactions uses one ATP and one NADH for each PEP molecule metabolized.

4. Fructose 1,6-bisphosphate is dephosphorylated to form fructose 6-phosphate by the enzyme **fructose 1,6-bisphosphatase**, in the reaction:

$$\text{fructose 1,6-bisphosphate} + H_2O \longrightarrow \text{fructose 6-phosphate} + Pi$$

5. Fructose 6-phosphate is converted to glucose 6-phosphate by the glycolytic enzyme phosphoglucose isomerase.

6. Glucose 6-phosphate is converted to glucose by the enzyme **glucose 6-phosphatase**. This enzyme is bound to the smooth endoplasmic reticulum and catalyzes the reaction:

$$\text{glucose 6-phosphate} + H_2O \longrightarrow \text{glucose} + P_i$$

Energy used

As would be expected, the synthesis of glucose by gluconeogenesis needs the input of energy. Two pyruvate molecules are required to synthesize one molecule of glucose. Energy is required at the following steps:

pyruvate carboxylase	1 ATP ($\times$2) = 2 ATP
PEP carboxykinase	1 GTP ($\times$2) = 2 ATP
phosphoglycerate kinase	1 ATP ($\times$2) = 2 ATP
	Total = 6 ATP

This compares with only two ATPs as the net ATP yield from glycolysis. Thus an extra four ATPs per glucose are required to reverse glycolysis.

In fact, the glyceraldehyde 3-phosphate dehydrogenase reaction also consumes NADH, equivalent to two molecules of NADH for each molecule of glucose synthesized. Since each cytosolic NADH would normally be used to generate approximately two ATP molecules via the glycerol 3-phosphate shuttle and oxidative phosphorylation (Section L2), this is equivalent to the input of another four ATPs per glucose synthesized.

Transport of oxaloacetate

Pyruvate carboxylase is a mitochondrial matrix enzyme whereas the other enzymes of gluconeogenesis are located outside the mitochondrion. Thus oxaloacetate, produced by pyruvate carboxylase, needs to exit the mitochondrion. However, the inner mitochondrial membrane is not permeable to this compound. Thus oxaloacetate is converted to malate inside the mitochondrion by **mitochondrial malate dehydrogenase**, the malate is transported through the mitochondrial membrane by a special transport protein and then the malate is converted back to oxaloacetate in the cytoplasm by a **cytosolic malate dehydrogenase** (Figure 2).

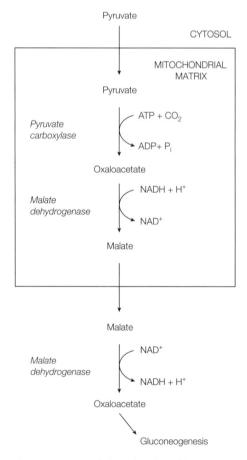

Figure 2. Transport of oxaloacetate out of the mitochondrion.

Pyruvate carboxylase activation

Oxaloacetate has two main roles. It is an intermediate that is consumed in gluconeogenesis and it is also a key intermediate in the citric acid cycle where it fuses with acetyl CoA to form citrate, eventually being regenerated by the cycle. So pyruvate carboxylase generates oxaloacetate for gluconeogenesis but also must maintain oxaloacetate levels for citric acid cycle function. For the latter reason, the activity of pyruvate carboxylase depends absolutely on the presence of acetyl CoA; the biotin prosthetic group of the enzyme cannot be carboxylated unless acetyl CoA is bound to the enzyme. This allosteric activation by acetyl CoA ensures that more oxaloacetate is made when excess acetyl CoA is present. In this role of maintaining the level of citric acid cycle intermediates, the pyruvate carboxylase reaction is said to be **anapleurotic**, that is 'filling up'.

Reciprocal regulation of glycolysis and gluconeogenesis

The conversion of fructose 6-phosphate to fructose 1,6-bisphosphate by PFK and the reverse reaction catalyzed by fructose 1,6-bisphosphatase is an example of a **substrate cycle** (Figure 3). If both reactions occur simultaneously, the net reaction is the hydrolysis of ATP. For this reason, such reactions have also been called **futile cycles**.

Figure 3. The fructose 6-phosphate–fructose 1,6-bisphosphate substrate cycle

Looking at the complete pathways, glycolysis generates two ATPs net per glucose whereas gluconeogenesis uses four ATPs and two GTPs per glucose. Therefore, if both glycolysis and gluconeogenesis were allowed to operate simultaneously, converting glucose to pyruvate and back again, the only net result would be the utilization of two ATPs and two GTPs. This is prevented by tight coordinate regulation of glycolysis and gluconeogenesis. Since many of the steps of the two pathways are common, the steps that are distinct in each pathway are the sites of this regulation, in particular the interconversions between fructose 6-phosphate and fructose 1,6-bisphosphate and between PEP and pyruvate. The situation is summarized in Figure 4 and described in detail below.

Regulation of PFK and fructose 1,6-bisphosphatase

When the level of AMP is high, this indicates the need for more ATP synthesis. AMP stimulates PFK, increasing the rate of glycolysis, and inhibits fructose 1,6-bisphosphatase, turning off gluconeogenesis. Conversely, when ATP and citrate levels are high, this signals that no more ATP need be made. ATP and citrate inhibit PFK, decreasing the rate of glycolysis, and citrate stimulates fructose 1,6-bisphosphatase, increasing the rate of gluconeogenesis.

Glycolysis and gluconeogenesis are made responsive to starvation by the level of the regulatory molecule **fructose 2,6-bisphosphate (F-2,6-BP)**. F-2,6-BP is synthesized from fructose 6-phosphate and hydrolyzed back to fructose 6-phosphate by a single

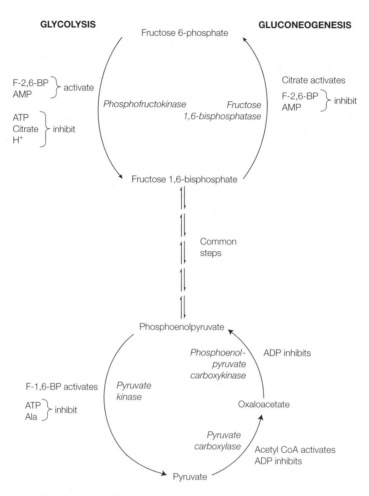

Figure 4. Reciprocal regulation of glycolysis and gluconeogenesis.

polypeptide with two enzymatic activities (PFK2 and FBPase2; Section J3). The level of F-2,6-BP is under hormonal control. During starvation, when the level of blood glucose is low, the hormone glucagon is released into the bloodstream and triggers a cAMP cascade (Section J7), eventually causing phosphorylation of the PFK2/FBPase2 polypeptide. This activates FBPase2 and inhibits PFK2, lowering the level of F-2,6-BP (Section J3). In the fed state, when blood glucose is at a high level, the hormone insulin is released and has the opposite effect, causing an elevation in the level of F-2,6-BP. Since F-2,6-BP strongly stimulates PFK and inhibits fructose 1,6-bisphosphatase (Figure 4), glycolysis is stimulated and gluconeogenesis is inhibited in the fed animal. Conversely, during starvation, the low level of F-2,6-BP allows gluconeogenesis to predominate.

Regulation of pyruvate kinase, pyruvate carboxylase and PEP carboxykinase

- In liver, pyruvate kinase is inhibited by high levels of ATP and alanine so that glycolysis is inhibited when ATP and biosynthetic intermediates are already plentiful (Section J3). Acetyl CoA is also abundant under these conditions and activates pyruvate carboxylase, favoring gluconeogenesis. Conversely, when the energy status of the cell is low, the ADP concentration is high and this inhibits both pyruvate carboxylase and PEP carboxykinase, switching off gluconeogenesis. At this time, the ATP level will be low so pyruvate kinase is not inhibited and glycolysis will operate.

- Pyruvate kinase is also stimulated by fructose 1,6-bisphosphate (Section J3; feedforward activation) so that its activity rises when needed, as glycolysis speeds up.

- During starvation, the priority is to conserve blood glucose for the brain and muscle so under these conditions pyruvate kinase in the liver is switched off. This occurs because the hormone glucagon is secreted into the bloodstream and activates a cAMP cascade (Section J7) that leads to the phosphorylation and inhibition of this enzyme.

The Cori cycle

Under the limiting oxygen conditions experienced during vigorous exercise, the formation of NADH by glycolysis exceeds the ability of the respiratory chain to oxidize it back to NAD^+. The pyruvate produced by glycolysis in muscle is then converted to lactate by lactate dehydrogenase, a reaction that regenerates NAD^+ and so allows glycolysis to continue to produce ATP (Section J3).

However, lactate is a metabolic 'dead-end' in that it cannot be metabolized further until it is converted back to pyruvate. Lactate diffuses out of the muscle and is carried in the bloodstream to the liver. Here it diffuses into liver cells and is converted back to pyruvate by lactate dehydrogenase. The pyruvate is then converted to glucose by gluconeogenesis and the glucose is released back into the bloodstream ready to be taken up by skeletal muscle (and brain). This cycle of reactions (Figure 5) is called **the Cori cycle**.

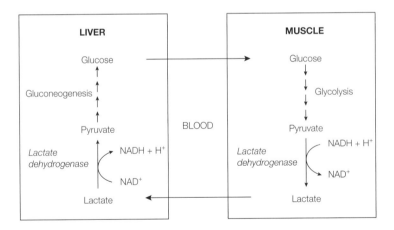

Figure 5. The Cori cycle.

J5 Pentose phosphate pathway

Key Notes

Overview

The two major products of the pathway are nicotinamide adenine dinucleotide phosphate (reduced form; NADPH) and ribose 5-phosphate. Ribose 5-phosphate and its derivatives are components of important cellular molecules such as RNA, DNA, NAD^+, FAD, ATP and CoA. NADPH is required for many biosynthetic pathways and particularly for synthesis of fatty acids and steroids. Hence the pathway is very active in tissues such as adipose tissue, mammary gland and the adrenal cortex.

Main reactions of the pathway

The reactions of the pathway can be grouped into three stages. In the first stage, oxidative reactions convert glucose 6-phosphate into ribulose 5-phosphate, generating two NADPH molecules. In the second stage, ribulose 5-phosphate is converted to ribose 5-phosphate by isomerization. The third stage of reactions, catalyzed by transketolase and transaldolase, converts ribose 5-phosphate into fructose 6-phosphate and glyceraldehyde 3-phosphate and hence links the pentose phosphate pathway with glycolysis.

Control of the pathway

The transketolase and transaldolase reactions are reversible and so allow either the conversion of ribose 5-phosphate into glycolytic intermediates when it is not needed for other cellular reactions, or the generation of ribose 5-phosphate from glycolytic intermediates when more is required. The rate of the pentose phosphate pathway is controlled by $NADP^+$ regulation of the first step, catalyzed by glucose 6-phosphate dehydrogenase.

Related topics

(J1) Monosaccharides and disaccharides
(J3) Glycolysis

(K3) Fatty acid synthesis

Overview

Reducing power is available in a cell both as NADH and NADPH but these have quite distinct roles. NADH is oxidized by the respiratory chain to generate ATP via oxidative phosphorylation (Section L2). NADPH is used for biosynthetic reactions that require reducing power. Despite their similar structures (Section D1), NADH and NADPH are not metabolically interchangeable and so the cell must carry out a set of reactions that specifically create NADPH. This set of reactions is the pentose phosphate pathway (also known as the **hexose monophosphate shunt** or the **phosphogluconate pathway**). It takes place in the cytosol and is particularly important in tissues such as adipose tissue, mammary gland and the adrenal cortex that synthesize fatty acids and steroids from acetyl CoA (Section K3). The activity of the pathway is very low in skeletal muscle, for example, which does not synthesize fatty acids or steroids.

The core set of reactions of the pathway oxidize glucose 6-phosphate to ribose 5-phosphate and generate NADPH. Thus, as well as generating NADPH, the pathway has a second important role in converting hexoses into pentoses, in particular ribose 5-phosphate. Ribose 5-phosphate or derivatives of it are required for the synthesis of RNA, DNA, NAD$^+$, FAD, ATP, CoA and other important molecules. Thus the two main products of the pathway are NADPH and ribose 5-phosphate.

Main reactions of the pathway

The core reactions of the pathway can be summarized as:

$$\text{glucose 6-phosphate} + 2\text{NADP}^+ + \text{H}_2\text{O} \longrightarrow \text{ribose 5-phosphate} + 2\,\text{NADPH} + 2\text{H}^+ + \text{CO}_2$$

The pathway has three stages:

Stage 1 – Oxidative reactions that convert glucose 6-phosphate into ribulose 5-phosphate, generating two NADPH molecules

Glucose 6-phosphate is oxidized by **glucose 6-phosphate dehydrogenase** to 6-phosphoglucono-δ-lactone (producing NADPH) and this is then hydrolyzed by **lactonase** to 6-phosphogluconate. The 6-phosphogluconate is subsequently converted by **6-phosphogluconate dehydrogenase** to ribulose 5-phosphate. This is an **oxidative decarboxylation** (i.e. the 6-phosphogluconate is oxidized and a carbon is removed as CO$_2$). These reactions are shown below:

Glucose 6-phosphate   6-Phosphoglucono-δ-lactone   6-Phosphogluconate   Ribulose 5-phosphate

Stage 2 – Isomerization of ribulose 5-phosphate to ribose 5-phosphate

The ribulose 5-phosphate is now converted to ribose 5-phosphate by isomerization, a reaction catalyzed by **phosphopentose isomerase**:

Ribulose 5-phosphate   Ribose 5-phosphate

Stage 3 – Linkage of the pentose phosphate pathway to glycolysis via transketolase and transaldolase

If at any time only a little ribose 5-phosphate is required for nucleic acid synthesis and other synthetic reactions, it will tend to accumulate and is then converted to fructose 6-phosphate and glyceraldehyde 3-phosphate by the enzymes **transketolase** and **transaldolase**. These two products are intermediates of glycolysis. Therefore, these reactions provide a link between the pentose phosphate pathway and glycolysis. The outline reactions are shown below.

Xylulose 5-phosphate + Ribose 5-phosphate
(C5)   (C5)

Transketolase

Glyceraldehyde 3-phosphate + Sedoheptulose 7-phosphate
(C3)   (C7)

Transaldolase

Fructose 6-phosphate + Erythrose 4-phosphate
(C6)   (C4)

Xylulose 5-phosphate ———   *Transketolase*
(C5)

Glyceraldehyde 3-phosphate + Fructose 6-phosphate
(C3)   (C6)

Details of these reactions, showing the structures of the molecules involved, are given in Figure 1. These reactions require xylulose 5-phosphate as well as ribose 5-phosphate. Xylulose 5-phosphate is an epimer (Section J1) of ribulose 5-phosphate and is made by **phosphopentose epimerase**:

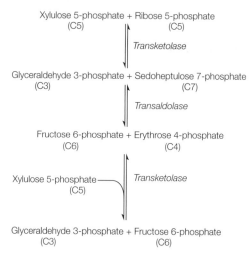

Ribulose 5-phosphate   Xylulose 5-phosphate

Overall the reactions in this stage can be summarized as:

2 xylulose 5-phosphate + ribose 5-phosphate ⇌ 2 fructose 6-phosphate + glyceraldehyde 3-phosphate

Control of the pathway

The transketolase and transaldolase reactions are reversible, so the final products of the pentose phosphate pathway can change depending on the metabolic needs of the cell. Therefore when the cell needs NADPH but not ribose 5-phosphate, the latter is converted to glycolytic intermediates and enters glycolysis. At the other extreme, when the need for ribose 5-phosphate exceeds that for NADPH, fructose 6-phosphate and glyceraldehyde 3-phosphate can be taken from glycolysis and converted into ribose 5-phosphate by reversal of the transketolase and transaldolase reactions.

Figure 1. Details of the transaldolase and transketolase reactions.

The first reaction of the pathway, the oxidation of glucose 6-phosphate by glucose 6-phosphate dehydrogenase, is rate limiting and irreversible. The enzyme is regulated by $NADP^+$. As the cell uses NADPH, the concentration of $NADP^+$ rises, stimulating glucose 6-phosphate dehydrogenase and so increasing the rate of the pathway and NADPH regeneration.

J6 Glycogen metabolism

Key Notes

Roles of glycogen metabolism

Glycogen is stored mainly by the liver and skeletal muscle as an energy reserve. The role of stored glycogen in muscle is to provide a source of energy upon prolonged muscle contraction. In contrast, glycogen stored in the liver is used to maintain blood glucose levels.

Glycogen degradation

Glycogen degradation is carried out by glycogen phosphorylase and glycogen-debranching enzyme. Phosphorylase removes glucose units sequentially from the nonreducing ends of a glycogen molecule, producing glucose 1-phosphate as the product. It breaks only α1–4 glycosidic bonds and cannot break the α1–6 branchpoints. The glucose 1-phosphate is converted to glucose 6-phosphate by phosphoglucomutase. In liver, this is further converted to glucose by glucose 6-phosphatase and the glucose enters the bloodstream. Muscle lacks glucose 6-phosphatase. Rather, here the glucose 6-phosphate enters glycolysis and is oxidized to yield energy for muscle contraction.

Glycogen synthesis

UDP-glucose is synthesized by UDP-glucose pyrophosphorylase from UTP and glucose 1-phosphate. Glycogen synthase then uses the UDP-glucose as a substrate to synthesize glycogen, adding one residue at a time to the nonreducing end of the glycogen molecule, forming α1–4 bonds between neighboring glucosyl residues. The enzyme can only extend chains and therefore requires a primer, called glycogenin, in order to begin synthesis. Glycogenin is a protein with eight glucose units joined by α1–4 bonds. The branches in glycogen are created by branching enzyme that breaks an α1–4 bond in the glycogen chain and moves about seven residues to an internal location, joining them to the main chain by an α1–6 bond.

Related topics

(J1) Monosaccharides and disaccharides
(J3) Glycolysis

(J4) Gluconeogenesis
(J7) Control of glycogen metabolism

Roles of glycogen metabolism

Glycogen is a large polymer of glucose residues linked by α1–4 glycosidic bonds with branches every 10 residues or so via α1–6 glycosidic bonds (Section J2 for structure). Glycogen provides an important energy reserve for the body. The two main storage sites are the liver and skeletal muscle where the glycogen is stored as granules in the cytosol. The granules contain not only glycogen but also the enzymes and regulatory proteins that are required for glycogen degradation and synthesis. Glycogen metabolism is important because it enables the blood glucose level to be maintained between meals (via glycogen stores in the liver) and also provides an energy reserve for muscular activity. The maintenance of blood glucose is essential in order to supply tissues with an easily metabolizable energy source, particularly the brain, which uses only glucose except after a long starvation period.

Glycogen degradation

Glycogen degradation requires two enzymes; **glycogen phosphorylase** and **glycogen-debranching enzyme**.

Glycogen phosphorylase (often called simply **phosphorylase**) degrades glycogen by breaking α1–4 glycosidic bonds to release glucose units one at a time from the nonreducing end of the glycogen molecule (the end with a free 4′-OH group; Section J2) as glucose 1-phosphate. The other substrate required is inorganic phosphate (P_i). The reaction is an example of **phosphorolysis**, which is breakage of a covalent bond by the addition of a phosphate group. The (reversible) reaction is as follows:

$$\text{glycogen} + P_i \rightleftharpoons \text{glycogen} + \text{glucose 1-phosphate}$$
$$(n\text{ residues}) \qquad\qquad (n{-}1\text{ residues})$$

However, glycogen phosphorylase can remove only those glucose residues that are more than five residues from a branchpoint. Glycogen-debranching enzyme removes the α1–6 branches and so allows phosphorylase to continue degrading the glycogen molecule. The glucose 1-phosphate produced is converted to glucose 6-phosphate by the enzyme **phosphoglucomutase**:

$$\text{glucose 1-phosphate} \rightleftharpoons \text{glucose 6-phosphate}$$

The fate of the glucose 6-phosphate depends on the tissue. Liver contains the enzyme glucose 6-phosphatase, which converts the glucose 6-phosphate to glucose, which then diffuses out into the bloodstream and so maintains the blood glucose concentration:

$$\text{glucose 6-phosphate} + H_2O \rightarrow \text{glucose} + P_i$$

During glycogen degradation in muscle, the main aim is to produce energy quickly and so the glucose 6-phosphate is metabolized immediately via glycolysis. This tissue does not contain glucose 6-phosphatase.

Glycogen synthesis

Three enzymes are needed to synthesize glycogen:

1. **UDP-glucose pyrophosphorylase** catalyzes the synthesis of UDP-glucose (Figure 1) from UTP and glucose 1-phosphate:

    $$\text{UTP} + \text{glucose 1-phosphate} \longrightarrow \text{UDP-glucose} + PP_i$$

    The pyrophosphate (PP_i) is immediately hydrolyzed by inorganic pyrophosphatase, releasing energy. Thus the overall reaction is very exergonic and essentially irreversible.

2. **Glycogen synthase** now transfers the glucosyl residue from UDP-glucose to the C-4 OH group at the nonreducing end of a glycogen molecule, forming an α1–4 glycosidic bond (Figure 2). Interestingly, glycogen synthase can only extend an existing chain. Thus it needs a primer; this is a protein called **glycogenin**. Glycogenin contains eight

Figure 1. The UDP-glucose pyrophosphorylase reaction.

glucosyl units linked via α1–4 linkages, which are added to the protein by itself (i.e. autocatalysis). It is this molecule that glycogen synthase then extends. Each glycogen granule contains only a single glycogenin molecule at its core. The fact that glycogen synthase is fully active only when in contact with glycogenin limits the size of the glycogen granule.

3.   **Branching enzyme [amylo-(1–4→1–6) transglycosylase]** is a different enzyme from glycogen-debranching enzyme. After a number of glucose units have been joined as a straight chain with α1–4 linkages, branching enzyme breaks one of the α1–4 bonds and transfers a block of residues (usually about seven) to a more interior site in the glycogen molecule, reattaching these by creating an α1–6 bond. The branches are important because the enzymes that synthesize and degrade glycogen (glycogen synthase and glycogen phosphorylase, respectively) work only at the ends of the glycogen molecule. Thus the existence of many termini allows a far more rapid rate of synthesis and degradation than would be possible with a nonbranched polymer.

Figure 2. Synthesis of glycogen by glycogen synthase.

J7 Control of glycogen metabolism

Key Notes

Overview

Glycogen degradation and glycogen synthesis are controlled both by allosteric regulation and by hormonal control.

Allosteric control and covalent modification

Phosphorylase exists in a phosphorylated active *a* form and a dephosphorylated normally inactive *b* form. The two forms are interconverted by phosphorylase kinase and protein phosphatase I. In muscle, phosphorylase *b* is activated by the high concentrations of AMP generated by strenuous exercise and thus degrades glycogen, but the AMP stimulation is opposed by high concentrations of ATP and glucose 6-phosphate and so the enzyme is inactive in resting muscle. In liver, phosphorylase *b* is not responsive to AMP but phosphorylase *a* is deactivated by glucose so that glycogen degradation to produce glucose occurs only when glucose levels are low. Glycogen synthase exists as a phosphorylated normally inactive *b* form and a dephosphorylated active *a* form. A high concentration of glucose 6-phosphate can activate synthase *b* in resting muscle, stimulating glycogen synthesis, but the enzyme is inactive in contracting muscle where the glucose 6-phosphate concentration is low.

Hormonal control by epinephrine and glucagon

Epinephrine (adrenaline) stimulates glycogen degradation in skeletal muscle. Epinephrine and glucagon stimulate glycogen degradation in liver. The hormone binds to a plasma membrane receptor and activates adenylate cyclase via a G protein. Adenylate cyclase synthesizes cAMP from ATP, which in turn activates protein kinase A. Protein kinase A phosphorylates phosphorylase kinase, which activates it. The phosphorylase kinase then converts inactive phosphorylase *b* to active phosphorylase *a* by phosphorylation. The same active protein kinase A inactivates glycogen synthase by phosphorylation, converting glycogen synthase *a* to glycogen synthase *b*. When hormone levels fall, stimulation of glycogen degradation is turned off by degradation of cAMP to 5′ AMP by phosphodiesterase and by dephosphorylation of the phosphorylated forms of phosphorylase and synthase by protein phosphatase I.

Hormonal control by insulin

Insulin is released into the bloodstream when the blood glucose concentration is high and it stimulates glycogen synthesis. It binds to and activates a receptor protein kinase in the plasma membrane of target cells. This leads to activation of an insulin-responsive protein kinase that

	then activates protein phosphatase I by phosphorylation. Activated protein phosphatase I ensures that phosphorylase and glycogen synthase are dephosphorylated, thus inhibiting glycogen degradation and activating glycogen synthesis.
Calcium control of glycogen metabolism	During muscle contraction, Ca^{2+} ions released from the sarcoplasmic reticulum partially activate dephosphorylated phosphorylase kinase and this in turn activates phosphorylase, stimulating glycogen degradation.
Related topics	(A2) Eukaryotic cells (J3) Glycolysis (E5) Signal transduction (J4) Gluconeogenesis (J1) Monosaccharides and (J6) Glycogen metabolism disaccharides

Overview

If glycogen synthesis and glycogen degradation were allowed to occur simultaneously (Figure 1), the net effect would be hydrolysis of UTP via a substrate cycle (also called a futile cycle; Section J4). However, both pathways are tightly controlled and avoid unnecessary UTP hydrolysis. This control is carried out via allosteric regulation and covalent modification of both the glycogen synthase and phosphorylase. In addition, the covalent modification is under close hormonal control.

Figure 1. Simultaneous operation of glycogen synthesis and degradation would result in net hydrolysis of UTP.

Allosteric control and covalent modification

Phosphorylase exists in two interchangeable forms; active **phosphorylase *a*** and a *normally* inactive **phosphorylase *b***. Phosphorylase *b* is a dimer and is converted into phosphorylase *a* by phosphorylation of a single serine residue on each subunit by the enzyme **phosphorylase kinase**. The process can be reversed and phosphorylase inactivated by removal of the phosphate group by **protein phosphatase I** (Figure 2a).

In skeletal muscle, high concentrations of AMP can activate phosphorylase *b* (by acting at an **allosteric site**) but this is opposed by the concentrations of ATP and glucose 6-phosphate found in resting muscle so that in this condition phosphorylase *b* is indeed inactive. Since most of the phosphorylase in resting muscle is phosphorylase *b*, significant glycogen degradation does not occur under these conditions. However, during exercise,

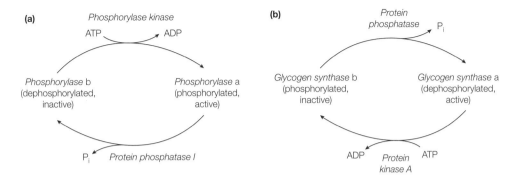

Figure 2. Regulation of (a) glycogen phosphorylase activity and (b) glycogen synthase activity by phosphorylation (covalent modification).

the concentrations of ATP and glucose 6-phosphate fall and the concentration of AMP rises. Thus phosphorylase *b* becomes activated and this leads to the rapid degradation of glycogen to yield energy as required. Phosphorylase *a* is unaffected by ATP, AMP or glucose 6-phosphate and so remains active under all conditions.

In liver, phosphorylase *b* is not activated by AMP and is therefore always inactive. Unlike muscle, therefore, glycogen degradation in liver is not responsive to the energy status of the cell. Rather, phosphorylase *a* is deactivated by glucose. This fits with the different role of glycogen storage in liver, namely to maintain blood levels of glucose. Thus as glucose levels rise, glycogen degradation by liver phosphorylase *a* is shut off and degradation starts again only as the glucose level falls.

Glycogen synthase is also regulated by covalent modification and allosteric interactions. The enzyme exists as an active glycogen synthase *a* and a *normally* inactive glycogen synthase *b*. However, in contrast to phosphorylase, it is the active form of glycogen synthase (synthase *a*) that is dephosphorylated whereas the inactive synthase *b* form is the phosphorylated form (Figure 2b).

A high concentration of glucose 6-phosphate can activate glycogen synthase *b*. During muscle contraction, glucose 6-phosphate levels are low and therefore glycogen synthase *b* is inhibited. This is at the time when phosphorylase *b* is most active (see above). Thus glycogen degradation occurs and glycogen synthesis is inhibited, preventing the operation of a futile cycle. When the muscle returns to the resting state and ATP and glucose 6-phosphate levels rise, phosphorylase *b* is inhibited (see above), turning off glycogen degradation, whereas glycogen synthase is activated to rebuild glycogen reserves. The synthase *a* form is active irrespective of the concentration of glucose 6-phosphate.

Hormonal control by epinephrine and glucagon

Glycogen metabolism is tightly controlled by hormones. When blood glucose levels fall, glucagon is secreted by the α cells of the pancreas and acts on the liver to stimulate glycogen breakdown to glucose, which is then released into the bloodstream to boost blood glucose levels again. Muscular contraction or nervous stimulation (the 'fight or flight' response) causes the release of epinephrine (adrenaline) from the adrenal medulla and this acts on muscle to increase glycogen breakdown ready to supply the energy needs of the cells.

Consider first the activation of glycogen degradation by epinephrine in the liver. The hormone binds to a receptor, called the β-**adrenergic receptor**, in the plasma membrane of the target cell (Figure 3). Binding of the hormone to the receptor causes a conformational change in the protein, which in turn activates an enzyme called **adenylate cyclase**. The receptor does not activate adenylate cyclase directly but rather by activating a **G protein** as an intermediary in the signaling process (see Section E5 for details). Activated adenylate cyclase converts ATP to 3′, 5′-cyclic AMP (cAMP). The cAMP binds to **protein kinase A (PKA)**, also known as **cAMP-dependent protein kinase**. This enzyme consists of two regulatory subunits (R) and two catalytic subunits (C), making a complex, R_2C_2, that is normally inactive (Figure 3). The binding of two molecules of cAMP to each of the regulatory subunits leads to dissociation of the complex into an R_2 complex and two free C subunits that are now catalytically active. The active protein kinase A phosphorylates phosphorylase kinase, which can exist as an inactive dephosphorylated form and an active phosphorylated form. Thus phosphorylase kinase is now also activated and in turn phosphorylates phosphorylase *b*, converting it to the active phosphorylase *a* that now carries out a rapid degradation of glycogen.

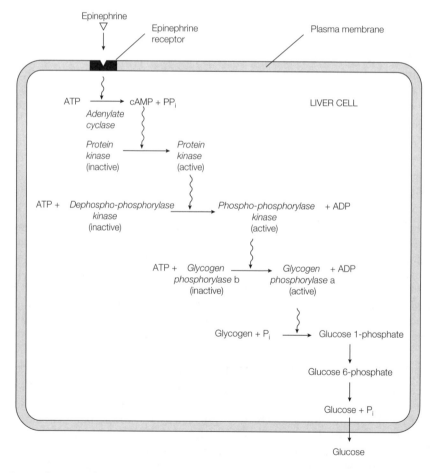

Figure 3. Mechanism of action of epinephrine.

This set of reactions is called a **cascade** and is organized to amplify the original signal of a small number of hormone molecules. For example, each bound hormone molecule causes the production of many cAMP molecules inside the cell; the activated protein kinase A in turn activates many molecules of phosphorylase kinase; each active phosphorylase kinase activates many molecules of phosphorylase. Thus a small hormonal signal can cause a major shift in cell metabolism.

To prevent the operation of a futile cycle, glycogen synthesis is switched off during epinephrine or glucagon stimulation of glycogen breakdown. This is achieved by the activated protein kinase A that, as well as phosphorylating phosphorylase kinase, also phosphorylates glycogen synthase *a*, converting it to the inactive synthase *b* form (Figure 4). Thus protein kinase A activates glycogen degradation and inhibits glycogen synthesis.

When epinephrine and glucagon levels in the bloodstream fall again, the hormone dissociates from the receptor, no more cAMP is made and existing cAMP is converted to 5′ AMP (i.e. 'normal' AMP not the cyclic form) by cAMP phosphodiesterase, an enzyme that is constantly active in the cell. This decline in the cAMP level shuts off the activation cascade. The enzymes that had been phosphorylated are now dephosphorylated by protein phosphatase I, restoring them to their original condition.

Hormonal control by insulin

Insulin is released into the bloodstream by the β cells of the pancreas when blood glucose levels are high after feeding, and stimulates glycogen synthesis to store excess glucose as glycogen. This control is also achieved via phosphorylation events. Insulin binds to its receptor in the plasma membrane and activates it. This receptor has tyrosine kinase activity (i.e. it will phosphorylate selected tyrosine residues on target proteins; Section E5). Its activation leads to the activation of an insulin-responsive protein kinase that then phosphorylates protein phosphatase I, thus activating it. This enzyme now ensures that glycogen synthase is dephosphorylated (and hence active) and that phosphorylase kinase is also dephosphorylated (and hence inactive). The net effect is to stimulate glycogen synthesis.

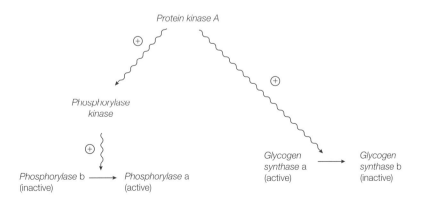

Figure 4. Dual control of glycogen degradation and glycogen synthesis by protein kinase A.

Calcium control of glycogen metabolism

As we have seen, during epinephrine or glucagon hormonal control, dephosphorylated phosphorylase kinase is activated by being phosphorylated by protein kinase. This then activates phosphorylase and stimulates glycogen degradation. However, there is also another way to activate dephosphorylated phosphorylase kinase, at least partially, and that is by a high concentration of Ca^{2+} ions. This is important in muscle contraction, which is triggered when Ca^{2+} is released from the internal store in the sarcoplasmic reticulum (Section B5). Thus, as well as allosteric control and hormonal control during muscle contraction, both of which stimulate glycogen degradation, there is also calcium control.

K1 Fatty acid structure and function

Key Notes

Structure and properties
Fatty acids consist of a long hydrocarbon chain with a terminal carboxylic acid group. Most fatty acids have an even number of carbon atoms in an unbranched chain. Saturated fatty acids have no double bonds between the carbon atoms, whereas mono- and polyunsaturated fatty acids have one or more double bonds. The properties of a fatty acid depend on the chain length and the number of double bonds.

Nomenclature
Fatty acids are named according to the number of carbon atoms in the chain and the number and position of any double bonds. Some of the more common fatty acids are palmitate (C16:0), stearate (C18:0), oleate (C18:1), linoleate (C18:2), linolenate (C18:3) and arachidonate (C20:4). The double bonds in a fatty acid are usually in the *cis* configuration.

Roles
Fatty acids have four major biological roles:

1. they act as energy stores (triacylglycerols) and fuel molecules;
2. they are components of membranes (glycerophospholipids and sphingolipids);
3. several proteins are covalently modified by fatty acids;
4. fatty acid derivatives serve as hormones and intracellular second messengers.

Prostaglandins
Prostaglandins and the other eicosanoids (prostacyclins, thromboxanes and leukotrienes) are derived from arachidonate. These compounds all act as local hormones. Aspirin reduces inflammation by inhibiting prostaglandin H2 synthase, the enzyme that catalyzes the first step in prostaglandin synthesis.

Related topics

(E1) Membrane lipids	(K2) Fatty acid breakdown
(E2) Membrane structure	(K3) Fatty acid synthesis
(E5) Signal transduction	(K4) Triacylglycerols

Structure and properties

A fatty acid consists of a **hydrocarbon chain** and a **terminal carboxylic acid group** (Figure 1). Most fatty acids found in biology have an **even number of carbon atoms** arranged in an **unbranched chain**. Chain length usually ranges from 14 to 24 carbon atoms, with the most common fatty acids containing 16 or 18 carbon atoms. A **saturated**

fatty acid has all of the carbon atoms in its chain saturated with hydrogen atoms (Figure 1a). This gives the general formula $CH_3(CH_2)_nCOOH$, where n is an even number. **Monounsaturated fatty acids** have one double bond in their structure (Figures 1b and 1c), while **polyunsaturated fatty acids** have two or more double bonds (Figure 1d). The double bonds in polyunsaturated fatty acids are separated by at least one methylene group.

The properties of fatty acids depend on their **chain length** and the **number of double bonds**. Shorter chain length fatty acids have lower melting temperatures than those with longer chains. Unsaturated fatty acids have lower melting temperatures than saturated fatty acids of the same chain length, whilst the corresponding polyunsaturated fatty acids have even lower melting temperatures (Section E1).

Nomenclature

Fatty acids are named according to the total number of carbon atoms, and to the number and position of any double bonds. The systematic names for fatty acids are made by adding '-oic acid' on to the name of the parent hydrocarbon. However, as fatty acids are ionized at physiological pH they are usually written as RCOO⁻, and have names ending in '-ate' rather than '-oic acid'. Thus, a C18 saturated fatty acid would be called octadecanoate. However, many nonsystematic names are still in use (Table 1).

(a)

Palmitate (hexadecanoate) C16:0

(b)

Palmitoleate (*cis*-Δ^9-hexadecenoate) C16:1

(c)

(*Trans*-Δ^9-octadecenoate) C18:1

(d)

Linoleate (*cis, cis*-Δ^9, Δ^{12}-octadecadienoate) C18:2

Figure 1. Structures of (a) a saturated fatty acid (palmitate, C16:0); (b) a monounsaturated fatty acid with the double bond in the *cis* configuration (palmitoleate, C16:1); (c) a monounsaturated fatty acid with the double bond in the *trans* configuration (C18:1); and (d) a polyunsaturated fatty acid (linoleate, C18:2).

Table 1. The names and formulae of some common fatty acids

Fatty acid	Formula	No. of double bonds	No. of carbon atoms
Palmitate	$CH_3(CH_2)_{14}COO^-$	None	16
Stearate	$CH_3(CH_2)_{16}COO^-$	None	18
Oleate	$CH_3(CH_2)_7CH=CH(CH_2)_7COO^-$	1	18
Linoleate	$CH_3(CH_2)_4(CH=CHCH_2)_2(CH_2)_6COO^-$	2	18
Linolenate	$CH_3CH_2(CH=CHCH_2)_3(CH_2)_6COO^-$	3	18
Arachidonate	$CH_3(CH_2)_4(CH=CHCH_2)_4(CH_2)_2COO^-$	4	20

There is also a shorthand notation to show the number of carbon atoms and the number of any double bonds in the structure. A fatty acid with 18 carbons and no double bonds is designated 18:0, while one with 18 carbons and two double bonds is 18:2. The carbon atoms in fatty acids are numbered from the carboxylic acid residue, and so the position of double bonds can be described using the number of the first carbon involved in the bond (e.g. Δ^9 shows a double bond between carbons 9 and 10 of the fatty acid chain; Figure 1b). The configuration of the double bonds in most unsaturated fatty acids is *cis*; so called because the two hydrogens on the carbon atoms either side of the double bond are on the same side of the molecule (Figure 1b) (Latin, *cis*=on this side of). Thus, the full systematic name of linoleate (Table 1) is *cis*, *cis*-Δ^9,Δ^{12}-octadecadienoate (Figure 1d). During the degradation of fatty acids (Section K2) some ***trans***-isomers are formed (Figure 1c), where the hydrogens on the carbon atoms either side of the double bond are on opposite sides of the molecule (Latin, *trans*=across).

Roles

Fatty acids have four major biological roles:

1. Fatty acids act as **fuel molecules**, being stored as **triacylglycerols**, and broken down to generate energy (Sections K2 and K4)

2. They are used to make **glycerophospholipids** and **sphingolipids** that are essential components of biological membranes (Section E1)

3. Numerous proteins are **covalently modified** by fatty acids (Section E2). Myristate (C14:0) and palmitate (C16:0) are directly attached to some proteins, while phosphatidylinositol is covalently linked to the C-terminus of other proteins via a complex glycosylated structure

4. Derivatives of fatty acids serve as **hormones** (such as the prostaglandins) and **intracellular second messengers** (such as DAG and IP$_3$) (Section E5)

Prostaglandins

Prostaglandins, and the structurally related molecules **prostacyclins**, **thromboxanes** and **leukotrienes**, are called **eicosanoids** because they contain 20 carbon atoms (Greek *eikosi*=20). These hormones are relatively short-lived and hence act locally near to their site of synthesis in the body. They are derived from the common precursor **arachidonate** (Figure 2). This polyunsaturated fatty acid is a derivative of linoleate (Table 1). Prostaglandins stimulate inflammation, modulate synaptic transmission between nerve cells and induce sleep. Although **aspirin** (acetylsalicylic acid) has been used for centuries to decrease inflammation, pain and fever, it was not until 1974 that John Vane discovered how aspirin works. Aspirin inhibits the synthesis of prostaglandins by irreversibly inhibiting **prostaglandin H2 synthase**. This enzyme catalyzes the first step in the synthesis of prostaglandins, prostacyclins and thromboxanes (Figure 2). Other **nonsteroidal anti-inflammatory drugs** (**NSAIDs**), such as **ibuprofen**, also inhibit prostaglandin H2 synthase.

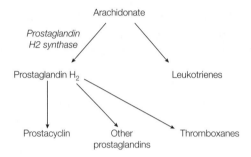

Figure 2. Biosynthetic relationship of the eicosanoids.

K2 Fatty acid breakdown

Key Notes

Overview

Fatty acid breakdown (also called β-oxidation) brings about the oxidation of long-chain fatty acids with the production of energy in the form of ATP. The fatty acids are converted into their acyl CoA derivatives and then metabolized by the removal of two-carbon acetyl CoA units from the end of the acyl chain.

Activation

Fatty acid breakdown occurs in the cytosol of prokaryotes and in the mitochondrial matrix of eukaryotes. The fatty acid is activated by forming a thioester link with CoA before entering the mitochondria.

Transport into mitochondria

The inner mitochondrial membrane is not permeable to long-chain acyl CoA derivatives and so these are transported into the mitochondria as carnitine derivatives by carnitine/ acylcarnitine translocase.

β-Oxidation pathway

Fatty acid breakdown involves a repeating sequence of four reactions:

1. Oxidation of the acyl CoA by FAD to form a $trans$-Δ^2-enoyl CoA.
2. Hydration to form 3-hydroxyacyl CoA.
3. Oxidation by NAD^+ to form 3-ketoacyl CoA.
4. Thiolysis by a second CoA molecule to form acetyl CoA and an acyl CoA shortened by two carbon atoms.

The $FADH_2$ and NADH produced feed directly into oxidative phosphorylation, while the acetyl CoA feeds into the citric acid cycle where further $FADH_2$ and NADH are produced.

Oxidation of unsaturated fatty acids

Unsaturated fatty acids require the action of additional enzymes in order to be completely degraded by β-oxidation.

Oxidation of odd-chain fatty acids

Fatty acids having an odd number of carbon atoms give rise to acetyl CoA and propionyl CoA (three carbon atoms) in the final round of fatty acid degradation.

Regulation

The rate of fatty acid degradation is controlled by the availability of free fatty acids in the blood, which arise from the breakdown of triacylglycerols.

Energy yield

Complete degradation of palmitate (C16:0) in β-oxidation generates 28 ATP molecules from oxidation of the NADH and $FADH_2$ produced directly and 80 ATPs from the breakdown of the acetyl CoA molecules in the citric acid cycle. However, two ATP equivalents are required to activate the palmitate to

	its acyl CoA derivative prior to oxidation. Thus the net yield is 106 ATPs.
Ketone bodies	When in excess, acetyl CoA produced from the β-oxidation of fatty acids is converted into acetoacetate and D-3-hydroxybutyrate. Together with acetone, these compounds are collectively termed ketone bodies.
Related topics	(K1) Fatty acid structure and (K5) Cholesterol function (L1) Citric acid cycle (K3) Fatty acid synthesis (L2) Electron transport and (K4) Triacylglycerols oxidative phosphorylation

Overview

Fatty acid breakdown brings about the oxidation of long-chain fatty acids. The fatty acids are first converted to their **acyl coenzyme A (CoA) derivatives** and then degraded by the successive removal of two-carbon units from the end of the fatty acid as **acetyl CoA**. The pathway produces $FADH_2$ and NADH directly. The acetyl CoA produced can also enter the citric acid cycle and produce further $FADH_2$ and NADH (Section L1). The $FADH_2$ and NADH are then oxidized by the respiratory electron transport chain to yield energy in the form of ATP (Section L2).

Activation

Fatty acid breakdown occurs in the cytosol of prokaryotes, in peroxisomes in plants and in the mitochondrial matrix of all other eukaryotes. Before entering the **mitochondrial matrix**, the fatty acid is **activated** by forming a **thioester link** with **CoA** (Figure 1). This reaction is catalyzed by **acyl CoA synthetase** (also called **fatty acid thiokinase**), which is present on the outer mitochondrial membrane, and uses a molecule of ATP. The overall reaction is irreversible due to the subsequent hydrolysis of PP_i to two molecules of P_i.

Transport into mitochondria

Small- and medium-chain acyl CoA molecules (up to 10 carbon atoms) are readily able to cross the **inner mitochondrial membrane** by diffusion. However, longer chain acyl CoAs do not readily cross the inner mitochondrial membrane, and require a specific transport mechanism. To achieve this, the longer chain acyl CoAs are conjugated to the polar **carnitine** molecule, which is found in both plants and animals. This reaction, catalyzed by an enzyme on the outer face of the inner mitochondrial membrane (**carnitine acyltransferase I**), removes the CoA group and substitutes it with a carnitine molecule (Figure 2). The acylcarnitine is then transported across the inner mitochondrial membrane by a **carnitine/acylcarnitine translocase**. This integral membrane transport

Figure 1. Activation of a fatty acid.

Figure 2. Transport of fatty acids across the inner mitochondrial membrane.

protein (Section E3) transports acylcarnitine molecules into the mitochondrial matrix and free carnitine molecules out. Once inside the mitochondrial matrix the acyl group is transferred back on to CoA, releasing free carnitine, by the enzyme **carnitine acyltransferase II**, which is located on the matrix side of the inner mitochondrial membrane (Figure 2).

β-Oxidation pathway

The individual reactions involved in the degradation of fatty acids by β-oxidation are as follows (Figure 3):

1. **Oxidation** of the fatty acyl CoA to enoyl CoA forming a *trans*-Δ^2-double bond on the fatty acyl chain and producing $FADH_2$ (catalyzed by **acyl CoA dehydrogenase**).

2. **Hydration** of the *trans*-Δ^2-enoyl CoA to form 3-hydroxyacyl CoA (catalyzed by **enoyl CoA hydratase**).

3. **Oxidation** of 3-hydroxyacyl CoA to 3-ketoacyl CoA producing NADH (catalyzed by **hydroxyacyl CoA dehydrogenase**).

4. Cleavage, or **thiolysis**, of 3-ketoacyl CoA by a second CoA molecule, giving acetyl CoA and an acyl CoA shortened by two carbon atoms (catalyzed by β**-ketothiolase**).

Thus, the breakdown of individual fatty acids occurs as a repeating sequence of four reactions: **oxidation** (by FAD), **hydration**, **oxidation** (by NAD+) and **thiolysis**. These four reactions form one 'round' of fatty acid degradation (Figure 3) and their overall effect is to remove two-carbon units sequentially in the form of acetyl CoA from the fatty acid chain. The cleavage of the Δ^2 (or β) bond of the fatty acyl chain (see Figure 3, top structure, for nomenclature) gives fatty acid breakdown its alternative name, β**-oxidation**. The shortened acyl CoA then undergoes further cycles of β-oxidation until the last cycle, when the acyl CoA with four carbon atoms is split into two molecules of acetyl CoA. Thus a C16 saturated acyl CoA, such as palmitoyl CoA, would be completely degraded into eight molecules of acetyl CoA by seven rounds of degradation, leading to the overall equation:

$$\text{Palmitoyl CoA} + 7\text{ FAD} + 7\text{ NAD}^+ + 7\text{ CoA} + 7\text{ H}_2\text{O} \longrightarrow$$
$$8\text{ acetyl CoA} + 7\text{ FADH}_2 + 7\text{ NADH} + 7\text{ H}^+$$

Mitochondria contain three **acyl CoA dehydrogenases**, which act on short-, medium- and long-chain acyl CoAs, respectively. In contrast, there is just one each of the enzymes

Figure 3. Summary of the reactions involved in the degradation of fatty acids.

enoyl CoA hydratase, hydroxyacyl CoA dehydrogenase and β-ketothiolase, which all have a broad specificity with respect to the length of the acyl chain.

In animals the acetyl CoA produced from fatty acid degradation cannot be converted into pyruvate or oxaloacetate. Although the two carbon atoms from acetyl CoA enter the citric acid cycle, they are both oxidized to CO_2 in the reactions catalyzed by isocitrate dehydrogenase and α-ketoglutarate dehydrogenase (Section L1). Thus, **animals cannot convert fatty acids into glucose**. In contrast, plants have two additional enzymes, isocitrate lyase and malate synthase, that enable them to convert the carbon atoms of acetyl CoA into oxaloacetate. This is accomplished via the **glyoxylate pathway**, a route involving enzymes of both the mitochondrion and the glyoxysome, a specialized membranous plant organelle.

Oxidation of unsaturated fatty acids

Unsaturated fatty acids require some additional processing before they can be degraded completely by β-oxidation. Unsaturated fatty acyl CoAs with double bonds at odd-numbered carbon atoms (i.e. between, for example, C-9 and C-10 as in palmitoleate; Section K1, Figure 1b) are acted on in the normal way by the degradation mechanism until the acyl CoA dehydrogenase encounters the *cis*-Δ^3-enoyl CoA formed at the end of the third round. The presence of the double bond between C-3 and C-4 prevents the formation of another double bond between C-2 and C-3. To overcome this problem an **isomerase** converts the *cis*-Δ^3 bond into a *trans*-Δ^2 double bond, and the resulting *trans*-Δ^2-enoyl CoA can then continue down the β-oxidation pathway (Figure 4).

Another enzyme, in addition to the isomerase, is required for the oxidation of **polyunsaturated fatty acids**, which have a double bond at an even-numbered carbon atom. In this case the 2,4-dienoyl intermediate resulting from the action of acyl CoA dehydrogenase is acted on by **2,4-dienoyl CoA reductase** to form *cis*-Δ^3-enoyl CoA (Figure 4). This is then converted by the isomerase into the *trans* form, which continues down the pathway. These reactions are important since over half the fatty acids of plant and animal lipids are unsaturated (and often polyunsaturated).

Figure 4. Accessory enzymes required for the metabolism of unsaturated fatty acids.

Oxidation of odd-chain fatty acids

Fatty acids having an **odd number of carbon atoms** (which are relatively rare in nature) are also degraded by the β-oxidation pathway in the same way as those with an even number of carbon atoms. The only difference is that in the final round the five carbon acyl CoA intermediate is cleaved into one molecule of the C3 **propionyl CoA** and one molecule of the C2 acetyl CoA. The propionyl CoA is then converted into succinyl CoA, which enters the citric acid cycle (Section L1).

Regulation

The major point of control of β-oxidation is the availability of fatty acids. The major source of free fatty acids in the blood is from the breakdown of **triacylglycerol** stores in adipose tissue, which is regulated by the action of hormone-sensitive triacylglycerol lipase (Section K4). Fatty acid breakdown and fatty acid synthesis are coordinately controlled to prevent a futile cycle (Section K3).

Energy yield

For each round of degradation, one $FADH_2$, one NADH and one acetyl CoA molecule are produced. Each NADH generates 2.5 ATP molecules, and each $FADH_2$ generates 1.5 ATPs during oxidative phosphorylation (Section L2). In addition, each acetyl CoA yields 10 ATPs on oxidation by the citric acid cycle (Section L1). The total yield for each round of fatty acid degradation is therefore 14 ATP molecules.

The complete degradation of palmitoyl CoA (C16:0) requires seven rounds of degradation and hence produces $7 \times (2.5+1.5)=28$ ATP molecules. A total of eight acetyl CoA molecules are produced and hence another $8 \times 10=80$ ATP. Thus the total ATP yield per molecule of palmitate degraded is $28+80=108$ ATP. However, one ATP is hydrolyzed to AMP and PP_i in the activation of palmitate to palmitoyl CoA, resulting in two high-energy bonds being cleaved. Thus the net yield is **106 ATPs** (Table 1).

The yield of ATP is reduced slightly for unsaturated fatty acids, since the additional metabolic reactions, which enable them to be degraded by the β-oxidation pathway, either involve using NADPH or bypass an $FADH_2$-producing reaction (Figure 4).

Ketone bodies

When the level of acetyl CoA from β-oxidation increases in excess of that required for entry into the citric acid cycle, the acetyl CoA is converted into **acetoacetate** and **D-3-hydroxybutyrate** by a process known as **ketogenesis**. D-3-hydroxybutyrate, acetoacetate and its nonenzymic breakdown product **acetone** are referred to collectively as **ketone bodies** (Figure 5).

Table 1. Calculation of the ATP yield from the complete oxidation of palmitate

Degradative step	ATP yield
7×4 ATP for oxidation of NADH and $FADH_2$ produced by each round of degradation	28
8×10 ATP for the breakdown of acetyl CoA by the citric acid cycle	80
–2 ATP equivalents for the activation of palmitate	–2
Total=	106

Two molecules of acetyl CoA initially condense to form acetoacetyl CoA in a reaction, which is essentially the reverse of the thiolysis step in β-oxidation. The acetoacetyl CoA reacts with another molecule of acetyl CoA to form **3-hydroxy-3-methylglutaryl CoA (HMG CoA)** (Figure 5). This molecule is then cleaved to form acetoacetate and acetyl CoA. (HMG CoA is also the starting point for cholesterol biosynthesis; Section K5.) The acetoacetate is then either reduced to D-3-hydroxybutyrate in the mitochondrial matrix or undergoes a slow, spontaneous decarboxylation to acetone (Figure 5). In diabetes, acetoacetate is produced faster than it can be metabolized. Hence untreated diabetics have high levels of ketone bodies in their blood, and the smell of acetone can often be detected on their breath.

Acetoacetate and D-3-hydroxybutyrate are produced mainly in the liver and are not just degradation products of little physiological value. They are used in preference to glucose as an energy source by certain tissues such as the heart muscle and kidney cortex. Although glucose is normally the major fuel for the brain, under conditions of starvation or diabetes this organ can switch to using predominantly acetoacetate.

Figure 5. Conversion of acetyl CoA to the ketone bodies acetoacetate, acetone and D-3-hydroxybutyrate.

K3 Fatty acid synthesis

Key Notes

Overview

Fatty acid synthesis involves the condensation of two-carbon units, in the form of acetyl CoA, to form long hydrocarbon chains in a series of reactions. These reactions are carried out on the fatty acid synthase complex using NADPH as reductant. The fatty acids are covalently linked to acyl carrier protein (ACP) during their synthesis.

Transport into the cytosol

Since fatty acid synthesis takes place in the cytosol, the acetyl CoA produced from pyruvate has to be transported out of the mitochondria. As the inner mitochondrial membrane is not permeable to acetyl CoA, it is first combined with oxaloacetate to form citrate, which readily crosses the membrane. In the cytosol the citrate is cleaved to regenerate the acetyl CoA.

The pathway

The first committed step in fatty acid biosynthesis is the carboxylation of acetyl CoA to form malonyl CoA, which is catalyzed by the biotin-containing enzyme acetyl CoA carboxylase. Acetyl CoA and malonyl CoA are then converted into their ACP derivatives. The elongation cycle in fatty acid synthesis involves four reactions: condensation of acetyl-ACP and malonyl-ACP to form acetoacetyl-ACP releasing free ACP and CO_2, then reduction by NADPH to form D-3-hydroxybutyryl-ACP, followed by dehydration to crotonyl-ACP, and finally reduction by NADPH to form butyryl-ACP. Further rounds of elongation add more two-carbon units from malonyl-ACP on to the growing hydrocarbon chain, until the C16 palmitate is formed. Further elongation of fatty acids takes place on the cytosolic surface of the SER.

Formation of double bonds

The enzymes for introducing double bonds into the acyl chain are also present on the cytosolic surface of the SER. The polyunsaturated fatty acids linoleate and linolenate cannot be synthesized by mammals and are therefore termed essential fatty acids as they have to be ingested in the diet.

Regulation

The key control point of fatty acid synthesis is acetyl CoA carboxylase. This enzyme is inactivated by phosphorylation by an AMP-activated protein kinase. Thus when the energy charge of the cell is low (high AMP, low ATP) acetyl CoA carboxylase is inactive. It is reactivated by dephosphorylation by protein phosphatase 2A. Glucagon and epinephrine inhibit fatty acid synthesis by inhibiting protein phosphatase 2A, whereas insulin stimulates fatty acid synthesis by activating the phosphatase. Acetyl CoA carboxylase is also

	allosterically regulated: citrate activates the enzyme, whereas palmitoyl CoA inhibits it.	
Related topics	(D5) Regulation of enzyme activity	(K2) Fatty acid breakdown
		(K4) Triacylglycerols
	(J5) Pentose phosphate pathway	(L1) Citric acid cycle
	(K1) Fatty acid structure and function	

Overview

Fatty acids are synthesized by the condensation of two-carbon units. However, in terms of the enzymic steps involved, the process is not the reverse of β-oxidation (Section K2). **Fatty acid synthesis** involves a separate series of reactions to build up long-chain hydrocarbons from **acetyl CoA** units. The key differences between fatty acid synthesis and breakdown are:

- Fatty acid synthesis occurs in the **cytosol** of both prokaryotes and eukaryotes whereas their degradation occurs in the mitochondria of eukaryotes.

- Fatty acid synthesis uses **NADPH** as the reductant whereas NADH is produced in β-oxidation.

- During their synthesis, fatty acids are covalently linked to an **acyl carrier protein (ACP)** as opposed to CoA in their degradation.

- The enzyme activities of fatty acid synthesis in higher organisms are present in a single, multifunctional polypeptide chain (as a dimer) called **fatty acid synthase**, whereas in β-oxidation the individual activities are present on separate enzymes.

Transport into the cytosol

Fatty acids are synthesized in the cytosol, but acetyl CoA is produced from pyruvate in the mitochondria (Section L1). Thus the acetyl CoA must be transferred from the mitochondria into the cytosol to allow fatty acid synthesis to occur. However, the **inner mitochondrial membrane** is not readily permeable to this molecule. This problem is overcome by the condensation of acetyl CoA with oxaloacetate to form **citrate** (Figure 1). This is then transported into the cytosol where it is cleaved to regenerate acetyl CoA and oxaloacetate by **ATP–citrate lyase** in an energy-requiring process. The oxaloacetate, which also cannot cross the inner mitochondrial membrane, is returned to the mitochondrial matrix through conversion first to malate (catalyzed by **malate dehydrogenase**) and then to pyruvate (catalyzed by **NADP$^+$-linked malate enzyme**) (Figure 1). This latter decarboxylation reaction generates NADPH, which can be used in fatty acid synthesis. The remaining NADPH required for fatty acid synthesis is provided by the pentose phosphate pathway (Section J5). Once back in the mitochondrial matrix, pyruvate is carboxylated to form oxaloacetate by **pyruvate carboxylase** with the hydrolysis of a further molecule of ATP (Figure 1).

The pathway

The **first committed step** in fatty acid biosynthesis is the carboxylation of acetyl CoA to form **malonyl CoA** using CO_2 in the form of bicarbonate HCO_3^- (Figure 2). This reaction is

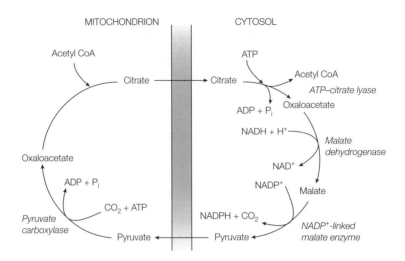

Figure 1. Transport of acetyl CoA from the mitochondrial matrix into the cytosol.

catalyzed by the enzyme **acetyl CoA carboxylase**, which has **biotin** as a prosthetic group, a common feature in CO_2-binding enzymes. One molecule of ATP is hydrolyzed in the reaction, which is irreversible. The **elongation steps** of fatty acid synthesis all involve intermediates linked to the terminal sulfhydryl group of the **phosphopantetheine** reactive unit in **ACP**; phosphopantetheine is also the reactive unit in CoA. Therefore, the next steps are the formation of acetyl-ACP and malonyl-ACP by the enzymes **acetyl transacylase** and **malonyl transacylase**, respectively (Figure 2). (For the synthesis of fatty acids with an odd number of carbon atoms, the three-carbon propionyl-ACP is the starting point instead of malonyl-ACP.)

The elongation cycle of fatty acid synthesis has four stages for each round of synthesis (Figure 3). For the first round of synthesis these are:

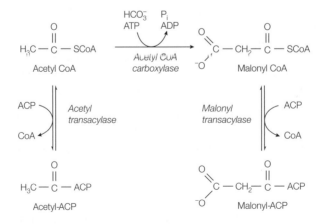

Figure 2. Formation of acetyl- and malonyl-ACP.

1. **Condensation** of acetyl-ACP and malonyl-ACP to form acetoacetyl-ACP, releasing free ACP and CO_2 (catalyzed by acyl-malonyl-ACP condensing enzyme).

2. **Reduction** of acetoacetyl-ACP to form D-3-hydroxybutyryl-ACP, using NADPH as reductant (catalyzed by β-ketoacyl-ACP reductase).

3. **Dehydration** of D-3-hydroxybutyryl-ACP to produce crotonyl-ACP (catalyzed by 3-hydroxyacyl-ACP dehydratase).

4. **Reduction** of crotonyl-ACP by a second NADPH molecule to give butyryl-ACP (catalyzed by enoyl-ACP reductase).

This first round of elongation produces the four-carbon butyryl-ACP. The cycle now repeats with malonyl-ACP adding two-carbon units in each cycle to the lengthening

Figure 3. The elongation cycle of fatty acid synthesis.

acyl-ACP chain. This continues until the 16-carbon **palmitoyl-ACP** is formed. This molecule is not accepted by the acyl-malonyl-ACP condensing enzyme, and so cannot be elongated further by this process. Instead it is hydrolyzed by a thioesterase to give palmitate and ACP.

The overall stoichiometry for the synthesis of palmitate is:

$$8 \text{ acetyl CoA} + 7 \text{ ATP} + 14 \text{ NADPH} + 6 \text{ H}^+ \rightarrow$$
$$\text{palmitate} + 14 \text{ NADP}^+ + 8 \text{ CoA} + 6 \text{ H}_2\text{O} + 7 \text{ ADP} + 7 \text{ P}_i$$

For each of the seven rounds of fatty acid elongation, one ATP is used in the synthesis of malonyl-CoA and two NADPH are used in the reduction reactions.

In eukaryotes, the elongation of fatty acids beyond C16 palmitate is carried out by enzymes located on the cytosolic surface of the SER. Malonyl CoA is used as the two-carbon donor, and the fatty acid is elongated as its CoA derivative rather than its ACP derivative.

In prokaryotes, each of the reactions of fatty acid synthesis is catalyzed by a separate enzyme. However, in eukaryotes, the enzymes of the fatty acid synthesis elongation cycle are present in a single polypeptide chain, multifunctional enzyme complex, called **fatty acid synthase**. The fatty acid synthase complex exists as a dimer, with the ACP moiety shuttling the fatty acyl chain between successive catalytic sites, and from one subunit of the dimer to the other. It is, in effect, a highly efficient production line for fatty acid biosynthesis.

Formation of double bonds

In eukaryotes, the SER has enzymes able to introduce **double bonds** into fatty acyl CoA molecules in an oxidation reaction that uses molecular oxygen. This reaction is catalyzed by a membrane-bound complex of three enzymes: NADH-cytochrome b_5 reductase, cytochrome b_5 and a desaturase. The overall reaction is:

$$\text{saturated fatty acyl CoA} + \text{NADH} + \text{H}^+ + \text{O}_2 \rightarrow$$
$$\text{monounsaturated acyl CoA} + \text{NAD}^+ + 2 \text{ H}_2\text{O}$$

The reaction may be repeated to introduce more than one double bond into a fatty acid.

Mammals lack the enzymes to insert double bonds at carbon atoms beyond C-9 in the fatty acid chain. Thus they cannot synthesize **linoleate** and **linolenate**, both of which have double bonds later in the chain than C-9 (linoleate has *cis, cis*-Δ^9, Δ^{12} double bonds, and linolenate has all *cis*-$\Delta^9,\Delta^{12},\Delta^{15}$ double bonds). Hence, in mammals linoleate and linolenate are called **essential fatty acids** since they have to be supplied in the diet. These two unsaturated fatty acids are also the starting points for the synthesis of other unsaturated fatty acids, such as **arachidonate**. This C20:4 fatty acid is the precursor of several biologically important molecules, including the prostaglandins, prostacyclins, thromboxanes and leukotrienes (Section K1).

Regulation

The synthesis of fatty acids takes place when carbohydrate and energy are plentiful and when fatty acids are scarce. The key enzyme in the regulation of fatty acid synthesis is **acetyl**

CoA carboxylase, which synthesizes malonyl CoA. This is a good example of **control at the committed step** of a metabolic pathway. Acetyl CoA carboxylase is inactivated by the **phosphorylation** of a single serine residue by an **AMP-activated protein kinase** (Figure 4) (Section D5). Unlike cAMP-dependent protein kinase (protein kinase A) (Section K4), this kinase is not affected by cAMP, but instead is stimulated by AMP and inhibited by ATP. Thus when the energy charge of the cell is low (i.e. there is a high AMP:ATP ratio) fatty acid synthesis is switched off. **Protein phosphatase 2A** removes the phosphate group from inactivated acetyl CoA carboxylase (Figure 4), thereby reactivating it.

Acetyl CoA carboxylase is also subject to hormonal regulation. When energy is required, **glucagon** and **epinephrine** inhibit protein phosphatase 2A, thus keeping acetyl CoA carboxylase in the inactive form (Figure 4) and blocking fatty acid synthesis. In the well-fed state, when blood glucose levels are high, **insulin** stimulates acetyl CoA carboxylase, possibly by activating protein phosphatase 2A (Figure 4), thereby leading to an increase in fatty acid synthesis.

As well as its control by phosphorylation/dephosphorylation, acetyl CoA carboxylase is also **allosterically regulated** (for a fuller description of allosteric regulation see Section D5). The citric acid cycle intermediate **citrate**, the level of which is high when both acetyl CoA and ATP are abundant, allosterically stimulates acetyl CoA carboxylase. This results in the conversion of the inactive phosphorylated form into a partially active form that is still phosphorylated (Figure 4), thereby activating fatty acid synthesis so that the excess acetyl CoA is 'stored' as fatty acid residues within triacylglycerol in adipose tissue. In contrast, high levels of **palmitoyl CoA**, which is abundant when there is an excess of fatty acids, antagonize the effect of citrate on acetyl CoA carboxylase, reducing its activity (Figure 4) and switching off further fatty acid synthesis.

Figure 4. Summary of the control of acetyl CoA carboxylase by phosphorylation and allosteric regulation.

K4 Triacylglycerols

Key Notes

Structure and function

Triacylglycerols (fats or triglycerides) consist of three fatty acid chains esterified to a glycerol backbone. Triacylglycerols are the major energy store and the major dietary lipid in humans. They are insoluble in water and are stored in specialized adipose (fat) cells.

Synthesis

Triacylglycerols are synthesized from glycerol 3-phosphate, which is derived from the glycolytic intermediate dihydroxyacetone phosphate, and fatty acyl CoAs. Acyl CoA molecules are added on to glycerol 3-phosphate to form first lysophosphatidic acid and then phosphatidic acid. The phosphate group is then removed to form DAG, which is further acylated to triacylglycerol. The energy for the synthesis of triacylglycerols comes from the hydrolysis of the high-energy thioester bond in acyl CoA.

Breakdown

The fatty acids in triacylglycerols are released from the glycerol backbone by the action of lipases. The free fatty acids can then be degraded by β-oxidation to produce energy. The glycerol is converted into dihydroxyacetone phosphate, which enters glycolysis.

Regulation

The concentration of free fatty acids in the blood is controlled by the rate at which hormone-sensitive triacylglycerol lipase hydrolyzes the triacylglycerols stored in adipose tissue. Glucagon, epinephrine and norepinephrine cause an increase in the intracellular level of cAMP, which allosterically activates cAMP-dependent protein kinase. The kinase in turn phosphorylates hormone-sensitive lipase, activating it, and leading to the release of fatty acids into the blood. Insulin has the opposite effect; it decreases the level of cAMP, which leads to the dephosphorylation and inactivation of hormone-sensitive lipase.

Related topics

(E1) Membrane lipids	(K3) Fatty acid synthesis
(E5) Signal transduction	(K5) Cholesterol
(K2) Fatty acid breakdown	(K6) Lipoproteins

Structure and function

Triacylglycerols (also called **fats** or **triglycerides**) consist of three fatty acid chains esterified to a glycerol backbone. **Simple triacylglycerols** have three identical fatty acids esterified to the glycerol backbone, while **mixed triacylglycerols** have two or three different fatty acid chains (Figure 1). Triacylglycerols constitute the **major fuel store** and the **major dietary lipid** in humans. Triacylglycerols are a highly concentrated **energy**

store. The energy yield from the complete oxidation of fatty acids is about 38 kJ g^{-1}, compared with an energy yield of 17 kJ g^{-1} of carbohydrate or protein. The hydrophobic properties of fats make them insoluble in water, and they are stored in specialized cells called **adipose cells** (fat cells), which consist almost entirely of triacylglycerol. These cells are specialized for the synthesis and storage of triacylglycerols and for their mobilization into fuel molecules. Triacylglycerols are transported round the body in large lipid–protein particles called **lipoproteins** (Section K6).

Figure 1. Structure of (a) a simple triacylglycerol (1,2,3-tripalmitoyl-glycerol) and (b) a mixed triacylglycerol (1-palmitoyl-2,3-dioleoyl-glycerol).

Synthesis

Triacylglycerols are synthesized from **fatty acyl CoAs** and **glycerol 3-phosphate** (Figure 2). The glycolytic intermediate **dihydroxyacetone phosphate** is first reduced to glycerol 3-phosphate, which is, in turn, acylated by glycerol-3-phosphate acyltransferase to form **lysophosphatidic acid**. This is then reacted with a further acyl CoA molecule to form **phosphatidic acid**. Removal of the phosphate group from phosphatidic acid generates DAG, which is further acylated with a third acyl CoA molecule to form **triacylglycerol** (Figure 2). ATP is not involved in the biosynthesis of triacylglycerols. Instead the reactions are driven by the cleavage of the high-energy thioester bond between the acyl moiety and CoA. Both phosphatidic acid (phosphatidate) and DAG are also used in the synthesis of membrane phospholipids (Section E1) and DAG is also used as a second messenger in cell signaling (Section E5).

Breakdown

The initial event in the utilization of both stored fat and dietary fat as energy sources is the hydrolysis of triacylglycerol by **lipases**. These enzymes release the three fatty acid chains from the glycerol backbone (Figure 3). The fatty acids can then be broken down in **β-oxidation** to generate energy (Section K2). The glycerol backbone is also utilized, being transformed into dihydroxyacetone phosphate, an intermediate in glycolysis (Figure 4). This requires two enzymes, glycerol kinase, which uses ATP to phosphorylate glycerol, producing L-glycerol 3-phosphate, and glycerol 3-phosphate dehydrogenase, which produces dihydroxyacetone phosphate.

In the intestine, dietary fats are hydrolyzed by **pancreatic lipase** and the released fatty acids taken up into the **intestinal cells**. Both the digestion and uptake processes are aided by the detergent-like properties of the **bile salts** (Section K5).

Figure 2. Synthesis of triacylglycerols.

Regulation

The breakdown of fatty acids in **β-oxidation** (Section K2) is controlled mainly by the concentration of free fatty acids in the blood, which is, in turn, controlled by the hydrolysis rate of triacylglycerols in adipose tissue by **hormone-sensitive triacylglycerol**

Figure 3. Breakdown of triacylglycerols.

lipase. This enzyme is regulated by **phosphorylation** and **dephosphorylation** (Figure 5) in response to hormonally controlled levels of the intracellular second messenger **cAMP** (Section E5). The catabolic hormones **glucagon**, **epinephrine** and **norepinephrine**, which are released in times of metabolic need, bind to receptor proteins on the cell

Figure 4. Conversion of glycerol into the glycolytic intermediate dihydroxyacetone phosphate.

surface and increase the levels of cAMP in adipose cells through activation of **adenylate cyclase** (see Section E5 for details of the signal transduction pathway). The cAMP allosterically activates **cAMP-dependent protein kinase** (otherwise known as **protein kinase A**), which phosphorylates various intracellular enzymes including hormone-sensitive lipase. Phosphorylation of hormone-sensitive lipase activates it, thereby stimulating the hydrolysis of triacylglycerols, raising the levels of fatty acids in the blood, and subsequently activating β-oxidation in tissues such as muscle and liver. Glucagon and epinephrine also prevent the dephosphorylation, and therefore activation, of **acetyl CoA carboxylase**, so that fatty acid synthesis is inhibited (Section K3).

The anabolic hormone **insulin** has the opposite effect to glucagon and epinephrine. It stimulates the formation of triacylglycerols through decreasing the level of cAMP, which promotes the dephosphorylation and inactivation of hormone-sensitive lipase (Figure 5). Insulin also stimulates the dephosphorylation of acetyl CoA carboxylase, thereby activating fatty acid synthesis (Section K3). Thus fatty acid synthesis and degradation are **coordinately controlled** so as to prevent a futile cycle.

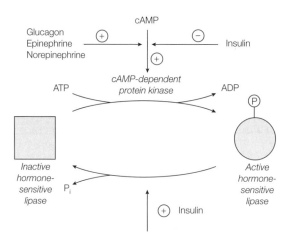

Figure 5. Summary of the control of hormone-sensitive triacylglycerol lipase.

K5 Cholesterol

Key Notes

Functions of cholesterol

Cholesterol is a component of cell membranes and is the precursor of steroid hormones and the bile salts.

Biosynthesis of cholesterol

All 27 carbon atoms of cholesterol are derived from acetyl CoA. First acetyl CoA and acetoacetyl CoA combine to form 3-hydroxy-3-methylglutaryl CoA (HMG CoA), which, in turn, is reduced to mevalonate by HMG CoA reductase. Mevalonate is converted into the five-carbon isoprene compounds 3-isopentenyl pyrophosphate and its isomer dimethylallyl pyrophosphate. These two compounds condense to form the C10 geranyl pyrophosphate, which is elongated to the C15 farnesyl pyrophosphate by the addition of another molecule of isopentenyl pyrophosphate. Two molecules of farnesyl pyrophosphate condense to form the C30 squalene, which then cyclizes and is converted to cholesterol.

Regulation of cholesterol biosynthesis

Cholesterol can either be obtained in the diet or synthesized in the liver. High levels of cholesterol and its metabolites decrease the amount and inhibit the activity of HMG CoA reductase, the enzyme that catalyzes the committed step in cholesterol biosynthesis. This enzyme can also be inhibited therapeutically by the statin compounds.

Bile salts

Bile salts (bile acids) are the major excretory form of cholesterol. These polar compounds are formed in the liver by converting cholesterol into either glycocholate or taurocholate. The detergent-like bile salts are secreted into the intestine where they aid the digestion and uptake of dietary lipids.

Vitamin D

Vitamin D is derived via cholesterol in a series of reactions, one of which requires the action of UV light to break the bond between two carbon atoms. Deficiency of vitamin D causes rickets in children and osteomalacia in adults.

Steroid hormones

The steroid hormones, including progestagens, androgens, estrogens, glucocorticoids and mineralocorticoids, are derived from cholesterol by a series of reactions that involve the heme-containing cytochrome P450 enzymes. These monooxygenases require both O_2 and NADPH to function and are involved in drug metabolism.

Related topics

(D5) Regulation of enzyme activity
(E1) Membrane lipids
(K2) Fatty acid breakdown
(K3) Fatty acid synthesis

(K4) Triacylglycerols
(K6) Lipoproteins
(M4) Hemes and chlorophylls

Functions of cholesterol

Cholesterol is a **steroid**. It is an important constituent of **cell membranes**, where, in mammals, it modulates their fluidity (Section E1). Cholesterol is also the precursor of **steroid hormones** such as progesterone, testosterone and cortisol, and the **bile salts** (see below). Thus, cholesterol is clearly an essential molecule, yet its deposition in arteries can give rise to cardiovascular disease and stroke, two of the leading causes of death in humans.

Biosynthesis of cholesterol

Animals are able to synthesize cholesterol *de novo* by an elegant series of reactions in which all 27 carbon atoms of cholesterol are derived from **acetyl CoA**. The acetate units are first converted into **C5 isoprene units**, that are then condensed to form a linear precursor to the cyclic cholesterol.

Figure 1. Synthesis of isopentenyl pyrophosphate.

The first stage in the synthesis of cholesterol is the formation of **isopentenyl pyrophosphate** (Figure 1). Acetyl CoA and acetoacetyl CoA combine to form **3-hydroxy-3-methylglutaryl CoA (HMG CoA)**. This process takes place in the liver, where the HMG CoA in the mitochondria is used to form ketone bodies during starvation (Section K2), whereas that in the cytosol is used to synthesize cholesterol in the fed state (under the influence of cholesterol). HMG CoA is then reduced to **mevalonate** by **HMG CoA reductase** (Figure 1). This is the committed step in cholesterol biosynthesis and is a key control point. Mevalonate is converted into **3-isopentenyl pyrophosphate** by three consecutive reactions each involving ATP, with CO_2 being released in the last reaction (Figure 1).

The C5 isoprene units in isopentenyl pyrophosphate are then condensed to form the C30 compound squalene (Figure 2). First, isopentenyl pyrophosphate isomerizes to **dimethylallyl pyrophosphate** (Figure 2a), which reacts with another molecule of isopentenyl pyrophosphate to form the C10 compound **geranyl pyrophosphate** (Figure 2b). Another molecule of isopentenyl pyrophosphate then reacts with geranyl pyrophosphate to form the C15 compound **farnesyl pyrophosphate**. Next, two molecules of farnesyl pyrophosphate condense to form **squalene** (Figure 2b).

Squalene is then converted into **squalene epoxide** in a reaction that uses O_2 and NADPH (Figure 2b). The squalene epoxide cyclizes to form **lanosterol**, and finally cholesterol is formed from lanosterol in a series of reactions by the removal of three methyl groups, the reduction of one double bond by NADPH, and the migration of the other double bond (Figure 2b).

Farnesyl pyrophosphate and the C20 compound **geranylgeranyl pyrophosphate** (which is formed by the condensation of another isopentenyl pyrophosphate with farnesyl pyrophosphate) are covalently linked to cysteine residues in a number of proteins, giving rise to **prenylated proteins**, promoting their association with membranes (Section E2). **Dolichol**, which contains some 20 isoprene units, is used to carry the biosynthetic precursor of the N-linked oligosaccharides that are subsequently attached to proteins (Section H5).

Regulation of cholesterol biosynthesis

Cholesterol can be obtained either from the diet or it can be synthesized *de novo*, mainly in the liver. Cholesterol is transported round the body in **lipoprotein** particles (Section K6). The rate of synthesis of cholesterol is dependent on the cellular level of cholesterol. High levels of cholesterol and its metabolites control cholesterol biosynthesis at the level of **HMG CoA reductase**, the enzyme that catalyzes the committed step in cholesterol biosynthesis. Short-term control of this enzyme is brought about by competitive inhibition, allosteric effects and covalent modification involving reversible phosphorylation. Like acetyl CoA carboxylase in fatty acid synthesis (Section K3), HMG CoA reductase is inactivated by phosphorylation by an AMP-activated protein kinase, retained in this form under the influence of glucagon during starvation. However, the primary regulatory mechanism for HMG CoA reductase is long-term feedback control of the amount of enzyme. The rate of synthesis of its mRNA is controlled by **the sterol regulatory element binding protein** (SREBP), which senses the amount of cholesterol in the cell.

HMG CoA reductase can be inhibited therapeutically by administering drugs called **statins** (e.g. **lovastatin** and **simvastatin**), which competitively inhibit the enzyme and hence decrease the rate of cholesterol biosynthesis. Therefore, these compounds are routinely used for the treatment of **hypercholesterolemia** (high levels of blood cholesterol) (Section K6).

(a)

Isopentenyl pyrophosphate   Dimethylallyl pyrophosphate

(b)

Figure 2. Synthesis of squalene and cholesterol from isopentenyl pyrophosphate:
(a) isomerization of isopentenyl pyrophosphate to dimethylallyl pyrophosphate; (b) synthesis
of cholesterol.

Figure 3. Structures of the bile salts (a) glycocholate and (b) taurocholate.

Bile salts

Bile salts (or bile acids) are polar derivatives of cholesterol and constitute the major pathway for the excretion of cholesterol in mammals. In the liver, cholesterol is converted into the activated intermediate **cholyl CoA**, which then reacts either with the amino group of glycine to form **glycocholate** (Figure 3a), or with the amino group of taurine (H_2N-CH_2-CH_2-SO_3^-, a derivative of cysteine) to form **taurocholate** (Figure 3b). After synthesis in the liver, the bile salts glycocholate and taurocholate are stored and concentrated in the gall bladder, before release into the small intestine. Since they contain both polar and nonpolar regions (that is are amphipathic molecules), the bile salts are very effective detergents and act to solubilize dietary lipids. The resulting increase in the surface area of the lipids aids their hydrolysis by lipases and their uptake into intestinal cells (Section K4). The intestinal absorption of the **lipid-soluble vitamins** A, D, E and K also requires the action of the bile salts.

Vitamin D

Vitamin D is derived from 7-dehydrocholesterol by the action of the **UV** component of sunlight on the skin. UV light brings about photolysis of 7-dehydrocholesterol between C-9 and C-10, leading to a rearrangement of the double bonds of the molecule to form **previtamin D$_3$** (Figure 4). This molecule spontaneously isomerizes to form **vitamin D$_3$** (**cholecalciferol**). Subsequent hydroxylation reactions take place in the liver and kidneys to produce 1,25-dihydroxycholecalciferol (1,25(OH)D$_3$), the active hormone (Figure 4). **Rickets**, which is caused by a deficiency of vitamin D, was historically a common disease of childhood in Britain due to the low vitamin D content of the national diet, and lack of exposure to sunlight. Even today, people whose cultures require the body to be clothed so that no skin is exposed to sunlight have problems in maintaining an adequate vitamin D level. In adults, this takes the form of **osteomalacia** – the softening or weakening of the bones.

Steroid hormones

Cholesterol is the precursor of the five major classes of **steroid hormones** (Table 1). The synthesis of steroid hormones is initiated by the removal of a six-carbon unit from carbon 20 of the cholesterol side chain to form **pregnenolone**, the common precursor of all steroid hormones (Figure 5). A series of reactions catalyzed by **cytochrome P450s** modify pregnenolone to give rise to the individual hormones (Figure 5).

Figure 4. Formation of vitamin D.

Table 1. Classes of steroid hormone

Class	Site of synthesis	Hormone	Action
Progestagens	Corpus luteum	Progesterone	Prepares uterine lining for egg implantation; maintenance of pregnancy
Androgens	Testis	Testosterone	Development of male secondary sex characteristics
Estrogens	Ovary	Estrone	Development of female secondary sex characteristics
Glucocorticoids	Adrenal cortex	Cortisol	Promotes gluconeogenesis and glycogen formation; enhances fat and protein degradation
Mineralocorticoids	Adrenal cortex	Aldosterone	Increases reabsorption of Na^+ and excretion of K^+ and H^+ by kidney tubules

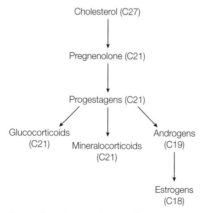

Figure 5. Biosynthetic pathway for the synthesis of the steroid hormones.

The cytochrome P450s are a group of **heme-containing enzymes** (Section M4) that get their name from the wavelength maximum of their absorbance spectra when bound to carbon monoxide. They are present in both the mitochondria and the SER of many cells, and consist of a family of structurally related enzymes with different substrate specificities. In humans, the cytochrome P450s are involved in **drug metabolism**, for example, oxidatively metabolizing drugs such as caffeine and ibuprofen. Although their action is generally beneficial, playing a protective role in removing foreign chemicals, some of the most powerful carcinogens are generated from harmless compounds by the action of the cytochrome P450s. The enzymes all catalyze so-called **monooxygenase** reactions, in which one oxygen atom from molecular oxygen is inserted into the substrate molecule, and the other oxygen atom forms water. The electrons required to bring about the reduction of oxygen to form water are supplied by specialized electron transport chains, which are functionally linked to the P450 enzymes. These electron transport chains usually have NADPH as the ultimate electron donor, so a cytochrome P450-catalyzed reaction is often characterized by the involvement of both O_2 and NADPH.

K6 Lipoproteins

Key Notes

Structure and function

Lipoproteins are globular particles consisting of a hydrophobic core of triacylglycerols and cholesterol esters surrounded by a coat of protein, phospholipid and cholesterol. The apoproteins on the surface help to solubilize the lipids and target the lipoproteins to the correct tissues. There are five different types of lipoprotein, classified according to their functional and physical properties: chylomicrons, very-low-density lipoproteins (VLDLs), intermediate-density lipoproteins (IDLs), low-density lipoproteins (LDLs), and high-density lipoproteins (HDLs). The major function of lipoproteins is to transport triacylglycerols, cholesterol and phospholipids around the body.

Chylomicrons

Chylomicrons are synthesized in the intestine and transport dietary triacylglycerols to skeletal muscle and adipose tissue, and dietary cholesterol to the liver. At these target tissues, the triacylglycerols are hydrolyzed by lipoprotein lipase on the surface of the cells. The resulting cholesterol-rich chylomicron remnants are transported in the blood to the liver where they are taken up by receptor-mediated endocytosis.

VLDLs, IDLs and LDLs

VLDLs are synthesized in the liver and transport triacylglycerols, cholesterol and phospholipids to other tissues, where lipoprotein lipase hydrolyzes the triacylglycerols and releases the fatty acids for uptake. The VLDL remnants are transformed first to IDLs and then to LDLs as all of their apoproteins other than apoB-100 are removed and their cholesterol esterified. The LDLs bind to the LDL receptor protein on the surface of target cells and are internalized by receptor-mediated endocytosis. The cholesterol, which is released from the lipoproteins by the action of lysosomal lipases, is either incorporated into the cell membrane or re-esterified for storage.

HDLs

HDLs are synthesized in the blood and extract cholesterol from cell membranes, converting it into cholesterol esters. Most HDLs are taken up into the liver cells by receptor-mediated endocytosis and the cholesterol disposed of in the form of bile salts.

Atherosclerosis

Atherosclerosis is characterized by cholesterol-rich arterial thickenings (atheromas) that narrow the arteries and cause blood clots to form. If these blood clots block the coronary arteries supplying the heart, the result is a myocardial infarction, or heart attack.

Familial hyper-cholesterolemia	This is an inherited disorder in which individuals have a lack of functional LDL receptors preventing cholesterol from being taken up by the tissues. The resulting high blood cholesterol level leads to an increase in the formation of atheromas and can cause death from myocardial infarction during childhood.	
Related topics	(E1) Membrane lipids (E4) Membrane transport: macromolecules	(K4) Triacylglycerols (K5) Cholesterol

Structure and function

Triacylglycerols (Section K4), phospholipids (Section E1) and cholesterol (Section K5) are relatively insoluble in aqueous solution. Hence, they are transported around the body in the blood as components of **lipoproteins**. These globular, micelle-like particles consist of a hydrophobic core of triacylglycerols and cholesterol esters surrounded by an amphipathic coat of protein, phospholipid and cholesterol. The protein components of lipoproteins are called **apolipoproteins** (or **apoproteins**). At least 10 different apoproteins are found in the different human lipoproteins. Their functions are to help solubilize the hydrophobic lipids and to act as cellular targeting signals. Lipoproteins are classified into five groups on the basis of their physical and functional properties (Table 1).

- **Chylomicrons** are the largest and least dense lipoproteins. They transport dietary (exogenous) triacylglycerols and cholesterol from the intestines to other tissues in the body.

- **Very-low-density lipoproteins** (VLDLs), **intermediate-density lipoproteins** (IDLs) and **low-density lipoproteins** (LDLs) are a group of related lipoproteins that transport internally produced (endogenous) triacylglycerols and cholesterol from the liver to the tissues.

- **High-density lipoproteins** (HDLs) transport endogenous cholesterol from the tissues to the liver.

Chylomicrons

Chylomicrons, the largest of the lipoproteins, are synthesized in the intestine. They transport ingested triacylglycerols to other tissues, mainly skeletal muscle and adipose

Table 1. Characteristics of the five classes of lipoproteins

Lipoprotein	Molecular mass (kDa)	Density (g ml^{-1})	% Protein	Major lipids	Apoproteins
Chylomicrons	>400 000	<0.95	1.5–2.5	TG	A, B-48, C, E
VLDLs	10 000–80 000	<1.006	5–10	TG, PL, CE	B-100, C, E
IDLs	5000–10 000	1.006–1.019	15–20	CE, TG, PL	B-100, C, E
LDLs	2300	1.019–1.063	20–25	CE, PL	B-100
HDLs	175–360	1.063–1.210	40–55	PL, CE	A, C, D, E

C, cholesterol; CE, cholesterol ester; TG, triacylglycerol; PL, phospholipid.

tissue, and transport ingested cholesterol to the liver (Figure 1). At the target tissues, the triacylglycerols are hydrolyzed by the action of **lipoprotein lipase**, an enzyme located on the outside of the cells that is activated by **apoC-II**, one of the apoproteins on the chylomicron surface. The released fatty acids and monoacylglycerols are taken up by the tissues, and either used for energy production or re-esterified to triacylglycerol for storage. As their triacylglycerol content is depleted, the chylomicrons shrink and form cholesterol-rich **chylomicron remnants**, which are transported in the blood to the liver (Figure 1). Here they bind to a specific cell-surface remnant receptor and are taken up into the liver cells by **receptor-mediated endocytosis** (Section E4).

VLDLs, IDLs and LDLs

VLDLs are synthesized in the liver and transport a variety of lipids (Table 1) to other tissues, again mainly adipose tissue and skeletal muscle. As with chylomicrons, the triacylglycerols in VLDLs are acted on by **lipoprotein lipase** and the released fatty acids taken up by the tissues (Figure 1). The VLDL remnants remain in the blood, first as IDLs and then as LDLs. In the transformation to LDLs, much of the cholesterol is esterified on its hydroxyl group on C-3 by the addition of a fatty acid chain from phosphatidylcholine (lecithin) by the enzyme **lecithin–cholesterol acyl transferase (LCAT)**. In addition, all of the apoproteins other than **apoB-100** are removed.

LDLs are then taken up by target cells through **receptor-mediated endocytosis** (Section E4). The **LDL receptor**, a transmembrane glycoprotein on the surface of the target cells, specifically binds apoB-100 in the LDL coat. The receptors then cluster into

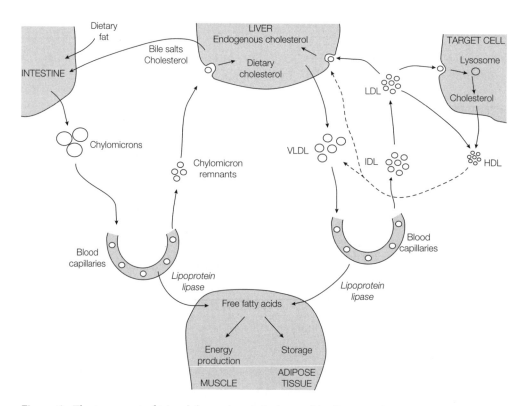

Figure 1. The transport of triacylglycerol and cholesterol by lipoproteins.

clathrin-coated pits and are internalized (Section E4, Figure 3). Once in the lysosomes, the LDLs are digested by lysosomal enzymes, with the cholesterol esters being hydrolyzed by a lysosomal lipase to release the cholesterol (Figure 1). This is then incorporated into the cell membrane and any excess is re-esterified for storage by **acyl CoA cholesterol acyltransferase** (ACAT).

To prevent the build-up of cholesterol and its ester derivatives in the cell, high levels of cholesterol:

● Decrease the synthesis of the LDL receptor, whose gene is regulated by SREBP, thereby reducing the rate of uptake of cholesterol by receptor-mediated endocytosis

● Inhibit the cellular biosynthesis of cholesterol through inhibition of HMG CoA reductase (Section K5)

HDLs

HDLs have the opposite function to that of LDLs in that they remove cholesterol from the tissues. The HDLs are synthesized in the blood mainly from components derived from the degradation of other lipoproteins. HDLs then acquire their cholesterol by extracting it from cell membranes and converting it into cholesterol esters by the action of LCAT (Figure 1). The HDLs are then either taken up directly by the liver or transfer their cholesterol esters to VLDLs, of which about half are taken up by the liver by receptor-mediated endocytosis (Figure 1). The liver is the only organ that can dispose of significant quantities of cholesterol, primarily in the form of **bile salts** (Section K5).

Atherosclerosis

Atherosclerosis, the most common type of **hardening of the arteries**, is characterized by the presence of cholesterol-rich arterial thickenings (**atheromas**). This progressive disease begins with the intracellular deposition of lipids, mainly cholesterol esters, in the smooth muscle cells of the arterial wall. These lesions become fibrous, calcified plaques that narrow and can eventually block the arteries. **Blood clots** are also more likely to occur, which may stop the blood flow and deprive the tissues of oxygen. If these blockages occur in the coronary arteries, those supplying the heart, the result is a **myocardial infarction** or **heart attack**, which is the most common cause of death in Western industrialized countries. Blood clots in cerebral arteries cause **stroke**, while those in peripheral blood vessels in the limbs can lead to possible gangrene and amputation.

Familial hypercholesterolemia

Familial hypercholesterolemia is an inherited disorder in which homozygotes have a markedly elevated level of cholesterol in their blood, whilst in heterozygotes the level is twice that of normal individuals. Not only does this result in the deposition of cholesterol in the skin as yellow nodules known as **xanthomas**, but also in the formation of **atheromas** that can cause death from **myocardial infarction** during childhood. The molecular defect in most cases of familial hypercholesterolemia is the lack of functional **LDL receptors**. Thus, the LDL cholesterol cannot be taken up by the tissues and results in a high concentration in the blood. Homozygotes can be treated by liver transplantation, while heterozygotes can be treated by inhibiting HMG CoA reductase with statins (Section K5) and reducing the intestinal reabsorption of bile salts, thereby decreasing the blood cholesterol level.

L1 Citric acid cycle

Key Notes

Role
The cycle oxidizes pyruvate (formed during the glycolytic breakdown of glucose) to CO_2 and H_2O, with the concomitant production of energy. Acetyl CoA from fatty acid breakdown and amino acid degradation products are also oxidized. In addition, the cycle has a role in producing precursors for biosynthetic pathways.

Location
The citric acid cycle occurs within the mitochondria of eukaryotes and the cytosol of prokaryotes.

The cycle
The citric acid cycle has eight stages:

1. Production of citrate from oxaloacetate and acetyl CoA (catalyzed by citrate synthase).
2. Isomerization of citrate to isocitrate (catalyzed by aconitase).
3. Oxidation of isocitrate to α-ketoglutarate (catalyzed by isocitrate dehydrogenase; the reaction requires NAD^+).
4. Oxidation of α-ketoglutarate to succinyl CoA (catalyzed by the α-ketoglutarate dehydrogenase complex; the reaction requires NAD^+).
5. Conversion of succinyl CoA to succinate [catalyzed by succinyl CoA synthetase; the reaction requires inorganic phosphate and GDP (or ADP)].
6. Oxidation of succinate to fumarate (catalyzed by succinate dehydrogenase; the reaction involves FAD).
7. Hydration of fumarate to malate (catalyzed by fumarase).
8. Oxidation of malate to oxaloacetate (catalyzed by malate dehydrogenase; the reaction requires NAD^+).

Energy yield
For each turn of the cycle, about 10 ATP molecules are produced, one directly from the cycle and about nine from the reoxidation of the three NADH and one $FADH_2$ molecules produced by the cycle by oxidative phosphorylation.

Regulation
The citric acid cycle is regulated at the steps catalyzed by citrate synthase, isocitrate dehydrogenase and α-ketoglutarate dehydrogenase via feedback inhibition by ATP, citrate, NADH and succinyl CoA, and stimulation of isocitrate dehydrogenase by ADP. Pyruvate dehydrogenase, which converts pyruvate to acetyl CoA to enter the cycle, is inhibited by acetyl CoA and NADH. In addition, this enzyme is inactivated by phosphorylation, a reaction catalyzed by pyruvate dehydrogenase kinase. A high ratio of $NADH/NAD^+$,

	acetyl CoA/CoA or ATP/ADP stimulates phosphorylation of pyruvate dehydrogenase and so inactivates this enzyme. Pyruvate inhibits the kinase. Removal of the phosphate group (dephosphorylation) by a phosphatase reactivates pyruvate dehydrogenase.
Biosynthetic pathways	Amino acids, purines and pyrimidines, porphyrins, fatty acids and glucose are all synthesized by pathways that use citric acid intermediates as precursors.
Related topics	(J3) Glycolysis (J4) Gluconeogenesis (K2) Fatty acid breakdown	(K3) Fatty acid synthesis (L2) Electron transport and oxidative phosphorylation (M2) Amino acid metabolism

Role

The citric acid cycle, also known as the **TCA (tricarboxylic acid) cycle** or **Krebs cycle** (after its discoverer in 1937), is used to oxidize the pyruvate formed during the glycolytic breakdown of glucose into CO_2 and H_2O. It also oxidizes acetyl CoA arising from fatty acid degradation (Section K2), and amino acid degradation products (Section M2). In addition, the cycle provides precursors for many biosynthetic pathways.

Location

The citric acid cycle operates in the mitochondria of eukaryotes and in the cytosol of prokaryotes. Succinate dehydrogenase, the only membrane-bound enzyme in the citric acid cycle, is embedded in the inner mitochondrial membrane in eukaryotes and in the plasma membrane in prokaryotes.

The cycle

The citric acid cycle forms the central part of a three-step process, which oxidizes organic fuel molecules into CO_2 with the concomitant production of ATP.

Step 1 – Oxidation of fuel molecules to acetyl CoA

A major source of energy is glucose, which is converted by glycolysis (Section J3) into pyruvate. **Pyruvate dehydrogenase**, a complex of three enzymes and five coenzymes located in the mitochondrial matrix, is the link between glycolysis and the citric acid cycle. It oxidizes the pyruvate to form acetyl CoA and CO_2.

$$\text{Pyruvate} + \text{CoA} + \text{NAD}^+ \rightarrow \text{acetyl CoA} + CO_2 + \text{NADH} + H^+$$

This irreversible reaction involves both an oxidation and a loss of CO_2 and so the process is called **oxidative decarboxylation**.

Step 2 – The citric acid cycle

The cycle carries out the oxidation of acetyl groups from acetyl CoA to CO_2 with the production of four pairs of electrons, stored initially in the reduced electron carriers NADH and $FADH_2$. The molecular structures of the metabolic intermediates are given in Figure 1.

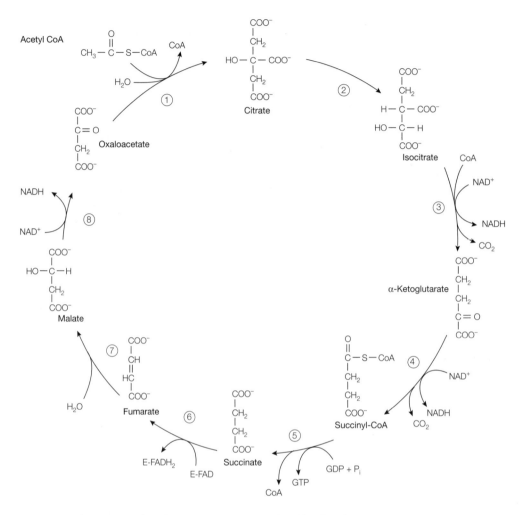

Figure 1. The citric acid cycle (reactions 1–8 are described in the text).

The cycle has eight stages:

1. Citrate (6C) is formed from the irreversible condensation of acetyl CoA (2C) and oxaloacetate (4C) – catalyzed by **citrate synthase**.

$$\text{Acetyl CoA} + \text{oxaloacetate} + H_2O \rightarrow \text{citrate} + \text{CoA}$$

2. Citrate is converted to isocitrate (6C) by an isomerization catalyzed by **aconitase**. This is actually a two-step reaction during which *cis*-aconitate is formed as an intermediate. It is the *cis*-aconitate that gives the enzyme its name.

$$\text{Citrate} \underset{}{\overset{H_2O}{\rightleftharpoons}} \textit{cis}\text{-aconitate} \underset{}{\overset{H_2O}{\rightleftharpoons}} \text{isocitrate}$$

3. Isocitrate is oxidized to α-ketoglutarate (5C) and CO_2 by **isocitrate dehydrogenase**. This mitochondrial enzyme requires NAD^+, which is reduced to NADH.

$$\text{Isocitrate} + NAD^+ \rightarrow \text{α-ketoglutarate} + CO_2 + \text{NADH}$$

4. α-Ketoglutarate is oxidized to succinyl CoA (4C) and CO_2 by the α-**ketoglutarate dehydrogenase complex**. Like pyruvate dehydrogenase, this is a complex of three enzymes and uses NAD^+ as a cofactor.

$$\text{α-Ketoglutarate} + \text{CoA} + NAD^+ \rightarrow \text{succinyl CoA} + CO_2 + \text{NADH}$$

5. Succinyl CoA is converted to succinate (4C) by **succinyl CoA synthetase**. The reaction uses the energy released by cleavage of the succinyl–CoA bond to synthesize either GTP (mainly in animals) or ATP (exclusively in plants) from P_i and, respectively, GDP or ADP, e.g.

$$\text{Succinyl CoA} + P_i + \text{GDP} \rightarrow \text{succinate} + \text{CoA} + \text{GTP}$$

6. Succinate is oxidized to fumarate (4C) by **succinate dehydrogenase**. FAD is tightly bound to the enzyme (indicated in the equation below as E-FAD) and is reduced to produce $FADH_2$.

$$\text{Succinate} + \text{E-FAD} \rightarrow \text{fumarate} + \text{E-FADH}_2$$

7. Fumarate is converted to malate (4C) by **fumarase**; this is a hydration reaction requiring the addition of a water molecule.

$$\text{Fumarate} + H_2O \rightarrow \text{malate}$$

8. Malate is oxidized to oxaloacetate (4C) by **malate dehydrogenase**. NAD^+ is again required by the enzyme as a cofactor to accept the free pair of electrons and produce NADH.

$$\text{Malate} + NAD^+ \rightleftharpoons \text{oxaloacetate} + \text{NADH} + H^+$$

Step 3 – Oxidation of NADH and FADH$_2$ produced by the citric acid cycle

The NADH and $FADH_2$ produced by the citric acid cycle are reoxidized and the energy released is used to synthesize ATP by **oxidative phosphorylation** (Section L2).

Energy yield

The overall reaction for the citric acid cycle is as follows:

$$\text{Acetyl CoA} + 3NAD^+ + \text{E-FAD} + \text{GDP} + P_i + 2H_2O \rightarrow 2CO_2 + 3\text{NADH}$$
$$+ \text{E-FADH}_2 + \text{GTP} + 2H^+ + \text{CoA}$$

Each of the three NADH molecules produced per turn of the cycle yields about 2.5 ATPs and the single $E\text{-FADH}_2$ yields about 1.5 ATPs by oxidative phosphorylation (Section L2). One GTP (or ATP) is synthesized directly during the conversion of succinyl CoA to succinate. Thus the oxidation of a single molecule of glucose via the citric acid cycle produces approximately 10 ATP molecules.

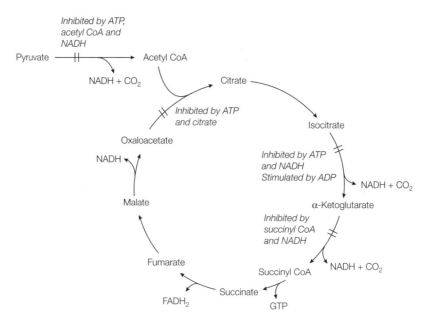

Figure 2. Regulation points of the citric acid cycle.

Regulation

Regulation of the cycle is governed by substrate availability, inhibition by accumulating products, and allosteric feedback inhibition by subsequent intermediates in the cycle. Three enzymes in the cycle itself are regulated (citrate synthase, isocitrate dehydrogenase and α-ketoglutarate dehydrogenase) and so is the enzyme that converts pyruvate to acetyl CoA to enter the cycle, namely pyruvate dehydrogenase (Figure 2):

- Citrate synthase is inhibited by citrate and also by ATP (the K_m for acetyl CoA is raised as the level of ATP rises).

- Isocitrate dehydrogenase is inhibited by NADH and ATP but activated by ADP.

- α-ketoglutarate dehydrogenase is inhibited by NADH and succinyl CoA.

- Pyruvate dehydrogenase is inhibited by NADH and acetyl CoA (i.e. product inhibition). However, in eukaryotes the enzyme is also controlled by phosphorylation/dephosphorylation via **pyruvate dehydrogenase kinase** and a **phosphatase**. The kinase catalyzes the phosphorylation of a specific Ser residue in pyruvate dehydrogenase, using ATP as the phosphate donor, and this inactivates the enzyme. Removal of the phosphate group by the phosphatase reactivates the enzyme. At any one time, the activity of pyruvate dehydrogenase is determined by the relative balance between the kinase and phosphatase reactions. Increasing the NADH/NAD$^+$, acetyl CoA/CoA or ATP/ADP ratio stimulates phosphorylation and hence inactivates pyruvate dehydrogenase. As pyruvate builds up, it inhibits the kinase and hence allows the phosphatase to reactivate pyruvate dehydrogenase, thus stimulating pyruvate conversion to acetyl CoA.

Overall, the cycle speeds up when cellular energy levels are low (high ADP concentration, low ATP and NADH) and slows down as ATP (and then NADH, succinyl CoA and citrate) accumulates.

Biosynthetic pathways

The intermediates in the cycle provide precursors for many biosynthetic pathways. For example:

- Synthesis of fatty acids from citrate (Section K3)

- Amino acid synthesis following transamination of α-ketoglutarate (Section M2)

- Synthesis of purine and pyrimidine nucleotides from α-ketoglutarate and oxaloacetate

- Oxaloacetate can be converted to glucose by gluconeogenesis (Section J4)

- Succinyl CoA is a central intermediate in the synthesis of the porphyrin ring of heme groups (Section M4)

L2 Electron transport and oxidative phosphorylation

Key Notes

Overview

Electron transport and oxidative phosphorylation reoxidize NADH and $FADH_2$ and trap the energy released as ATP. In eukaryotes, electron transport and oxidative phosphorylation occur in the inner membrane of mitochondria whereas in prokaryotes the process occurs in the plasma membrane.

Redox potential

The oxidation–reduction potential, E (or redox potential), of a substance is a measure of its affinity for electrons. The standard redox potential (E_0') is measured under standard conditions, at pH 7, and is expressed in volts. The standard free energy change of a reaction at pH 7, $\Delta G^{0\prime}$, can be calculated from the change in redox potential $\Delta E_0'$ of the substrates and products. A reaction with a positive $\Delta E_0'$ has a negative $\Delta G^{0\prime}$ (i.e. is exergonic).

Electron transport from NADH

Electrons are transferred from NADH to oxygen along the electron transport chain (also called the respiratory chain). NADH passes electrons to NADH-Q oxidoreductase (Complex I), a large protein complex that contains FMN and two types of iron–sulfur (FeS) clusters in iron–sulfur proteins. The electrons are accepted by the FMN to produce $FMNH_2$ and then passed to the iron atoms of the FeS clusters, which accept and donate electrons by alternating between Fe^{3+} and Fe^{2+} states. Electrons from NADH-Q oxidoreductase are passed to ubiquinone (coenzyme Q, Q), converting it to ubiquinol (or QH_2), and then to Q-cytochrome c oxidoreductase (Complex III). This contains cytochrome b and cytochrome c_1, as well as an FeS protein. A cytochrome contains a heme group with a central iron atom, which changes from the Fe^{3+} state to the Fe^{2+} state on accepting an electron. When the electron is donated to another component, the iron atom changes back to the Fe^{3+} state. The Q-cytochrome c oxidoreductase passes the electrons to cytochrome c, which in turn passes them to cytochrome c oxidase (Complex IV), a complex that contains two cytochromes (cytochrome a and a_3) paired with copper atoms (Cu_A and Cu_B, respectively). During electron transfer, the copper atoms cycle between the Cu^{2+} and Cu^+ states. Finally, cytochrome c oxidase passes four electrons to molecular oxygen to form two molecules of water.

Formation of an H⁺ gradient

The change in redox potential along the chain is a measure of the free energy change at each step. At the steps involving

NADH-Q oxidoreductase, Q-cytochrome *c* oxidoreductase and cytochrome *c* oxidase, the free energy change is large enough to pump H^+ ions across the inner mitochondrial membrane, from the mitochondrial matrix into the intermembrane space, to create an H^+ gradient. Therefore, each of these complexes is an H^+ pump driven by electron transport.

Electron transport from FADH$_2$

$FADH_2$ is reoxidized to FAD by donating two electrons to succinate-Q reductase (Complex II), a protein complex that contains FeS clusters. It passes the electrons on to ubiquinone in the main electron transport chain where their further transport leads to the formation of an H^+ gradient and ATP synthesis. However, succinate-Q reductase does not itself pump H^+ ions.

Electron transport inhibitors

Rotenone and amytal inhibit electron transport at NADH-Q oxidoreductase, antimycin A inhibits Q-cytochrome *c* oxidoreductase, and cyanide (CN^-), azide (N_3^-) and carbon monoxide (CO) all inhibit cytochrome *c* oxidase.

Oxidative phosphorylation

Oxidative phosphorylation is ATP synthesis linked to the oxidation of NADH and $FADH_2$ by electron transport through the respiratory chain. This occurs via a mechanism originally proposed as the chemiosmotic hypothesis. Energy liberated by electron transport is used to pump H^+ ions out of the mitochondrion to create an electrochemical proton (H^+) gradient. The protons flow back into the mitochondrion through the ATP synthase located in the inner mitochondrial membrane, and this drives ATP synthesis.

ATP synthase as a rotatory engine

ATP synthase is located in the inner mitochondrial membrane. It consists of two major components, F_1 ATPase attached to component F_0 (coupling factor 0). Hence, ATP synthase is also known as F_0F_1 ATPase. In mitochondria, this complex uses the energy released by electron transport to drive ATP synthesis. The F_1 ATPase has a subunit structure of $\alpha_3\beta_3\gamma\delta\epsilon$ whereas F_0 consists of a ring of 10–14 c subunits. During ATP synthesis, protons flow through a channel between the α subunits and c ring, which causes the c ring to rotate. The rotating c ring turns the γ subunit of F_1 ATPase within a ring formed by the α and β subunits. Conformational changes to the β subunits now lead to ATP synthesis.

Flow of ATP and ADP across the inner mitochondrial membrane

ATP exits the mitochondrial matrix and ADP enters via an ATP-ADP translocase protein that is located in the inner mitochondrial membrane; one ATP exits for one ADP entering. Likewise, P_i enters in exchange for OH^- exiting via a phosphate carrier protein also located in the inner mitochondrial membrane. These two transport proteins are associated with ATP synthase to form the ATP synthasome.

ATP yield

Approximately 2.5 ATP molecules are synthesized per NADH

	oxidized and approximately 1.5 ATPs are synthesized per $FADH_2$ oxidized.
Coupling and respiratory control	Electron transport is normally tightly coupled to ATP synthesis; electrons do not flow through the electron transport chain to oxygen unless ADP is simultaneously phosphorylated to ATP. If ADP is available, electron transport proceeds and ATP is made; as the ADP concentration falls, electron transport slows down. This process, called respiratory control, ensures that electron flow occurs only when ATP synthesis is required.
Uncouplers	Some chemicals (e.g. 2,4-dinitrophenol; DNP) are uncoupling agents; they allow electron transport to proceed without ATP synthesis. They uncouple mitochondria by carrying H^+ ions across the inner mitochondrial membrane and hence dissipate the proton gradient. The energy derived from uncoupled electron transport is released as heat. Uncoupling also occurs naturally in some tissues (e.g. the mitochondria of brown adipose tissue are uncoupled by a protein called thermogenin). The resulting production of heat (nonshivering thermogenesis) by the adipose tissue serves to protect sensitive body tissues in newborn animals and to maintain body temperature during hibernation.
Reoxidation of cytosolic NADH	Cytosolic NADH cannot cross the inner mitochondrial membrane and enter mitochondria to be reoxidized. However, it can be reoxidized via the glycerol 3-phosphate shuttle. Cytosolic glycerol 3-phosphate dehydrogenase oxidizes the NADH and reduces dihydroxyacetone phosphate to glycerol 3-phosphate. The glycerol 3-phosphate enters the mitochondrion and is converted back to dihydroxyacetone phosphate by mitochondrial glycerol 3-phosphate dehydrogenase (an FAD-linked enzyme). The dihydroxyacetone phosphate diffuses back to the cytosol. The enzyme-linked $FADH_2$ is reoxidized by transferring its electrons to ubiquinone in the electron transport chain. Since the electrons enter the electron transport chain from $FADH_2$, only about two ATPs are synthesized per molecule of cytosolic NADH. In heart and liver, cytosolic NADH can be reoxidized via the malate–aspartate shuttle. Oxaloacetate in the cytosol is reduced to malate by NADH and enters the mitochondrion via a malate–α-ketoglutarate carrier. In the matrix, the malate is reoxidized to oxaloacetate by NAD^+, which is converted to NADH, resulting in a net transfer of electrons from cytosolic NADH to matrix NADH. The oxaloacetate is converted to aspartate by transamination, leaves the mitochondrion and is reconverted to oxaloacetate in the cytosol, again by transamination.
Related topics	(J3) Glycolysis (L3) Photosynthesis (L1) Citric acid cycle

Overview

In eukaryotes, electron transport and oxidative phosphorylation occur in the inner membrane of mitochondria. These processes reoxidize the NADH and $FADH_2$ that arise from the citric acid cycle (located in the mitochondrial matrix; Section L1), glycolysis (located in the cytoplasm; Section J3) and fatty acid oxidation (located in the mitochondrial matrix; Section K2) and trap the energy released as ATP. In prokaryotes, the components of electron transport and oxidative phosphorylation are located in the plasma membrane (Section A1). Oxidative phosphorylation is by far the major source of ATP in the cell.

Redox potential

The oxidation of a molecule involves the loss of electrons. The reduction of a molecule involves the gain of electrons. Since electrons are not created or destroyed in a chemical reaction, if one molecule is oxidized, another must be reduced (i.e. it is an **oxidation–reduction reaction**). Thus, by definition, oxidation–reduction reactions involve the transfer of electrons. In the oxidation–reduction reaction:

$$NADH + H^+ + ½ O_2 \rightleftharpoons NAD^+ + H_2O$$

when the NADH is oxidized to NAD^+, it loses electrons. When the molecular oxygen is reduced to water, it gains electrons.

The **oxidation–reduction potential**, E (or **redox potential**), is a measure of the affinity of a substance for electrons and is measured relative to hydrogen. A positive redox potential means that the substance has a higher affinity for electrons than does hydrogen and so would accept electrons from hydrogen. A substance with a negative redox potential has a lower affinity for electrons than does hydrogen and would donate electrons to H^+, forming hydrogen. In the example above, NADH is a strong reducing agent with a negative redox potential and has a tendency to donate electrons. Oxygen is a strong oxidizing agent with a positive redox potential and has a tendency to accept electrons.

For biological systems, the **standard redox potential** for a substance (E_0') is measured under standard conditions, at pH 7, and is expressed in volts. In an oxidation–reduction reaction, where electron transfer is occurring, the total voltage change of the reaction (change in electric potential, ΔE) is the sum of the voltage changes of the individual oxidation–reduction steps. The standard free energy change of a reaction at pH 7, $\Delta G^{0\prime}$, can be readily calculated from the change in redox potential $\Delta E_0'$ of the substrates and products:

$$\Delta G^{0\prime} = -nF \Delta E_0'$$

where n is the number of electrons transferred, $\Delta E_0'$ is in volts (V), $\Delta G^{0\prime}$ is in kilocalories per mole (kcal mol^{-1}) and F is a constant called the Faraday (23.06 kcal V^{-1} mol^{-1}). Note that a reaction with a **positive** $\Delta E_0'$ has a **negative** $\Delta G^{0\prime}$ (i.e. is exergonic).

Thus for the reaction:

$$NADH + H^+ + ½ O_2 \rightleftharpoons NAD^+ + H_2O$$

$$\Delta E_0' = +1.14 \, V$$

$$\Delta G^{0\prime} = -52.6 \, kcal \, mol^{-1}$$

Electron transport from NADH

Comparing the energetics of the oxidation of NADH:

$$NADH + H^+ + \tfrac{1}{2} O_2 \rightleftharpoons NAD^+ + H_2O \qquad\qquad \Delta G^{0\prime} = -52.6 \text{ kcal mol}^{-1}$$

and the synthesis of ATP:

$$ADP + P_i + H^+ \rightleftharpoons ATP + H_2O \qquad\qquad \Delta G^{0\prime} = +7.3 \text{ kcal mol}^{-1}$$

it is clear that the oxidation of NADH releases sufficient energy to drive the synthesis of several molecules of ATP. However, NADH oxidation and ATP synthesis do not occur in a single step. Electrons are not transferred from NADH to oxygen directly. Rather the electrons are transferred from NADH to oxygen along a chain of electron carriers collectively called the **electron transport chain** (also called the **respiratory chain**).

Most of the protein components of the electron transport chain exist as four large protein complexes embedded in the inner mitochondrial membrane called:

● NADH-Q oxidoreductase (Complex I)

● Succinate-Q reductase (Complex II)

● Q-cytochrome *c* oxidoreductase (Complex III)

● Cytochrome *c* oxidase (Complex IV)

Electrons flow from NADH to oxygen through three of these complexes as shown in Figure 1. Each complex contains several electron carriers (see below) that work sequentially to carry electrons down the chain. Two small electron carriers are also needed to link these large complexes; **ubiquinone**, which is also called **coenzyme Q** (abbreviated as **Q**), and **cytochrome *c*** (Figure 1).

The flow of electrons from NADH down the respiratory chain can be summarized as follows:

1. NADH-Q oxidoreductase (Complex I)

NADH-Q oxidoreductase (Complex I) is a massive complex of at least 46 polypeptides. It binds NADH and reoxidizes it to NAD⁺, passing the two electrons from NADH to a prosthetic group called **FMN** (Figure 2) to produce **FMNH₂** (see Section D1 for structure of FMN). Each electron is accepted together with a hydrogen ion, H⁺, such that two electrons and two H⁺ are accepted in total. The electrons are then transferred, within Complex I, to **iron–sulfur clusters** (FeS) in **iron–sulfur proteins** (also called **nonheme**

Figure 1. Overview of the electron transport chain (respiratory chain).

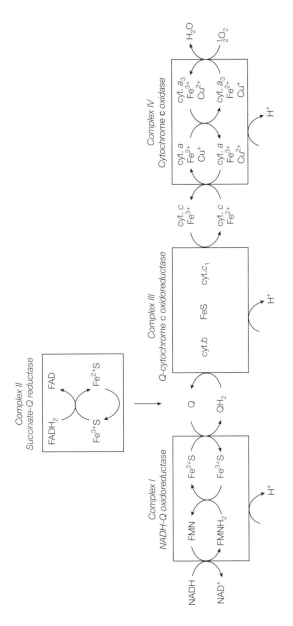

Figure 2. Details of electron transport.

iron proteins). Several types of FeS clusters exist but in each case the iron atoms are coordinated to inorganic sulfur atoms. Within an FeS cluster, an electron is carried by the iron atom, which, on accepting the electron, changes from the Fe^{3+} (ferric) state to the Fe^{2+} (ferrous) state (Figure 2). As the electron is passed to another electron carrier, the iron atom of the FeS cluster changes back again to the Fe^{3+} state.

2. Ubiquinone

Electrons from the FeS clusters of Complex I are passed on to ubiquinone (Q), a small lipid-soluble molecule in the inner mitochondrial membrane. This molecule can act as an electron carrier by accepting up to two electrons and two H^+ ions. In so doing, ubiquinone (Q) is converted to **ubiquinol** (QH_2).

3. Q-cytochrome c oxidoreductase (Complex III)

When ubiquinol (QH_2) donates its two electrons to the next carrier in the chain, the **Q-cytochrome *c* oxidoreductase** (**Complex III**), the H^+ ions are released once more. This complex (Figure 2) contains two types of cytochromes (**cytochrome *b*** and **cytochrome c_1**) as well as an iron–sulfur protein.

A cytochrome is a protein with a bound **heme group** that contains an iron atom (Section M4, Figure 1). Different cytochromes have different heme groups, but all cytochromes have the ability to act as electron carriers. As the electron is accepted, the iron atom of the heme group changes from the Fe^{3+} (ferric) state to the Fe^{2+} (ferrous) state. Cytochrome *b* has two hemes called heme b_L (L for low affinity) and heme b_H (H for high affinity). Cytochrome c_1 has a single heme.

The iron–sulfur protein has two iron ions and two inorganic sulfur atoms in a so-called **Rieske 2Fe2S center** (named after John Rieske, its discoverer) with one iron atom coordinated with two cysteine residues and one iron atom coordinated with two histidine residues (Figure 3).

Figure 2 shows the electrons passing from ubiquinol (QH_2) through the cytochrome *b*, FeS and cytochrome c_1 components of the Q-cytochrome *c* oxidoreductase complex to the next electron carrier, **cytochrome *c*.**

Figure 3. The Rieske iron–sulfur center showing two iron atoms, one coordinated to two cysteine residues and one coordinated to two histidine residues.

Since ubiquinol is a two-electron carrier whereas cytochromes are one-electron carriers, the pathway of electron transfer within the Q-cytochrome c oxidoreductase complex is complicated. The process involved, for coupling electron transfer from Q to cytochrome c and transporting protons across the membrane, is called the **Q cycle** (Figure 4):

● Two QH_2 molecules bind one after the other to the Q-cytochrome c oxidoreductase complex.

● The first QH_2 binds to a site in Complex III called a Q_o site (Figure 4a). It gives up two electrons and two H^+ ions (which are released to the intermembrane space i.e. between the two mitochondrial membranes). One electron is donated to the Rieske 2Fe2S center and then to cytochrome c_1 and finally to cytochrome c (reducing it). The other electron flows through the two heme groups of cytochrome b (first b_L and then b_H) and then to another Q bound to Complex III at a second site called a Q_i site (Figure 4a). When it receives the electron, this Q molecule becomes a semiquinone radical anion ($Q \cdot^-$).

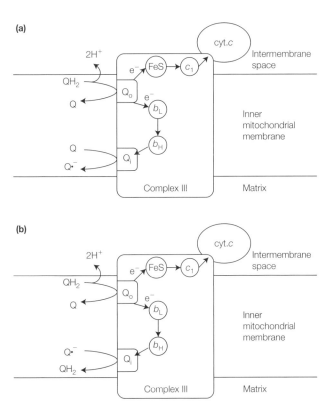

Figure 4. The Q cycle. (a) In the first phase, two electrons from a QH_2 bound at the Q_o site are transferred, one to cytochrome c and the other to a molecule of Q bound at the Q_i site to form the semiquinone $Q \cdot^-$. The newly formed Q dissociates from the Q_o site. (b) A second QH_2 binds to the Q_o site and donates two electrons, one to a second molecule of cytochrome c and the other to reduce the semiquinone $Q \cdot^-$ to QH_2, taking up two protons from the matrix in the process.

- The second QH_2 also binds to the Qo site and gives up two electrons and two H^+ ions (which are released to the intermembrane space). One electron flows via the Rieske center and cytochrome c_1 to another molecule of cytochrome c, reducing it (Figure 4b). The second electron flows through cytochrome b (first b_L and then b_H) and then to $Q_{\cdot}^-$ at the Q_i site, which takes up two protons (from the mitochondrial matrix) to form QH_2.

- Overall therefore, two QH_2 molecules are oxidized to two Q molecules and one Q molecule is reduced to QH_2. So the net result is that one QH_2 is converted to Q. In addition, two cytochrome c molecules each receive one electron and are reduced, four protons are released in the cytoplasmic side of the membrane and two protons are taken up from the mitochondrial matrix.

4. Cytochrome c

Cytochrome c is a peripheral membrane protein that is loosely bound to the outer surface of the inner mitochondrial membrane. It binds to the Q-cytochrome c oxidoreductase complex and accepts an electron via an Fe^{3+} to Fe^{2+} transition. Then it binds to the cytochrome c oxidase complex (Complex IV) and donates the electron, with the iron atom of the heme of cytochrome c then reverting to the Fe^{3+} state (Figure 2).

5. Cytochrome c oxidase (Complex IV)

Cytochrome c oxidase (Complex IV) contains two cytochromes (cytochrome a and a_3). Cytochrome a is paired with a copper atom, Cu_A, and cytochrome a_3 is paired with a different copper atom, Cu_B. During electron transfer, the iron atoms of the cytochromes cycle between the Fe^{3+} and Fe^{2+} states whilst the copper atoms cycle between Cu^{2+} and Cu^+. The cytochrome c oxidase reaction is complex; it transfers four electrons from four cytochrome c molecules and four H^+ ions to molecular oxygen to form two molecules of water:

$$\textit{Cytochrome c oxidase}$$
$$4 \text{ cyt. } c \text{ (Fe}^{2+}) + 4\,H^+ + O_2 \longrightarrow 4 \text{ cyt. } c \text{ (Fe}^{3+}) + 2\,H_2O$$

Formation of an H^+ gradient

All of the electron carriers in the electron transport chain interact according to their redox potentials. Every time that an electron transfer occurs, the accepting carrier has a higher affinity for electrons than the donating carrier. Thus there is a net flow of electrons from NADH (most negative redox potential, least affinity for electrons) to oxygen (most positive redox potential, highest affinity for electrons). This ensures a unidirectional flow of electrons. However, note that each cytochrome, each FeS center and each copper atom can carry only one electron but each NADH donates two electrons. Furthermore, each molecule of oxygen (O_2) needs to accept four electrons to be reduced to a molecule of water, H_2O. The various components are arranged in such a manner as to allow their different electron-handling properties to work in harmony.

The change in redox potential along the chain is a measure of the free energy change occurring (see above). The potential falls (i.e. becomes more positive) throughout the chain but mainly in three large steps that correspond to the three main protein complexes: Complex I, III and IV. The large free energy change at each of these three steps, and only these three steps, is large enough to pump H^+ ions from the mitochondrial matrix across the inner mitochondrial membrane and into the intermembrane space (Figure 5).

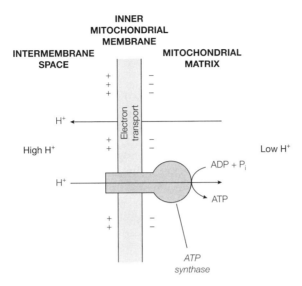

Figure 5. The mechanism of oxidative phosphorylation.

Thus, each of these three complexes (Complex I, III and IV) is an **H⁺ pump** driven by electron transport (Figures 1 and 2). Overall, therefore, electron transport along the chain from NADH releases energy that is used to create an **H⁺ gradient**.

Electron transport from FADH$_2$

Succinate dehydrogenase catalyzes the oxidation of succinate to fumarate in the citric acid cycle (Section L1). The succinate dehydrogenase contains bound FAD that is reduced to FADH$_2$ in the reaction. The reoxidation of the FADH$_2$ occurs via **succinate–Q reductase** (**Complex II**), an integral protein of the inner mitochondrial membrane. Succinate dehydrogenase is part of this complex but it also contains FeS clusters. During reoxidation of FADH$_2$, the two electrons pass from the FADH$_2$ to the FeS clusters and are then passed on to ubiquinone (Q; Figure 2). They then enter the main electron transport chain and cause H⁺ ions to be pumped out of the mitochondrion as for the oxidation of NADH. However, succinate–Q reductase itself is *not* an H⁺ pump because the free energy change of the overall reaction is too small. The FADH$_2$ of other **flavoproteins**, such as mitochondrial glycerol 3-phosphate dehydrogenase in the glycerol 3-phosphate shuttle (see below) and fatty acyl CoA dehydrogenase in fatty acid oxidation (Section K2), also feed their electrons into the electron transport chain at ubiquinone.

Electron transport inhibitors

Several inhibitors of specific electron carriers are known and were used in the original studies to determine the order of the components in the respiratory chain. For example:

- **rotenone** and **amytal** inhibit electron transport at NADH-Q oxidoreductase and so prevent NADH oxidation but the oxidation of FADH$_2$ can still occur since this feeds electrons into the chain at Q (Figure 1) (i.e. past the point of inhibition);

- **antimycin A** inhibits electron transport at the Q-cytochrome *c* oxidoreductase complex;

- **cyanide** (CN^-), **azide** (N_3^-) and **carbon monoxide** (CO) all inhibit cytochrome *c* oxidase.

Oxidative phosphorylation

Oxidative phosphorylation is the name given to the synthesis of ATP (*phosphorylation*) that occurs when NADH and $FADH_2$ are oxidized (hence *oxidative*) by electron transport through the respiratory chain. Unlike substrate level phosphorylation (Sections J3 and L1), it does not involve phosphorylated chemical intermediates. Rather, a very different mechanism was proposed by Peter Mitchell in 1961, the **chemiosmotic hypothesis**. This proposes that energy liberated by electron transport is used to create a proton gradient across the mitochondrial inner membrane and that it is this that is used to drive ATP synthesis. *Thus the proton gradient couples electron transport and ATP synthesis, not a chemical intermediate.* The evidence is overwhelming that this is indeed the way that oxidative phosphorylation works. The actual synthesis of ATP is carried out by an enzyme called **ATP synthase** located in the inner mitochondrial membrane (Figure 5).

In summary, the process is as follows. Electron transport down the respiratory chain from NADH oxidation causes H^+ ions to be pumped out of the mitochondrial matrix across the inner mitochondrial membrane into the intermembrane space (Figure 5) by the three H^+ pumps; Complex I, III and IV (see above). [Because $FADH_2$ is reoxidized via ubiquinone (Figures 1 and 2), its oxidation causes H^+ ions to be pumped out only by Complex III and IV and so the amount of ATP made from $FADH_2$ is less than from NADH.]

The free energy change in transporting an electrically charged ion across a membrane is related both to its electrical charge and the concentration of the species. The pumping out of the H^+ ions generates a higher concentration of H^+ ions in the intermembrane space and an electrical potential, with the side of the inner mitochondrial membrane facing the intermembrane space being positive (Figure 5). Thus, overall, an **electrochemical proton gradient** is formed. The protons flow back into the mitochondrial matrix through the ATP synthase and this drives ATP synthesis. The ATP synthase is driven by **proton-motive force**, which is the sum of the pH gradient (i.e. the chemical gradient of H^+ ions) and the membrane potential (i.e. the electrical charge potential across the inner mitochondrial membrane).

ATP synthase as a rotatory engine

The ATP synthase can be seen as spherical projections from the inner membrane (Figure 6). If mitochondria are subjected to sonic disruption, submitochondrial vesicles are formed in which the spheres of the ATP synthase point outward (Figure 6). In 1960, Racker showed that the spheres can be removed and that the isolated spheres *hydrolyze* ATP, that is, the spheres have ATPase activity (called F_1 **ATPase**; Figure 7). The stripped submitochondrial vesicles, devoid of the F_1 ATPase, can still transport electrons along the electron transport chain but cannot synthesize ATP. These stripped submitochondrial vesicles contain the other major part of the ATP synthase, called F_0 (coupling factor 0), which spans the inner mitochondrial membrane (Figure 7). Since it is composed of these two major component parts, ATP synthase is also known as F_0F_1 **ATPase**. The complete complex harnesses the energy released by electron transport to drive ATP synthesis whereas alone, without coupling to electron transport, the F_1 component hydrolyzes ATP.

The F_1 ATPase consists of five types of polypeptides in the following ratio: $\alpha 3$, $\beta 3$, γ, δ, ϵ. The six α and β subunits are arranged alternately in a ring, with a central stalk formed by the γ and ϵ subunits (Figure 7). F_0 consists of a ring of 10 to 14 **c subunits** embedded

Figure 6. Sonic disruption (sonication) of mitochondria produces submitochondrial vesicles.

in the inner mitochondrial membrane. This contacts a single **a subunit** that links to **two b subunits** and the single δ subunit to form a long column that connects to the head of the F_1 ATPase (Figure 7). This overall structure, the F_0F_1 ATPase, is a remarkable molecular motor. During ATP synthesis, protons flow through a channel created at the interface between the **a** subunit and **c** subunits and this causes the **c** subunit ring to spin relative to the static **a** subunit. Thus the **c** ring acts as a moving part or 'rotor' and the **a** subunit as a 'stator'. As the **c** subunit ring rotates, it turns the γε subunit stalk, which therefore turns rapidly inside the αβ ring. The αβ ring cannot rotate since it is held in place by the arm of the two **b** subunits and δ subunit (Figure 7).

The three β subunits in the αβ ring can bind ADP and inorganic phosphate. As the γ subunit rotates inside the αβ ring, the mechanical energy is used to drive ATP synthesis; three molecules of ATP are synthesized for each 360° rotation of the γ subunit (which equates to over 1000 molecules per second!).

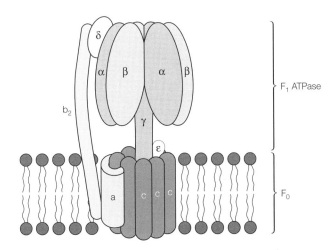

Figure 7. Schematic representation of the ATP synthase complex.

How is the ATP made? Isotope-exchange experiments showed that ATP is formed by ATP synthase even in the absence of a proton-motive force but the ATP does not leave the enzyme unless protons are flowing through it. Paul Boyer proposed the **binding-charge mechanism**, which accounts for this observation. Essentially, as the rotation causes the γ subunit to turn past the three β subunits in turn, the protein conformation of the nucleotide binding sites of each of the β subunits changes through three different states, which are ADP and P_i binding, ATP synthesis and then ATP release. So the mechanical energy of the rotating γ subunit drives the interconversion of these three different states in which ATP is not only synthesized but also released from the enzyme.

Flow of ATP and ADP across the inner mitochondrial membrane

The large amounts of ATP synthesized within mitochondria need to exit to the cytoplasm for use in numerous cellular reactions and ADP needs to enter the mitochondrial matrix to be phosphorylated. However, neither molecule can diffuse across the membrane. Their passage across this barrier is made possible by a very abundant transport protein, **ATP-ADP translocase**, that is sited in the inner mitochondrial membrane. It can bind ATP or ADP and alternately faces the matrix and the cytoplasmic side of the inner membrane; it binds an ADP on the cytoplasmic side and then releases this on the matrix side, binds an ATP and releases this on the cytoplasmic side and so on. Crucially for every ATP that exits, one ADP enters and vice versa.

The other component required for ATP synthesis inside the matrix is of course P_i. This enters via a **phosphate carrier** in exchange for OH^- leaving. In combination, the ATP-ADP translocase and the phosphate carrier allow one ADP and one P_i to enter in exchange for one ATP and one OH^- leaving the mitochondrion (at the cost of one H^+ entering). These two transport proteins in association with ATP synthase form a large complex called the **ATP synthasome**.

ATP yield

The exact number of ATP molecules generated by the oxidation of NADH and $FADH_2$ has been the subject of debate in recent years because its calculation depends on the stoichiometries of proton pumping.

Current data suggest that NADH-Q oxidoreductase, Q-cytochrome *c* oxidoreductase and cytochrome *c* oxidase pump 4, 2 and 4 electrons, respectively, out of the matrix and across the inner mitochondrial membrane. About three protons flowing through ATP synthase are enough to generate one ATP molecule and an additional proton is needed to transport ATP from the mitochondrial matrix to the cytoplasm (see above). So one NADH molecule reoxidized by the electron transport chain causes about 10 protons to be transported across the inner mitochondrial membrane and about four protons are needed to synthesize one ATP molecule; in other words about 2.5 ATP molecules are synthesized for every NADH oxidized. For the oxidation of $FADH_2$, the yield is about 1.5 molecules of ATP.

Coupling and respiratory control

Electron transport is normally tightly **coupled** to ATP synthesis (i.e. electrons do not flow through the electron transport chain to oxygen unless ADP is simultaneously phosphorylated to ATP). Clearly, it also follows that ATP is not synthesized unless electron transport is occurring to provide the proton gradient. Thus oxidative phosphorylation needs NADH or $FADH_2$, oxygen, ADP and inorganic phosphate. The actual rate of oxidative phosphorylation is set by the availability of ADP. If ADP is added to mitochondria, the rate of oxygen consumption rises as electrons flow down the chain and then the rate of oxygen utilization falls when all the ADP has been phosphorylated to ATP; a process called **respiratory control**. This mechanism ensures that electrons flow down the chain only when ATP synthesis is needed. If the level of ATP is high and the ADP level is low, no electron transport occurs, NADH and $FADH_2$ build up, as does excess citrate, and the citric acid cycle (Section L1) and glycolysis (Section J3) are inhibited.

Uncouplers

Some chemicals, such as 2,4-dinitrophenol (DNP), act as **uncoupling agents**, that is, when added to cells, they stop ATP synthesis but electron transport still continues and so oxygen is still consumed. The reason is that DNP and other uncoupling agents are lipid-soluble small molecules that can bind H^+ ions and transport them across membranes (i.e. they are **H^+ ionophores**). Electron transport occurs and pumps out H^+ ions across the inner mitochondrial membrane but DNP in the same membrane carries the H^+ ions back into the mitochondrion, preventing formation of a proton gradient. Since no proton gradient forms, no ATP can be made by oxidative phosphorylation. Rather the energy derived from electron transport is released as heat.

The production of heat by uncoupling is called **nonshivering thermogenesis**. It is important in certain biological situations. For example, uncoupling occurs naturally in brown adipose tissue. This tissue is rich in mitochondria, the inner mitochondrial membranes of which contain a protein called **thermogenin** (or **uncoupling protein**). Thermogenin allows H^+ ions to flow back into mitochondria without having to enter via the ATP synthase and so uncouples electron transport and oxidative phosphorylation, generating heat. The importance of this natural phenomenon is that brown adipose tissue is found in sensitive body areas of some newborn animals (including humans) where the heat production provides protection from cold conditions. In addition, thermogenesis by brown adipose tissue plays a role in maintaining body temperature in hibernating animals.

Reoxidation of cytosolic NADH

The inner mitochondrial membrane is impermeable to NADH. Therefore NADH produced in the cytoplasm during glycolysis must be reoxidized via a **membrane shuttle**, a combination of enzyme reactions that bypass this impermeability barrier. Figure 8 shows the **glycerol 3-phosphate shuttle**. Dihydroxyacetone phosphate in the cytosol is reduced to glycerol 3-phosphate, and NADH reoxidized to NAD^+, by cytosolic glycerol 3-phosphate dehydrogenase. The glycerol 3-phosphate diffuses across the inner mitochondrial membrane where it is converted back to dihydroxyacetone phosphate by mitochondrial glycerol 3-phosphate dehydrogenase, a transmembrane protein of the inner mitochondrial membrane. The dihydroxyacetone phosphate then diffuses back to the cytosol. The mitochondrial glycerol 3-phosphate dehydrogenase does not use NAD^+ but instead uses FAD. The enzyme-linked $FADH_2$ (E-$FADH_2$) is then reoxidized by transferring its electrons to ubiquinone in the same inner mitochondrial membrane (see above). Note that the shuttle does not allow cytoplasmic NADH to enter the mitochondrion but its operation effectively transports the two electrons from the NADH into the mitochondrion and feeds them into the electron transport chain. Since the electrons from cytoplasmic NADH actually enter the electron transport chain from $FADH_2$, only about 1.5 ATPs are synthesized instead of approximately 2.5 ATPs from each NADH that arises inside the mitochondrion from the citric acid cycle (Section L1) and fatty acid oxidation (Section K2).

Figure 8. The glycerol 3-phosphate shuttle.

A similar shuttle, the **malate–aspartate shuttle**, operates in heart and liver (Figure 9). Oxaloacetate in the cytosol is converted to malate by cytoplasmic malate dehydrogenase, reoxidizing NADH to NAD$^+$ in the process. The malate enters the mitochondrion via a **malate–α-ketoglutarate carrier** in the inner mitochondrial membrane. In the matrix, the malate is reoxidized to oxaloacetate by NAD$^+$ to form NADH. Oxaloacetate does not easily cross the inner mitochondrial membrane and so is transaminated to form aspartate, which then exits from the mitochondrion and is reconverted to oxaloacetate in the cytosol, again by transamination. The net result of this cycle of reactions is to transfer the electrons from NADH in the cytosol to NADH in the mitochondrial matrix, which is then reoxidized by the electron transport chain.

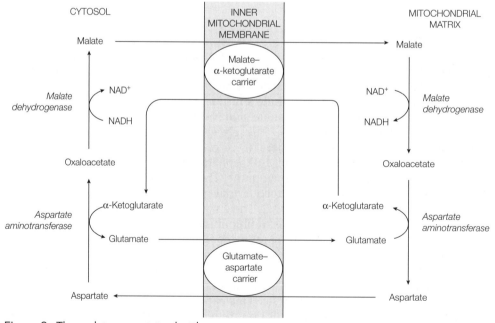

Figure 9. The malate–aspartate shuttle.

L3 Photosynthesis

Key Notes

Overview

Photosynthesis uses solar energy to synthesize carbohydrate from carbon dioxide and water. In the light reactions, the light energy drives the synthesis of NADPH and ATP. In the dark reactions (carbon-fixation reactions), the NADPH and ATP are used to synthesize carbohydrate from CO_2 and H_2O.

Location

In green plants and algae, photosynthesis takes place in chloroplasts. The light reactions occur in the thylakoid membranes and the dark reactions take place in the stroma. In photosynthetic bacteria, the light reactions take place in the bacterial plasma membrane, or in invaginations of it (chromatophores).

Light harvesting in green plants

Sunlight is absorbed by chlorophyll molecules, each of which is a magnesium porphyrin. Accessory pigments, such as carotenoids, absorb light at other wavelengths so maximizing light absorption. The pigments are arranged as photosystems, each photosystem consisting of an antenna complex and a photosynthetic reaction center. An antenna complex has several hundred chlorophyll molecules and accessory pigments clustered together in the thylakoid membrane. The absorption of a photon of light by a chlorophyll molecule raises an electron to a higher energy orbital. The excited chlorophyll can pass its extra energy on to another chlorophyll molecule in the complex by exciton transfer. The energy is channeled to two special chlorophyll molecules in the photosynthetic reaction center.

Photosystems I and II

Green plants and algae use two types of photosystem, photosystem I with chlorophyll P700 in its reaction center and photosystem II with P680 in its reaction center. The two photosystems are linked by a chain of electron carriers. When arranged in order of their redox potentials, the components form the so-called Z scheme. Light excites P680 of photosystem II to P680*. The excited P680* passes a high-energy electron to pheophytin, and is oxidized to P680+. The P680+ accepts an electron from water and returns to the ground state. Overall, the removal of four electrons from two molecules of water generates four H^+ ions and one molecule of O_2. The high-energy electrons accepted by pheophytin are passed in order to plastoquinone (Q), the cytochrome bf complex (also called cytochrome b_6f complex) and plastocyanin. Light excites P700 of photosystem I to P700*. The excited P700* passes a high-energy electron to ferredoxin, and becomes oxidized to P700+. The P700+ accepts an electron from plastocyanin and returns to the ground state. Finally, two electrons from two molecules of reduced ferredoxin are transferred to $NADP^+$ to form NADPH.

Noncyclic photo-phosphorylation

The cytochrome *bf* complex is a proton pump and, during electron transport, pumps H^+ ions from the stroma into the thylakoid space, creating an H^+ gradient. H^+ ions are also released into the thylakoid space when photosystem II oxidizes water to produce oxygen whilst the H^+ ions used to reduce $NADP^+$ to NADPH are taken up from the stroma. Both effects contribute to the H^+ gradient. The proton gradient drives ATP synthesis via an ATP synthase located in the thylakoid membrane (photophosphorylation). Since the electron transport involves a linear array of electron carriers, the system is called noncyclic photophosphorylation.

Cyclic photo-phosphorylation

When little $NADP^+$ is available to accept electrons, an alternative electron transport pathway is used. The high-energy electron donated by photosystem I passes to ferredoxin, then the cytochrome *bf* complex, then plastocyanin and back to the P700 of photosystem I. The resulting proton gradient generated by the cytochrome *bf* complex drives ATP synthesis (cyclic photophosphorylation) but no NADPH is made and no O_2 is produced.

Bacterial photosynthesis

Cyanobacteria use two photosystems as in green plants. The purple photosynthetic bacterium, *Rhodospirillum rubrum*, has only a single photosystem reaction center. This can carry out cyclic electron transport, synthesizing ATP (cyclic photophosphorylation). Alternatively, noncyclic electron transport can be used, producing NADH. Hydrogen sulfide (H_2S) can act as electron donor, generating sulfur (S). Hydrogen gas (H_2) and a variety of organic compounds can also be used as electron donors. Water is *not* used as electron donor and so no oxygen is produced.

The dark reactions

The dark reactions (carbon-fixation reactions) use the ATP and NADPH produced by the light reactions to 'fix' carbon dioxide as carbohydrate; sucrose and starch. The reactions form a cycle (the Calvin cycle) in which the enzyme ribulose bisphosphate carboxylase (rubisco), located in the stroma, condenses a CO_2 molecule with ribulose 1,5-bisphosphate to produce two molecules of 3-phosphoglycerate. Other reactions then regenerate the ribulose 1,5-bisphosphate. The fixation of three molecules of CO_2 requires six NADPH and nine ATP and leads to the net production of one molecule of glyceraldehyde 3-phosphate. For the synthesis of sucrose, glyceraldehyde 3-phosphate exits to the cytosol and is converted to fructose 6-phosphate and glucose 1-phosphate. The latter is then converted to UDP-glucose and reacts with fructose 6-phosphate to form sucrose 6-phosphate.

	Hydrolysis of the sucrose 6-phosphate yields sucrose. The glyceraldehyde 3-phosphate from the Calvin cycle is also used to synthesize glucose 1-phosphate, which generates ADP-glucose, cytidine diphosphate (CDP)-glucose or GDP-glucose as precursors for starch synthesis.
The C4 pathway	When the CO_2 concentration is low, rubisco can add O_2 to ribulose1,5-bisphosphate (oxygenase activity) instead of CO_2 (carboxylase activity) producing phosphoglycolate and 3-phosphoglycerate. Metabolism of phosphoglycolate releases CO_2 and NH_4^+ and wastes energy. This consumption of O_2 and release of CO_2 is called photorespiration. Plants in hot climates close their stomata to reduce water loss. This causes a drop in the CO_2 concentration in the leaf, favoring photorespiration. To avoid this problem, these plants carry out the Calvin cycle only in bundle-sheath cells that are protected from the O_2 in air by mesophyll cells. The CO_2 is transported from the air via the mesophyll cells to the bundle-sheath cells by combining with three-carbon molecules (C3) to produce four-carbon molecules (C4). This C4 pathway ensures a high CO_2 concentration for carbon fixation by rubisco in the bundle-sheath cells.
Related topics	(A2) Eukaryotic cells (M4) Hemes and chlorophylls (L2) Electron transport and oxidative phosphorylation

Overview

Photosynthesis occurs in green plants, algae and photosynthetic bacteria. Its role is to trap solar energy and use this to drive the synthesis of carbohydrate from carbon dioxide and water. Using (CH_2O) to represent carbohydrate, the overall reaction is:

$$\overset{Light}{H_2O + CO_2 \rightarrow (CH_2O) + O_2}$$

The reactions of photosynthesis occur in two distinct phases:

- **The light reactions**: which use light energy to synthesize NADPH and ATP.

- **The dark reactions**: that use the NADPH and ATP to synthesize carbohydrate from CO_2 and H_2O. In fact, the term 'dark reactions' is a misnomer; these **carbon-fixation reactions** should really be called light-independent reactions.

Location

In green plants and algae, photosynthesis takes place in chloroplasts (Section A2). Similar to a mitochondrion, a chloroplast has a highly permeable **outer membrane** and an **inner membrane** that is impermeable to most molecules and ions. Within each chloroplast lies the **stroma**, containing soluble enzymes (analogous to the matrix of a mitochondrion). However, whereas the inner membrane of a mitochondrion contains the electron transport chain and ATP synthase (Section L2), in a chloroplast these are located, together with photosynthetic light-absorbing systems, in stacks of flattened membranes within the stroma called **thylakoids** (Section A2). Thus the primary events of trapping solar energy in photosynthesis, the light reactions, occur in the thylakoid membranes. The dark reactions take place in the stroma. In photosynthetic bacteria, the light reactions take place in the bacterial plasma membrane, or in invaginations of it called **chromatophores**.

Light harvesting in green plants

Sunlight is absorbed by **chlorophyll** molecules. **Chlorophyll** is a porphyrin in which nitrogen atoms are coordinated to a magnesium ion (Section M4, Figure 1) (i.e. it is a **magnesium porphyrin**). This contrasts with a heme in which the nitrogen atoms are coordinated to an iron atom to form an iron porphyrin (Sections L2 and M4). Green plants contain two types of chlorophyll molecules, **chlorophyll *a*** and **chlorophyll *b***, that differ slightly in structure and in the wavelength of light they can absorb. Although light is trapped by **chlorophyll** molecules directly, several **accessory pigments** also exist that absorb light and pass the excitation energy on to chlorophyll molecules. Thus the **carotenoids** are important accessory pigments in green plants whilst **phycobilins** are accessory pigments in photosynthetic bacteria. These pigments absorb light at wavelengths different from that of chlorophyll and so act together to maximize the light harvested.

When a chlorophyll molecule is excited by a quantum of light (a **photon**), an electron is excited to a higher energy orbital. The excited chlorophyll can pass on its extra energy to a neighboring chlorophyll molecule by **exciton transfer** (also called **resonance energy transfer**) and so return to the unexcited state. Alternatively, the high-energy electron itself may be passed on, with the chlorophyll taking up a low-energy electron from another source.

The capture of solar energy occurs in **photosystems**. Each photosystem consists of an **antenna complex** and a **photosynthetic reaction center**. The antenna complex is composed of several hundred chlorophyll molecules and accessory pigments clustered together in the thylakoid membrane. When a chlorophyll molecule in the antenna complex absorbs light and is excited, the energy is passed by exciton transfer, from molecule to molecule, and is finally channeled to two special chlorophyll molecules in the photosynthetic reaction center. The reaction center passes on the energy as a high-energy electron to a chain of electron carriers in the thylakoid membrane (see below).

Photosystems I and II

Green plants and algae use two types of photosystem called **photosystem I** (**PSI**) and **photosystem II** (**PSII**). The chlorophyll in the reaction center of PSI has an absorption maximum at 700 nm and so is called **P700** (P for pigment) and that in the reaction center of PSII has an absorption maximum at 680 nm and so is called **P680**.

The two photosystems are linked by other electron carriers, in particular the cytochrome *bf* complex. When arranged according to their redox potentials (Section L2) the various components form the so-called **Z scheme** (Figure 1) because the overall shape of the redox diagram looks like a Z.

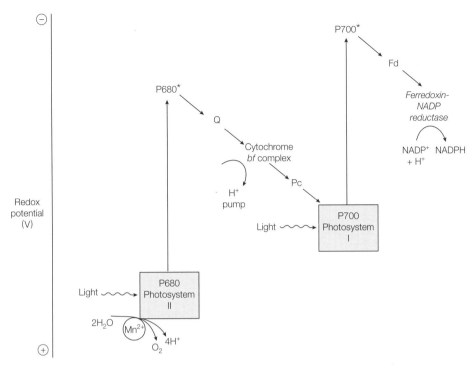

Figure 1. The Z scheme of noncyclic photophosphorylation in green plants. Note that the arrows in this diagram represent the sequence of events during photosynthesis, as described in the text, and are not metabolic interconversions.

The sequence of reactions during light absorption (Figure 1) is as follows:

1. Light is harvested by the antenna complex chlorophylls of PSII and the energy is channeled towards the reaction center at which P680 is located.

2. The excited P680 (P680*) emits a high-energy electron that passes to **plastoquinone** (Q), a mobile quinone in the thylakoid membrane that has a similar structure to ubiquinone in the mitochondrial electron transport chain (Section L2). This leaves P680 as the **P680$^+$** cation. Plastoquinone is reduced by accepting a total of two electrons and two H$^+$ ions to form **plastoquinol** (QH_2).

3. P680$^+$ extracts an electron from water, returning to its unexcited state. The removal of four electrons from two molecules of water requires four quanta of light to fall on PSII and leads to the production of four H$^+$ ions and one molecule of O$_2$:

$$\textit{4 photons}$$
$$2\,H_2O \rightarrow 4e^- + 4\,H^+ + O_2$$

This reaction is mediated by a cluster of four manganese ions (Mn^{2+}) in PSII.

The overall reaction catalyzed by photosystem II is:

$$\textit{Light}$$
$$2Q + 2H_2O \rightarrow O_2 + 2QH_2$$

4. The electrons are now passed from QH_2 via the **cytochrome *bf* complex** (also called the **cytochrome b_6f complex**) to **plastocyanin** (Pc). Pc is a copper-containing protein that accepts electrons by the copper cycling between Cu^{2+} and Cu^+ states:

    Cytochrome bf *complex*

    $$QH_2 + 2\ Pc\ (Cu^{2+}) \rightarrow Q + 2\ Pc\ (Cu^+) + 2H^+$$

    The cytochrome *bf* complex is analogous to Complex III in oxidative phosphorylation (Section L2) and similarly the reaction works via the Q cycle (Section L2) with plastoquinol and plastocyanin taking essentially the same roles as ubiquinol and cytochrome *c*. In the Complex III step in oxidative phosphorylation, protons are taken up from the matrix and released from the cytoplasmic side of the inner mitochondrial membrane whereas in the cytochrome *bf* complex step in photosynthesis, protons are taken up from the stroma and released into the thylakoid lumen.

5. Light energy falling onto the antenna complex of PSI is funneled to the reaction center. Here P700 is excited (to P700*) and emits a high-energy electron to **ferredoxin** (a protein that contains at least one FeS cluster; Section L2) itself becoming the **P700$^+$ cation**. The P700$^+$ receives the electron from the reduced Pc (from step 4 above) and so returns to the unexcited state.

    The overall reaction catalyzed by photosystem I is therefore:

    Light

    $$Pc\ (Cu^+) + Fd_{oxidized} \rightarrow Pc\ (Cu^{2+}) + Fd_{reduced}$$

6. Two high-energy electrons from two molecules of reduced ferredoxin are now transferred to NADP$^+$ to form NADPH. The reaction is carried out by **Ferredoxin-NADP reductase**.

    Ferredoxin-NADP reductase

    $$NADP^+ + 2\ e^- + H^+ \rightarrow NADPH$$

Taking account of the entire sequence of electron transport through PSI and PSII, the reaction can be written as follows, with electrons flowing from H_2O to NADP$^+$, reducing it to NADPH:

Light

$$2\ H_2O + 2\ NADP^+ \rightarrow 2\ NADPH + 2\ H^+ + O_2$$

Noncyclic photophosphorylation

During operation of the Z scheme, high-energy electrons are created by energy input via the two photosystems and the electrons then travel along a chain of carriers that decrease in redox potential (Figure 1). This is analogous to the passage of electrons along the respiratory chain in mitochondria (Section L2).

As described above, in a further analogy, the cytochrome *bf* complex is a **proton pump** (Figure 1) and pumps H$^+$ ions from the stroma into the thylakoid space (Figure 2) forming an H$^+$ gradient during electron transport. Because of the orientation of the various electron transport components in the thylakoid membrane (Figure 2), the H$^+$ ions released when PSII oxidizes water to produce oxygen are released into the thylakoid space whilst the H$^+$ ions used to reduce NADP$^+$ to NADPH by Ferredoxin-NADP reductase are taken up from the stroma. So these two reactions also contribute to the proton gradient.

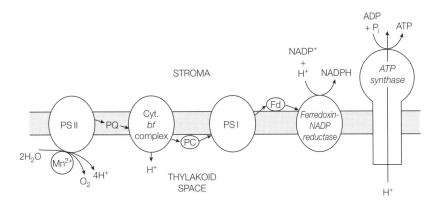

Figure 2. Formation of the proton gradient and ATP synthesis.

The proton gradient drives ATP synthesis via an ATP synthase located in the thylakoid membrane (Figure 2). This is called **photophosphorylation** and is analogous to ATP synthesis via a proton gradient during oxidative phosphorylation in mitochondria (Section L2).

The difference is that protons are pumped *out* of mitochondria but *into* a subcompartment, the thylakoid space, in chloroplasts. Because of the alternative ('cyclic') pathway for electron transport and ATP synthesis (see below), the formation of ATP via the joint operation of PSI and PSII (Figure 1; the Z scheme) is called **noncyclic photophosphorylation**.

Cyclic photophosphorylation

When the $NADPH/NADP^+$ ratio is high and little $NADP^+$ is available to accept electrons, an alternative electron transport pathway is used that involves only PSI and a few electron carriers (Figure 3). Here the high-energy electron is passed by ferredoxin to the cytochrome *bf* complex instead of to $NADP^+$. It then flows to plastocyanin and back to the P700 of PSI. The resulting proton gradient generated from the H^+ pump, cytochrome *bf* complex, then drives ATP synthesis. During this **cyclic photophosphorylation**, ATP is formed but no NADPH is made. Furthermore, since PSII is not involved, no O_2 is produced.

In summary, when electron transport is operating in noncyclic mode, via PSI and PSII, the products are NADPH and ATP. In cyclic electron transport, on the other hand, the sole product is ATP.

Bacterial photosynthesis

Cyanobacteria carry out photosynthesis using two photosystems as in green plants. However, other photosynthetic bacteria, such as the **purple photosynthetic bacterium** *Rhodospirillum rubrum*, have only a single photosystem reaction center. This can carry out cyclic electron transport, generating a proton gradient and hence synthesizing ATP (cyclic photophosphorylation).

Alternatively, a noncyclic pattern of electron transport can be carried out in which the electrons from the cytochromes pass to NAD^+ (rather than $NADP^+$ as in green plants)

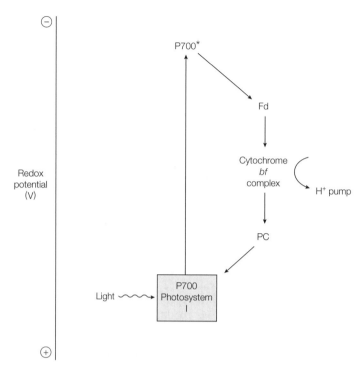

Figure 3. Cyclic photophosphorylation in green plants. Note that the arrows in this diagram represent the sequence of events during photosynthesis, as described in the text, and are not metabolic interconversions.

to produce NADH. The electron donor is, for example, hydrogen sulfide (H_2S), which generates sulfur (S). Hydrogen gas (H_2) and a variety of organic compounds can also be used as electron donors by certain photosynthetic bacteria. Since H_2O is *not* used as electron donor, no oxygen is produced.

The dark reactions

The dark reactions (also called the **carbon-fixation reactions**) use the ATP and NADPH produced by the light reactions to convert carbon dioxide into carbohydrate. The final products are sucrose and starch.

The key carbon fixation reaction is catalyzed by a large enzyme called **ribulose bisphosphate carboxylase** (often abbreviated to **rubisco**) that is located in the stroma. The reaction condenses a CO_2 molecule with **ribulose 1,5-bisphosphate** (a five-carbon molecule) to produce a transient six-carbon intermediate that rapidly hydrolyzes to two molecules of 3-phosphoglycerate (Figure 4).

Rubisco is a very slow enzyme, fixing only three molecules of its substrate every second and hence a large amount of this enzyme is needed by each plant. Typically, rubisco accounts for 50% or so of the total protein in a chloroplast. Indeed, it is probably the most abundant protein on Earth!

The rubisco reaction forms part of a cycle of reactions, called the **Calvin cycle**, that leads to the regeneration of ribulose 1,5-bisphosphate (ready to fix another CO_2) and the net

Figure 4. The rubisco reaction.

production of glyceraldehyde 3-phosphate for the synthesis of sucrose and starch. Three molecules of CO_2 must be fixed to generate one molecule of glyceraldehyde 3-phosphate (a three-carbon molecule). The overall reaction for this is:

$$3\ CO_2 + 6\ NADPH + 9\ ATP \rightarrow glyceraldehyde + 6\ NADP^+ + 9\ ADP + 8\ P_i$$
$$3\text{-phosphate}$$

The various reactions are shown in Figure 5. Following the formation of six molecules of 3-phosphoglycerate, six ATPs and six NADPHs are used to generate six molecules of glyceraldehyde 3-phosphate.

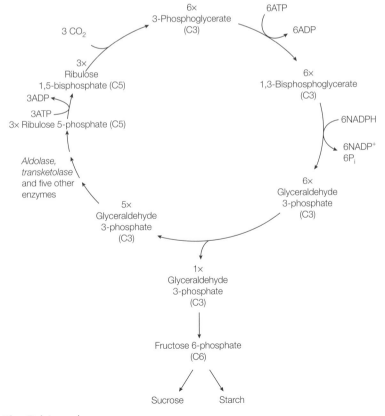

Figure 5. The Calvin cycle.

Only one of these molecules of glyceraldehyde 3-phosphate is then used to form fructose 6-phosphate and, eventually, starch. The other five molecules of glyceraldehyde 3-phosphate are converted in a series of steps (involving aldolase, transketolase and five other enzymes) to three molecules of ribulose 5-phosphate, which are then phosphorylated (using three molecules of ATP) to form three molecules of ribulose 1,5-bisphosphate, ready for another turn of the cycle. Hence one complete turn of the cycle, trapping three molecules of CO_2 as one molecule of glyceraldehyde 3-phosphate, requires 6+3=9 ATPs and 6 NADPHs.

Synthesis of sucrose

Much of the glyceraldehyde 3-phosphate produced by the Calvin cycle in chloroplasts is exported to the cytosol and used to produce the disaccharide, sucrose. First the glyceraldehyde 3-phosphate is converted to fructose 6-phosphate and glucose 1-phosphate. The chemical reactions involved are essentially a reversal of glycolysis (Section J3). The glucose 1-phosphate is then converted to UDP-glucose and this reacts with fructose 6-phosphate to synthesize sucrose 6-phosphate:

$$\text{UDP-glucose} + \text{fructose 6-phosphate} \rightarrow \text{sucrose 6-phosphate} + \text{UDP}$$

Hydrolysis of the sucrose 6-phosphate yields sucrose. This is the major sugar that is transported between plant cells, analogous to the supply of glucose via the bloodstream to animal tissues (Section J4).

Synthesis of starch

Whereas animals store excess carbohydrate as glycogen (Sections J2 and J6), plants do so in the form of starch (Section J2). Starch is produced in the stroma of chloroplasts and stored there as starch grains. Starch synthesis occurs from ADP-glucose, CDP-glucose or GDP-glucose (but *not* UDP-glucose). The pathway involves the conversion of glyceraldehyde 3-phosphate (from the Calvin cycle) to glucose 1-phosphate, which in turn is used to synthesize the nucleotide sugar derivatives.

The C4 pathway

Under normal atmospheric conditions, rubisco adds CO_2 to ribulose 1,5-bisphosphate. However, when the CO_2 concentration is low, it can add O_2 instead. This produces phosphoglycolate and 3-phosphoglycerate. The phosphoglycolate can be salvaged and used for biosynthetic reactions but the pathway for achieving this releases CO_2 and NH_4^+ and wastes metabolic energy. Because the net result of this process is to consume O_2 and release CO_2, it is known as **photorespiration**.

This is a major problem for plants in hot climates. The plants close the gas exchange pores in their leaves (stomata) to conserve water but this leads to a drop in the CO_2 concentration within the leaf, favoring photorespiration. In addition, as temperature rises, the oxygenase activity of rubisco (using O_2) increases more rapidly than the carboxylase activity (using CO_2), again favoring photorespiration. To avoid these problems, some plants adapted to live in hot climates, such as corn and sugar cane, have evolved a mechanism to maximize the carboxylase activity of rubisco. In these plants, carbon fixation using the Calvin cycle takes place only in **bundle-sheath cells** that are protected from the air by **mesophyll cells**. Since the bundle-sheath cells are not exposed to air, the O_2 concentration is low. The CO_2 is transported from the air via the mesophyll cells to the bundle-sheath cells by combining with three-carbon molecules (C3) to produce four-carbon molecules (C4).

These enter the bundle-sheath cells where they are broken down to C3 compounds, releasing CO_2. The C3 molecules return to the mesophyll cell to accept more CO_2.

This cycle ensures a high CO_2 concentration for the carboxylase activity of rubisco action in the bundle-sheath cells. Since it relies on CO_2 transport via four-carbon molecules, it is called the **C4 pathway** and plants that use this mechanism are called **C4 plants**. All other plants are called **C3 plants** since they trap CO_2 directly as the three-carbon compound 3-phosphoglycerate (Figure 4).

Details of the C4 pathway are shown in Figure 6. The steps involved are as follows:

- In the mesophyll cell, phosphoenolpyruvate (C3) accepts CO_2 to form oxaloacetate (C4); a reaction catalyzed by **phosphoenolpyruvate carboxylase**.

- Oxaloacetate is converted to malate (C4) by **$NADP^+$-linked malate dehydrogenase**.

- Malate enters the bundle-sheath cell and releases CO_2, forming pyruvate (C3); catalyzed by **$NADP^+$-linked malate enzyme**.

- Pyruvate returns to the mesophyll cell and is used to regenerate phosphoenolpyruvate. This reaction, catalyzed by **pyruvate-P_i dikinase**, is unusual in that it requires ATP and P_i and breaks a high-energy bond to generate AMP and pyrophosphate.

The pyrophosphate from the pyruvate-P_i dikinase is rapidly degraded so that, overall, the net price the plant pays for operation of this CO_2 pump is the hydrolysis of two high-energy phosphate bonds for every molecule of CO_2 transported:

$$CO_2 \text{ (in air)} + ATP \rightarrow CO_2 \text{ (bundle-sheath cell)} + AMP + 2\,P_i$$

Figure 6. The C4 pathway.

M1 Nitrogen fixation and assimilation

Key Notes

The nitrogen cycle   The nitrogen cycle is the movement of nitrogen through the food chain from simple inorganic compounds, mainly ammonia, to complex organic compounds.

Nitrogen fixation   Nitrogen fixation is the conversion of N_2 gas into ammonia, a process carried out by some soil bacteria, cyanobacteria and the symbiotic bacteria *Rhizobium* that invade the root nodules of leguminous plants. This process is carried out by the nitrogenase complex, which consists of a reductase and an iron–molybdenum-containing nitrogenase. At least 16 ATP molecules are hydrolyzed to form two molecules of ammonia. Leghemoglobin is used to protect the nitrogenase in the *Rhizobium* from inactivation by O_2.

Nitrogen assimilation   Ammonia is assimilated by all organisms into organic nitrogen-containing compounds (amino acids, nucleotides, etc.) by the action of glutamate dehydrogenase (to form glutamate) and glutamine synthetase (to form glutamine).

Related topics   (B3) Myoglobin and hemoglobin   (L3) Photosynthesis
(L1) Citric acid cycle   (M2) Amino acid metabolism
(L2) Electron transport and oxidative phosphorylation   (M4) Hemes and chlorophylls

The nitrogen cycle

The nitrogen cycle refers to the movement of nitrogen through the food chain of living organisms (Figure 1). This complex cycle involves bacteria, plants and animals. All organisms can convert **ammonia** (NH_3) to organic nitrogen compounds, that is compounds containing C–N bonds. However, only a few microorganisms can synthesize ammonia from nitrogen gas (N_2). Although N_2 gas makes up about 80% of the Earth's atmosphere, it is a chemically unreactive compound. The first stage in the nitrogen cycle is the reduction of N_2 gas to ammonia, a process called **nitrogen fixation**. Ammonia can also be obtained by reduction of nitrate ion (NO_3^-) that is present in the soil. **Nitrate reduction** can be carried out by most plants and microorganisms. The ammonia resulting from these two processes can then be assimilated by all organisms. Within the biosphere, there is a balance between total inorganic and total organic forms of nitrogen. The conversion of organic to inorganic nitrogen comes about through catabolism, denitrification and decay (Figure 1).

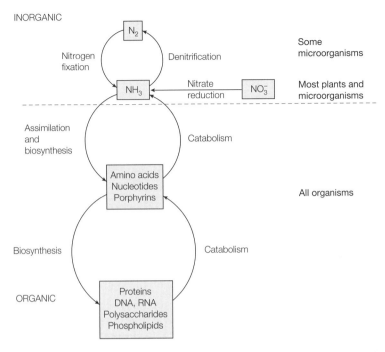

Figure 1. The interrelationships between inorganic and organic nitrogen metabolism.

Nitrogen fixation

The process of converting atmospheric N_2 gas into ammonia (nitrogen fixation) is carried out by only a few microorganisms, termed **diazatrophs**. These are some free-living soil bacteria such as *Klebsiella* and *Azotobacter*, cyanobacteria (blue–green algae), and the **symbiotic bacteria** (mainly *Rhizobium*). The symbiotic *Rhizobium* bacteria invade the roots of **leguminous green plants** (plants belonging to the pea family, e.g. beans, clover, alfalfa) and form **root nodules** where nitrogen fixation takes place. The amount of N_2 fixed by these diazatrophic microorganisms has been estimated to be in the order of 10^{11} kg per year, about 60% of the Earth's newly fixed nitrogen. Lightning and ultraviolet radiation fix another 15%, with the remainder coming from industrial processes.

The chemical unreactivity of the N≡N bond is clearly seen when one considers the industrial process of nitrogen fixation. This process, devised by Fritz Haber in 1910 and still used today in fertilizer factories, involves the reduction of N_2 in the presence of H_2 gas over an iron catalyst at a temperature of 500°C and a pressure of 300 atmospheres.

$$N_2 + 3\,H_2 \rightleftharpoons 2\,NH_3$$

Nitrogenase complex

Biological nitrogen fixation is carried out by the **nitrogenase complex**, which consists of two proteins: a **reductase**, which provides electrons with high reducing power, and a **nitrogenase**, which uses these electrons to reduce N_2 to NH_3 (Figure 2). The reductase is a 64 kDa dimer of identical subunits that contains one iron–sulfur cluster and two ATP binding sites. The nitrogenase is a larger protein of 220 kDa that consists of two α- and two β-subunits ($\alpha_2\beta_2$) and contains an iron–molybdenum complex. The transfer of electrons from the reductase to the nitrogenase protein is coupled to the hydrolysis of ATP by the reductase. Although the reduction of N_2 to NH_3 is only a six-electron process:

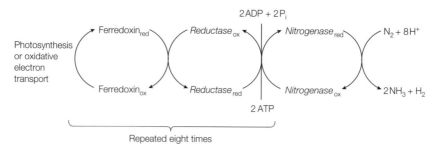

Figure 2. The flow of electrons in the nitrogenase-catalyzed reduction of N_2.

$$N_2 + 6\,e^- + 6\,H^+ \rightarrow 2\,NH_3$$

the reductase is imperfect and H_2 is also formed. Thus two additional electrons are also required:

$$N_2 + 8\,e^- + 8\,H^+ \rightarrow 2\,NH_3 + H_2$$

The eight high-potential electrons come from reduced **ferredoxin** that is produced either in chloroplasts by the action of photosystem I or in oxidative electron transport (Figure 2) (Sections L2 and L3). The overall reaction of biological nitrogen fixation:

$$N_2 + 8\,e^- + 8\,H^+ + 16\,ATP + 16\,H_2O \rightarrow 2\,NH_3 + H_2 + 16\,ADP + 16\,P_i$$

highlights that it is energetically very costly, with at least 16 ATP molecules being hydrolyzed.

Leghemoglobin

The nitrogenase complex is extremely sensitive to inactivation by O_2, so the enzyme must be protected from this reactive substance. In the root nodules of leguminous plants, protection is afforded by the symbiotic synthesis of **leghemoglobin**. The globin part of this monomeric oxygen-binding protein (Section B3) is synthesized by the plant, whereas the **heme** group (Section M4) is synthesized by the *Rhizobium*. The leghemoglobin has a very high affinity for O_2, so maintaining a low enough concentration to protect the nitrogenase.

Nitrogen assimilation

The next step in the nitrogen cycle is the assimilation of inorganic nitrogen, in the form of ammonia or the ammonium ion, into organic nitrogen-containing compounds. All organisms assimilate ammonia via two main reactions catalyzed by **glutamate dehydrogenase** and **glutamine synthetase** giving rise to the amino acids glutamate (Glu) and glutamine (Gln), respectively. The amino nitrogen in Glu and the amide nitrogen in Gln are then used in further biosynthetic reactions to give rise to other compounds.

Glutamate dehydrogenase

Glutamate dehydrogenase catalyzes the reductive amination of the citric acid cycle intermediate α-ketoglutarate (Figure 3a) (Section L1). Although the reaction is reversible, the reductant used in the biosynthetic reaction is NADPH. This enzyme is also involved in the catabolism of amino acids (Section M2).

(a)

$$\begin{array}{c}\text{COO}^-\\|\\\text{CH}_2\\|\\\text{CH}_2\\|\\\text{C}=\text{O}\\|\\\text{COO}^-\end{array} + \text{NH}_3 + \text{NADPH} + 2\text{H}^+ \rightleftharpoons \begin{array}{c}\text{COO}^-\\|\\\text{CH}_2\\|\\\text{CH}_2\\|\\\text{H}-\text{C}-\overset{+}{\text{N}}\text{H}_3\\|\\\text{COO}^-\end{array} + \text{H}_2\text{O} + \text{NADP}^+$$

α-Ketoglutarate   Glutamate

(b)

$$\begin{array}{c}\text{COO}^-\\|\\\text{CH}_2\\|\\\text{CH}_2\\|\\\text{H}-\text{C}-\overset{+}{\text{N}}\text{H}_3\\|\\\text{COO}^-\end{array} + \text{NH}_3 + \text{ATP} \longrightarrow \begin{array}{c}\text{O}\\||\\\text{C}-\text{NH}_2\\|\\\text{CH}_2\\|\\\text{CH}_2\\|\\\text{H}-\text{C}-\overset{+}{\text{N}}\text{H}_3\\|\\\text{COO}^-\end{array} + \text{ADP} + \text{P}_i$$

Glutamate   Glutamine

Figure 3. Assimilation of ammonia by (a) glutamate dehydrogenase and (b) glutamine synthetase.

Glutamine synthetase

Glutamine synthetase catalyzes the incorporation of ammonia into glutamine, deriving energy from the hydrolysis of ATP (Figure 3b). This enzyme is named a **synthetase**, rather than a synthase, because the reaction couples bond formation with the hydrolysis of ATP. In contrast, a **synthase** does not require ATP.

M2 Amino acid metabolism

Key Notes

Biosynthesis of amino acids

Some organisms can synthesize all of the 20 standard amino acids, others cannot. Nonessential amino acids are those that can be synthesized, essential amino acids have to be taken in the diet. The 20 standard amino acids can be grouped into six biosynthetic families depending on the metabolic intermediate from which their carbon skeleton is derived.

Amino acid degradation

Amino acids are degraded by the removal of the α-amino group and the conversion of the resulting carbon skeleton into one or more metabolic intermediates. Amino acids are termed glucogenic if their carbon skeletons can give rise to the net synthesis of glucose, and ketogenic if they can give rise to ketone bodies. Some amino acids are both glucogenic and ketogenic.

Transamination

The α-amino groups are removed from amino acids by a process called transamination. The acceptor for this reaction is usually the α-keto acid, α-ketoglutarate, which results in the formation of glutamate and the corresponding α-keto acid. The coenzyme of all transaminases is pyridoxal phosphate, which is derived from vitamin B_6.

Oxidative deamination of glutamate

The glutamate produced by transamination is oxidatively deaminated by glutamate dehydrogenase to produce ammonia. This enzyme is unusual in being able to use either NAD^+ or $NADP^+$, and is subject to allosteric regulation.

Amino acid oxidases

Small amounts of amino acids are degraded by L- and D-amino acid oxidases that utilize FMN or FAD as coenzyme, respectively.

Metabolism of phenylalanine

Phenylalanine is first converted to tyrosine by the monooxygenase phenylalanine hydroxylase; a reaction involving the coenzyme tetrahydrobiopterin. The tyrosine is then converted by transamination and a dioxygenase reaction to homogentisate, which in turn is further metabolized to fumarate and acetoacetate.

Inborn errors of metabolism

These are inherited metabolic disorders caused by the absence of an enzyme in a metabolic pathway. Alkaptonuria is caused by the lack of homogentisate oxidase and is harmless, whereas phenylketonuria, which is due to a lack of phenylalanine hydroxylase, can cause severe mental retardation.

Related topics	(B1) Amino acid structure	(K3) Fatty acid synthesis
	(J3) Glycolysis	(L1) Citric acid cycle
	(J4) Gluconeogenesis	(L2) Electron transport and
	(J5) Pentose phosphate	oxidative phosphorylation
	pathway	(M1) Nitrogen fixation and
	(K2) Fatty acid breakdown	assimilation
		(M3) The urea cycle

Biosynthesis of amino acids

Plants and microorganisms can synthesize all of the 20 standard amino acids. Mammals, however, cannot synthesize all 20 and must obtain some of them in their diet. Those amino acids that are supplied in the diet are referred to as **essential**, whereas the remainder that can be synthesized by the organism are termed **nonessential**. This designation refers to the needs of an organism under a particular set of conditions. In humans the nonessential amino acids are Ala, Arg, Asn, Asp, Cys, Glu, Gln, Gly, Pro, Ser and Tyr, while the essential ones are His, Ile, Leu, Lys, Met, Phe, Thr, Trp and Val (see Section B1 for the structures of the 20 standard amino acids). The pathways for the biosynthesis of amino acids are diverse and often vary from one organism to another. However, they all have an important feature in common: their carbon skeletons come from key intermediates in central metabolic pathways (glycolysis, Section J3; the citric acid cycle, Section L1; or the pentose phosphate pathway, Section J5) (Figure 1). The amino acids can be grouped together into six biosynthetic pathways depending on the intermediate from which they are derived (Figure 1). The primary amino group usually comes from transamination of glutamate.

Amino acid degradation

As there is no store for excess amino acids, and as proteins are constantly being turned over, amino acids have to be continually degraded. The α-amino group is removed first and the resulting carbon skeleton is converted into one or more major metabolic intermediates and used as metabolic fuel. The carbon skeletons of the 20 standard amino acids are funneled into only seven molecules: pyruvate, acetyl CoA, acetoacetyl CoA, α-ketoglutarate, succinyl CoA, fumarate and oxaloacetate (Figure 2). Amino acids that are degraded to pyruvate, α-ketoglutarate, succinyl CoA, fumarate and oxaloacetate are termed **glucogenic** as they can give rise to the net synthesis of glucose. This is because the citric acid cycle intermediates and pyruvate can be converted into phosphoenolpyruvate and then into glucose via gluconeogenesis (Sections J4 and L1). In contrast, amino acids that are degraded to acetyl CoA or acetoacetyl CoA are termed **ketogenic** because they give rise to ketone bodies (Section K2); the acetyl CoA or acetoacetyl CoA can also be used to synthesize lipids (Section K3). Of the standard set of 20 amino acids, only Leu and Lys are solely ketogenic. Ile, Phe, Trp and Tyr are both ketogenic and glucogenic as some of their carbon atoms end up in acetyl CoA or acetoacetyl CoA, whereas others end up in precursors of glucose. The remaining 14 amino acids are classified as solely glucogenic.

Transamination

Prior to the metabolism of their carbon skeletons into a major metabolic intermediate, the α-amino group of the amino acid has first to be removed by a process known as

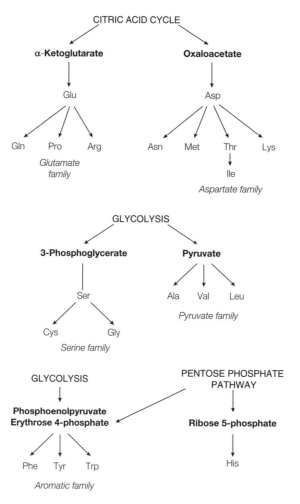

Figure 1. Biosynthetic families of amino acids.

transamination. In this process the α-amino group of most amino acids is transferred to α-**ketoglutarate** to form **glutamate** and the corresponding α-keto acid:

$$\alpha\text{-amino acid} + \alpha\text{-ketoglutarate} \rightleftharpoons \alpha\text{-keto acid} + \text{glutamate}$$

The enzymes that catalyze these reactions are called **transaminases** (aminotransferases) and in mammals are found predominantly in the liver. For example, aspartate transaminase catalyzes the transfer of the amino group of aspartate to α-ketoglutarate (Figure 3a), while alanine transaminase catalyzes the transfer of the amino group of alanine to α-ketoglutarate (Figure 3b).

Pyridoxal phosphate
The coenzyme (or prosthetic group) of all transaminases is **pyridoxal phosphate**, which is derived from pyridoxine (**vitamin B$_6$**), and which is transiently converted into

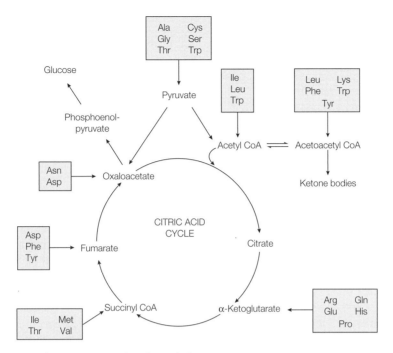

Figure 2. Fates of the amino acid carbon skeletons.

pyridoxamine phosphate during transamination (Figure 4). In the absence of substrate, the aldehyde group of pyridoxal phosphate forms a covalent **Schiff base linkage** (imine bond) with the amino group in the side-chain of a specific lysine residue in the active site of the enzyme. On addition of substrate, the α-amino group of the incoming amino acid displaces the amino group of the active site lysine and a new Schiff base linkage is

(a)

$$
\begin{array}{cccc}
\text{COO}^- & \text{COO}^- & \text{COO}^- & \text{COO}^- \\
| & | & | & | \\
\text{CH}_2 & \text{CH}_2 & \text{CH}_2 & \text{CH}_2 \\
| & | & | & | \\
\text{H}-\text{C}-\text{NH}_3^+ \;+\; \text{CH}_2 & \rightleftharpoons & \text{CH}_2 \;+\; \text{H}-\text{C}-\overset{+}{\text{N}}\text{H}_3 \\
| & | & | & | \\
\text{COO}^- & \text{C}=\text{O} & \text{C}=\text{O} & \text{COO}^- \\
& | & | & \\
& \text{COO}^- & \text{COO}^- &
\end{array}
$$

Aspartate   α-Ketoglutarate   Oxaloacetate   Glutamate

(b)

$$
\begin{array}{cccc}
& \text{COO}^- & & \text{COO}^- \\
& | & & | \\
& \text{CH}_2 & & \text{CH}_2 \\
& | & & | \\
\text{CH}_3 & \text{CH}_2 & \text{CH}_3 & \text{CH}_2 \\
| & | & | & | \\
\text{H}-\text{C}-\overset{+}{\text{N}}\text{H}_3 \;+\; \text{C}=\text{O} & \rightleftharpoons & \text{C}=\text{O} \;+\; \text{H}-\text{C}-\overset{+}{\text{N}}\text{H}_3 \\
| & | & | & | \\
\text{COO}^- & \text{COO}^- & \text{COO}^- & \text{COO}^-
\end{array}
$$

Alanine   α-Ketoglutarate   Pyruvate   Glutamate

Figure 3. Reactions catalyzed by (a) aspartate transaminase and (b) alanine transaminase.

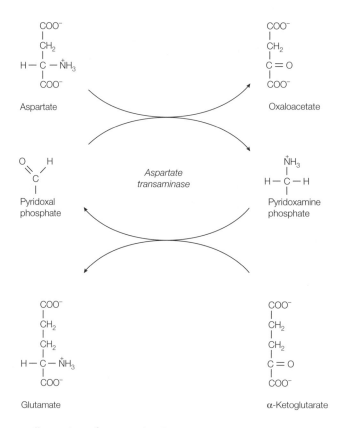

Figure 4. Structures of (a) pyridoxine (vitamin B$_6$), (b) pyridoxal phosphate and (c) pyridoxamine phosphate.

formed with the amino acid substrate. The resulting amino acid–pyridoxal phosphate–Schiff base that is formed remains tightly bound to the enzyme by multiple noncovalent interactions.

The amino acid is then hydrolyzed to form an α-keto acid and pyridoxamine phosphate, the α-amino group having been temporarily transferred from the amino acid substrate on to pyridoxal phosphate (Figure 5). These steps constitute one half of the overall transamination reaction. The second half occurs by a reversal of the above reactions with a second α-keto acid reacting with the pyridoxamine phosphate to yield a second amino acid and regenerate the enzyme–pyridoxal phosphate complex (Figure 5).

Figure 5. The overall reaction of transamination.

$$\begin{array}{l}
\text{COO}^- \\
| \\
\text{CH}_2 \\
| \\
\text{CH}_2 \\
| \\
\text{H}-\overset{+}{\underset{|}{\text{C}}}-\overset{+}{\text{NH}_3} \\
\text{COO}^-
\end{array} + \text{NAD}^+ + \text{H}_2\text{O} \rightleftharpoons \overset{+}{\text{NH}_4} + \begin{array}{l}
\text{COO}^- \\
| \\
\text{CH}_2 \\
| \\
\text{CH}_2 \\
| \\
\text{C}=\text{O} \\
| \\
\text{COO}^-
\end{array} + \text{NADH} + \text{H}^+$$

Glutamate   α-Ketoglutarate

Figure 6. Oxidative deamination of glutamate by glutamate dehydrogenase.

The reactions catalyzed by transaminases are **anergonic** as they do not require an input of metabolic energy. They are also freely reversible, the direction of the reaction being determined by the relative concentrations of the amino acid–keto acid pairs. Pyridoxal phosphate is not just used as the coenzyme in transamination reactions, but is also the coenzyme for several other reactions involving amino acids including decarboxylations, deaminations, racemizations and aldol cleavages.

Oxidative deamination of glutamate

The α-amino groups that have been funneled into glutamate from the other amino acids are then converted into **ammonia** by the action of **glutamate dehydrogenase** (Figure 6). This enzyme is unusual in being able to utilize either NAD$^+$ or NADP$^+$ (Section D1). In the biosynthesis of glutamate, the NADP$^+$ form of the coenzyme is used (Section M1), whereas NAD$^+$ is used in its degradation. Glutamate dehydrogenase consists of six identical subunits and is subject to **allosteric regulation** (see Section D5 for a detailed description of allosteric regulation). GTP and ATP are **allosteric inhibitors**, whereas GDP and ADP are **allosteric activators**. Hence, when the energy charge of the cell is low (i.e. there is more ADP and GDP than their triphosphate forms) glutamate dehydrogenase is activated and the oxidation of amino acids increases. The resulting carbon skeletons are then utilized as metabolic fuel, feeding into the citric acid cycle (Section L1) and ultimately giving rise to energy through oxidative phosphorylation (Section L2).

Amino acid oxidases

The major route for the deamination of amino acids is transamination followed by the oxidative deamination of glutamate. However, a minor route also exists that involves direct oxidation of the amino acid by L-**amino acid oxidase**. This enzyme utilizes **FMN** as its coenzyme (Section D1), with the resulting FMNH$_2$ being reoxidized by molecular O$_2$, a process that also generates the toxic H$_2$O$_2$ (Figure 7). The H$_2$O$_2$ is rendered harmless by the action of **catalase**. Kidney and liver are also rich in the **FAD**-containing D-**amino acid**

$$\begin{array}{l}
\text{R} \\
| \\
\text{H}-\overset{|}{\underset{|}{\text{C}}}-\overset{+}{\text{NH}_3} \\
\text{COO}^-
\end{array} + \text{FMN} + \text{H}_2\text{O} \longrightarrow \begin{array}{l}
\text{R} \\
| \\
\text{C}=\text{O} \\
| \\
\text{COO}^-
\end{array} + \text{FMNH}_2 + \overset{+}{\text{NH}_4}$$

Amino acid   α-Keto acid

$$\text{FMNH}_2 + \text{O}_2 \longrightarrow \text{FMN} + \text{H}_2\text{O}_2$$

Figure 7. Action of L-amino acid oxidase.

oxidase. However, the function of this enzyme in animals is unclear, since the D-isomers of amino acids are rare in nature, only occurring in bacterial cell walls (Section A1) and certain antibiotics.

Metabolism of phenylalanine

The metabolism of phenylalanine will now be considered in some detail, as two inborn errors of metabolism are known that affect this pathway. **Phenylalanine** is first hydroxylated by **phenylalanine hydroxylase** to form another aromatic amino acid **tyrosine** (Figure 8). The coenzyme for this reaction is the reductant **tetrahydrobiopterin**, which is oxidized to **dihydrobiopterin**. Phenylalanine hydroxylase is classified as a **monooxygenase** as one of the atoms of O_2 appears in the product and the other in H_2O. The tyrosine is then transaminated to p-hydroxyphenylpyruvate, which is in turn converted into **homogentisate** by p-hydroxyphenylpyruvate hydroxylase. This hydroxylase is an example of a **dioxygenase**, as both atoms of O_2 become incorporated into the product (Figure 8). The homogentisate is then cleaved by homogentisate oxidase, another dioxygenase, before fumarate and acetoacetate are produced in a final reaction. The fumarate can feed into the citric acid cycle (Section L1), whereas acetoacetate can be used to form ketone bodies (Section K2). Thus phenylalanine and tyrosine are each both glucogenic and ketogenic.

Figure 8. The metabolism of phenylalanine.

Inborn errors of metabolism

Two inborn errors of metabolism are known that affect phenylalanine metabolism. These are **inherited metabolic disorders** caused by the absence of one of the enzymes in the pathway. One of these disorders, **alkaptonuria**, is caused by the absence of **homogentisate oxidase**. This results in the accumulation of homogentisate that is subsequently excreted in the urine, and which oxidizes to a black color on standing. This defect is harmless, in contrast with the other disorder, **phenylketonuria**. In phenylketonuria there is a block in the hydroxylation of phenylalanine to tyrosine caused by an absence or deficiency of **phenylalanine hydroxylase** or, more rarely, of its tetrahydrobiopterin coenzyme. The result is a 20-fold increase in the levels of phenylalanine in the blood with its subsequent **transamination** to phenylpyruvate. If untreated, severe mental retardation occurs, with a life expectancy of on average 20 years. With an incidence of 1:20 000 this condition is now screened for at birth, with treatment being to restrict the intake of phenylalanine (by putting the individual on a **low phenylalanine diet**) and thus minimize the need to metabolize the excess. However, enough phenylalanine must be provided to meet the needs for growth and replacement. The artificial sweetener **aspartame** (**NutraSweet®**) is Asp-Phe-methyl ester and is often found in diet soft drinks and other dietetic food products. As this is a source of dietary phenylalanine, a warning label for phenyl-ketonuriacs appears on all products containing aspartame.

M3 The urea cycle

Key Notes

Ammonia excretion

Excess nitrogen is excreted as ammonia. Ammonotelic organisms excrete ammonia directly, uricotelic organisms excrete it as uric acid, and ureotelic organisms excrete it as urea.

The urea cycle

In the urea cycle, ammonia is first combined with CO_2 to form carbamoyl phosphate. This then combines with ornithine to form citrulline. Citrulline then condenses with aspartate, the source of the second nitrogen atom in urea, to form argininosuccinate. This compound is in turn split to arginine and fumarate, and the arginine then splits to form urea and regenerate ornithine. The first two reactions take place in the mitochondria of liver cells, the remaining three in the cytosol.

Link to the citric acid cycle

The fumarate produced in the urea cycle can enter directly into the citric acid cycle and be converted into oxaloacetate. Oxaloacetate can then be either transaminated to aspartate, which feeds back into the urea cycle, or be converted into citrate, pyruvate or glucose.

Hyperammonemia

Hyperammonemia is an increase in the levels of ammonia in the blood caused by a defect in an enzyme of the urea cycle. The excess ammonia is channeled into glutamate and glutamine with a deleterious effect on brain function.

Formation of nitric oxide

The gas nitric oxide (NO) is derived from arginine by the action of nitric oxide synthase. This short-lived metabolite triggers a signal transduction pathway that causes dilation of arteries.

Formation of creatine phosphate

The urea cycle intermediate arginine can be condensed with glycine to form guanidinoacetate, which in turn is methylated by the methyl donor *S*-adenosyl methionine to creatine. The creatine is then phosphorylated to form creatine phosphate, a high-energy store found in muscle.

The activated methyl cycle

S-Adenosyl methionine is the major methyl donor in biological reactions. It is regenerated via the intermediates *S*-adenosyl homocysteine, homocysteine and methionine in the activated methyl cycle.

Uric acid

Uric acid, the major nitrogenous waste product of uricotelic organisms, is also formed in other organisms from the breakdown of purine bases. Gout is caused by the deposition of excess uric acid crystals in the joints.

Related topics	(J4) Gluconeogenesis	(L1) Citric acid cycle
	(J6) Glycogen metabolism	(M2) Amino acid metabolism
	(K4) Triacylglycerols	

Ammonia excretion

There is no store for nitrogen-containing compounds as there is for carbohydrate (glycogen, Section J6) or lipids (triacylglycerol, Section K4). Thus nitrogen ingested in excess of what is required by the organism has to be excreted. The excess nitrogen is first converted into **ammonia** and is then excreted from living organisms in one of three ways. Many aquatic animals simply excrete the ammonia itself directly into the surrounding water. Birds and terrestrial reptiles excrete the ammonia in the form of **uric acid**, while most terrestrial vertebrates convert the ammonia into **urea** before excretion. These three classes of organisms are called: **ammonotelic**, **uricotelic** and **ureotelic**, respectively.

The urea cycle

Urea is synthesized in the liver by the **urea cycle**. It is then secreted into the bloodstream and taken up by the kidneys for excretion in the urine. The urea cycle was the first cyclic meta-bolic pathway to be discovered by Hans Krebs and Kurt Henseleit in 1932, 5 years before Krebs discovered the citric acid cycle (Section L1). The overall reaction of the pathway is:

$$NH_4^+ + HCO_3^- + H_2O + 3\,ATP + aspartate \rightarrow$$
$$urea + 2\,ADP + AMP + 2P_i + PP_i + fumarate$$

One of the nitrogen atoms of urea comes from ammonia, the other is transferred from the amino acid aspartate, while the carbon atom comes from CO_2 via bicarbonate (HCO_3^-). **Ornithine**, an amino acid that is not in the standard set of 20 amino acids and is not found in proteins, is the carrier of these nitrogen and carbon atoms. Five enzymatic reactions are involved in the urea cycle (Figure 1), the first two of which take place in mitochondria, the other three in the cytosol:

1. **Carbamoyl phosphate synthetase**, which is technically not a member of the urea cycle, catalyzes the condensation and activation of ammonia (from the oxidative deamination of glutamate by glutamate dehydrogenase; Section M2) and CO_2 (in the form of bicarbonate, HCO_3^-) to form **carbamoyl phosphate**. The hydrolysis of two ATP molecules makes this reaction essentially irreversible.

2. The second reaction also occurs in the mitochondria and involves the transfer of the carbamoyl group from carbamoyl phosphate to **ornithine** by **ornithine transcarbamoylase**. This reaction forms another nonstandard amino acid **citrulline**, which then has to be transported out of the mitochondrion into the cytosol where the remaining reactions of the cycle take place.

3. The citrulline is then condensed with aspartate, the source of the second nitrogen atom in urea, by the enzyme **argininosuccinate synthetase** to form **argininosuccinate**. This reaction is driven by the hydrolysis of ATP to AMP and PP_i, with subsequent hydrolysis of the pyrophosphate. Thus both of the high-energy bonds in ATP are ultimately cleaved.

4. **Argininosuccinase** then removes the carbon skeleton of aspartate from argininosuccinate in the form of **fumarate**, leaving the nitrogen atom on the other

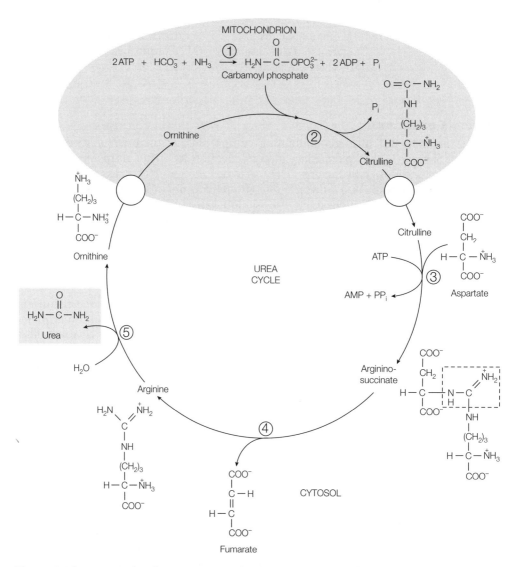

Figure 1. The urea cycle. The enzymes involved in this cycle are: ① carbamoyl phosphate synthetase; ② ornithine transcarbamoylase; ③ argininosuccinate synthetase; ④ argininosuccinase; and ⑤ arginase.

product **arginine**. As the urea cycle also produces arginine, this amino acid is classified as nonessential in ureotelic organisms. Arginine is the immediate precursor of urea.

5. The **urea** is then formed from arginine by the action of **arginase** with the regeneration of ornithine. The ornithine is then transported back into the mitochondrion ready to be combined with another molecule of carbamoyl phosphate.

Link to the citric acid cycle

The synthesis of fumarate by argininosuccinase links the urea cycle to the citric acid cycle (Figure 2). Fumarate is an intermediate of this latter cycle, which is then hydrated to malate, which in turn is oxidized to **oxaloacetate** (Section L1).

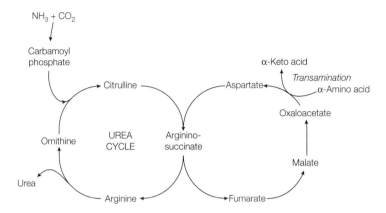

Figure 2. The urea cycle and the citric acid cycle are linked by fumarate and the transamination of oxaloacetate to aspartate.

Oxaloacetate has several possible fates:

- **Transamination** to aspartate (Section M2), which can then feed back into the urea cycle

- Condensation with acetyl CoA to form citrate, which then continues on round the citric acid cycle (Section L1)

- Conversion into glucose via gluconeogenesis (Section J4)

- Conversion into pyruvate

Hyperammonemia

Why do organisms need to detoxify ammonia in the first place? The answer to this question is obvious when one considers what happens if there is a block in the urea cycle due to a defective enzyme. A block in any of the urea cycle enzymes leads to an increase in the amount of ammonia in the blood, so-called **hyperammonemia**. The most common cause of such a block is a genetic defect that becomes apparent soon after birth, when the afflicted baby becomes lethargic and vomits periodically. If left untreated, coma and irreversible brain damage will follow. The reasons for this are not entirely clear but may be because the excess ammonia leads to the increased formation of **glutamate** and **glutamine** (Figure 3) (Section M1). These reactions result in depletion of the citric acid cycle intermediate α-ketoglutarate, which may then compromise energy production, especially in the brain. It also leads to an increase in the acidic amino acids glutamate and glutamine, which may directly cause damage to the brain.

$$\text{α-Ketoglutarate} \underset{\substack{Glutamate \\ dehydrogenase}}{\overset{NH_3}{\rightleftharpoons}} \text{Glutamate} \underset{\substack{Glutamine \\ synthetase}}{\overset{NH_3}{\rightleftharpoons}} \text{Glutamine}$$

Figure 3. Excess ammonia leads to the formation of glutamate and glutamine.

Formation of nitric oxide

The short-lived free radical gas **nitric oxide** (**NO**) is produced from **arginine** in a complex reaction catalyzed by **nitric oxide synthase** (NOS). NO acts by binding to and activating guanylate cyclase, which is involved in **signal transduction** (Section E5). NO is produced by endothelial cells in response to a range of agents and physiological conditions, induces the dilation of arteries, and is responsible for, amongst other things, penile erection.

Formation of creatine phosphate

The urea cycle is also the starting point for the synthesis of another important metabolite **creatine phosphate**. This **phosphagen** provides a reservoir of high-energy phosphate in muscle cells (Section B5) as the energy released upon its hydrolysis is greater than

Figure 4. Formation of creatine phosphate.

that released upon the hydrolysis of ATP (ΔG for creatine phosphate hydrolysis = -10.3 kcal mol^{-1} compared with -7.3 kcal mol^{-1} for ATP hydrolysis, Section D2). The first step in the formation of creatine phosphate is the condensation of arginine and glycine to form guanidinoacetate (Figure 4). Ornithine is released in this reaction and can then be re-utilized by the urea cycle. The guanidinoacetate is then methylated by the methyl group donor **S-adenosyl methionine** to form creatine, which is in turn phosphorylated by **creatine kinase** to form creatine phosphate (Figure 4).

The activated methyl cycle

S-Adenosyl methionine serves as a donor of **methyl groups** in numerous biological reactions [e.g. in the formation of creatine phosphate (see above) and in the synthesis of nucleic acids]. It is formed through the action of the activated methyl cycle (Figure 5). During donation of its methyl group to another compound, S-adenosyl methionine is converted into S-adenosyl homocysteine. To regenerate S-adenosyl methionine, the adenosyl group is removed from the S-adenosyl homocysteine to form homocysteine. This is then methylated by the enzyme homocysteine methyltransferase, one of only two **vitamin B$_{12}$**-containing enzymes found in eukaryotes, to form **methionine**. The resulting methionine is then activated to S-adenosyl methionine with the release of all three of the phosphates from ATP.

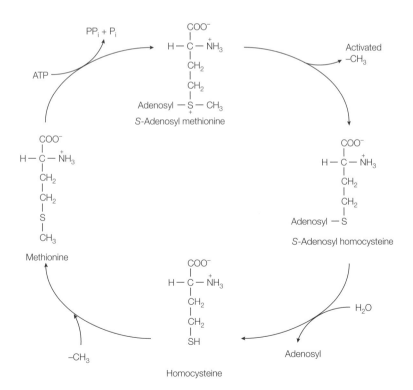

Figure 5. The activated methyl cycle.

Uric acid

Uric acid (Figure 6) is the main nitrogenous waste product of uricotelic organisms (reptiles, birds and insects), but is also formed in ureotelic organisms from the breakdown of the **purine bases** from DNA and RNA (Sections F1 and G1). Some individuals have a high serum level of sodium urate (the predominant form of uric acid at neutral pH), which can lead to crystals of this compound being deposited in the joints and kidneys, a condition known as **gout**, a type of arthritis characterized by extremely painful joints.

Figure 6. Uric acid.

M4 Hemes and chlorophylls

Key Notes

Tetrapyrroles

The tetrapyrroles are a family of pigments based on a common chemical structure that includes the hemes and chlorophylls. Hemes are cyclic tetrapyrroles that contain iron and are commonly found as the prosthetic group of hemoglobin, myoglobin and the cytochromes. The chlorophylls are modified tetrapyrroles containing magnesium that occur as light-harvesting and reaction center pigments of photosynthesis in plants, algae and photosynthetic bacteria.

Biosynthesis of hemes and chlorophylls

The starting point for heme and chlorophyll synthesis is aminolaevulinic acid (ALA), which is made in animals from glycine and succinyl CoA by the enzyme ALA synthase. Two molecules of ALA then condense to form porphobilinogen in a reaction catalyzed by ALA dehydratase. Porphobilinogen deaminase catalyzes the condensation of four porphobilinogens to form a linear tetrapyrrole. This compound then cyclizes to form uroporphyrinogen III, the precursor of hemes, chlorophylls and vitamin B_{12}. Further modifications take place to form protoporphyrin IX. The biosynthetic pathway then branches, and either iron is inserted to form heme, or magnesium is inserted to begin a series of conversions to form chlorophyll.

Heme degradation

Heme is broken down by heme oxygenase to the linear tetrapyrrole biliverdin. This green pigment is then converted to the red-orange bilirubin by biliverdin reductase. The lipophilic bilirubin is carried in the blood bound to serum albumin, and is then converted into a more water-soluble compound in the liver by conjugation to glucuronic acid. The resulting bilirubin diglucuronide is secreted into the bile, and finally excreted in the feces. Jaundice is due to a build-up of insoluble bilirubin in the skin and whites of the eyes. In higher plants heme is broken down to the phycobiliprotein phytochrome, which is involved in coordinating light responses, while in algae it is metabolized to the light-harvesting pigments phycocyanin and phycoerythrin.

Related topics

(B3) Myoglobin and hemoglobin
(D5) Regulation of enzyme activity
(L1) Citric acid cycle

(L2) Electron transport and oxidative phosphorylation
(L3) Photosynthesis
(M2) Amino acid metabolism

Tetrapyrroles

The red **heme** and green **chlorophyll** pigments, so important in the energy-producing mechanisms of respiration and photosynthesis, are both members of the family of pigments called **tetrapyrroles**. They share similar structures (Figure 1), and have some common steps in their synthesis and degradation. The basic structure of a tetrapyrrole is four pyrrole rings surrounding a central metal atom.

Hemes (Figure 1a) are a diverse group of tetrapyrrole pigments, being present as the prosthetic group of both the **globins** (hemoglobin and myoglobin; Section B3) and the **cytochromes** (including those involved in respiratory and photosynthetic electron transport; Sections L2 and L3) and the cytochrome P450s that are used in detoxification reactions. Some enzymes, including the catalases and peroxidases, contain heme. In all these hemoproteins the function of the heme is either to bind and release a ligand to its central **iron atom**, or for the iron atom to undergo a change in oxidation state, releasing or accepting an electron for participation in a redox reaction.

The **chlorophylls** are also a diverse family of pigments, existing in different forms in photosynthetic bacteria, algae and higher plants. They share a common function in all of these organisms to act as light-harvesting and reaction center pigments in **photosynthesis** (Section L3). This function is achieved by a number of modifications to the basic tetrapyrrole structure. These include: the insertion of **magnesium** as the central metal ion, the addition of a fifth ring to the tetrapyrrole structure, loss of a double bond from one or more of the pyrrole rings, and binding of one specific side-chain to a long fat-like molecule called **phytol** (Figure 1b). These changes give chlorophylls and bacteriochlorophylls a number of useful properties. For example, chlorophylls are membrane bound, absorb light at longer wavelengths than heme, and are able to respond to excitation by light. In this way, chlorophylls can accept and release light energy and drive photosynthetic electron transport (Section L3).

Figure 1. Structure of (a) heme and (b) chlorophyll.

Biosynthesis of hemes and chlorophylls

In animals, fungi and some bacteria, the first step in tetrapyrrole synthesis is the condensation of the amino acid **glycine** with **succinyl CoA** (an intermediate of the citric acid cycle; Section L1) to form **aminolaevulinic acid** (ALA). This reaction is catalyzed by the enzyme **ALA synthase** (Figure 2a), which requires the coenzyme **pyridoxal phosphate** (Section M2) and is located in the mitochondria of eukaryotes. This committed step in the pathway is subject to regulation. The synthesis of ALA synthase is feedback-inhibited by heme (Section D5). In plants, algae and many bacteria there is an alternative route for ALA synthesis that involves the conversion of the five-carbon skeleton of glutamate in a series of three steps to yield ALA. In all organisms, two molecules of ALA then condense to form **porphobilinogen** in a reaction catalyzed by **ALA dehydratase** (also called **porphobilinogen synthase**) (Figure 2a). Inhibition of this enzyme by lead is one of the major manifestations of acute **lead poisoning**.

Four porphobilinogens then condense head-to-tail in a reaction catalyzed by **porphobilinogen deaminase** to form a **linear tetrapyrrole** (Figure 2b). This enzyme-bound linear tetrapyrrole then cyclizes to form **uroporphyrinogen III**, which has an asymmetric arrangement of side-chains (Figure 2b). Uroporphyrinogen III is the common precursor of all hemes and chlorophylls, as well as of **vitamin B$_{12}$**. The pathway continues with a number of modifications to groups attached to the outside of the ring structure, finally forming **protoporphyrin IX** (Figure 2b). At this point either iron or magnesium is inserted into the central cavity, committing the porphyrin to either heme or chlorophyll synthesis, respectively. From here further modifications occur, and finally the specialized porphyrin prosthetic groups are attached to their respective **apoproteins** (the form of the protein consisting of just the polypeptide chain) to form the biologically functional **holoprotein**.

Figure 2. Pathway of the synthesis of heme and chlorophyll: (a) synthesis of porphobilinogen from glycine and succinyl CoA; (b) synthesis of protoporphyrin IX from porphobilinogen. A=CH$_2$COOH, M=CH$_3$, P=CH$_2$CH$_2$COOH.

Heme biosynthesis takes place primarily in immature erythrocytes (85% of the body's heme groups), with the remainder occurring in the liver. Several genetic defects in heme biosynthesis have been identified that give rise to the disorders called **porphyrias**.

Heme degradation

Bile pigments exist in both the plant and animal kingdoms, and are formed by breakdown of the cyclic tetrapyrrole structure of heme. In animals, this pathway is an excretory system by which the heme from the hemoglobin of aging red blood cells, and other hemoproteins, is removed from the body. In the plant kingdom, however, heme is broken down to form bile pigments, which have major roles to play in coordinating light responses in higher plants (the phycobiliprotein **phytochrome**), and in light harvesting in algae (the phycobiliproteins **phycocyanin** and **phycoerythrin**).

In all organisms, the degradation of heme begins with a reaction carried out by a single common enzyme. This enzyme, **heme oxygenase**, is present mainly in the spleen and liver of vertebrates, and carries out the oxidative ring opening of heme to produce the green bile pigment **biliverdin**, a linear tetrapyrrole (Figure 3). Heme oxygenase is a member of the **cytochrome P450** family of enzymes, and requires NADPH and O_2. In birds, reptiles and amphibians this water-soluble pigment is the final product of heme degradation and is excreted directly. In mammals, however, a further conversion to the red-orange **bilirubin** takes place; a reaction catalyzed by **biliverdin reductase** (Figure 3). The changing color of a bruise is a visible indicator of these degradative reactions. The bilirubin, like other lipophilic molecules such as free fatty acids, is then transported in the blood bound to **serum albumin**. In the liver, its water solubility is increased by conjugation

Figure 3. Degradation of heme to the bile pigments biliverdin and bilirubin. M=methyl (CH_3), V=vinyl $(CH=CH_2)$, P=propionyl $(CH_2CH_2CH_2OH)$.

to two molecules of **glucuronic acid**, a sugar residue that differs from glucose in having a COO^- group at C-6 rather than a CH_2OH group. The resulting **bilirubin diglucuronide** is secreted into the bile and then into the intestine, where it is further metabolized by bacterial enzymes and finally excreted in the feces.

When the blood contains excessive amounts of the insoluble bilirubin, it is deposited in the skin and the whites of the eyes, resulting in a yellow discoloration. This condition, called **jaundice**, is indicative either of impaired liver function, obstruction of the bile duct, or excessive breakdown of erythrocytes.

Further reading

There are many comprehensive textbooks of biochemistry and molecular biology and no one book that can satisfy all needs. Different readers subjectively prefer different textbooks and hence we do not feel it would be particularly helpful to recommend one book over another. Rather we have listed some of the leading books which we know from experience have served their student readers well.

General reading

Alberts, B., Johnson, A., Lewis, J., Raff, M., Roberts, K. and Walter, P. (2007) *Molecular Biology of the Cell*, 5th Edn. Garland Science, Taylor & Francis Group, New York.

Berg, J.M., Tymoczko, J.L. and Stryer, L. (2006) *Biochemistry: International Edition*, 6th Edn. W.H. Freeman and Company, New York.

Brown, T.A. (2006) *Genomes 3*, 3rd Edn. Garland Science, Taylor & Francis Group, New York.

Lodish, H., Berk, A., Kaiser, C.A., Krieger, M., Scott, M.P., Bretscher, A., Ploegh, H. and Matsudaira, P. (2007) *Molecular Cell Biology*, 6th Edn. W.H. Freeman and Company, New York.

Voet, D., Voet, J.G. and Pratt, C.W. (2008) *Principles of Biochemistry*, 3rd Edn. John Wiley and Sons, New York.

Watson, J.D., Baker, T.A., Bell, S.P., Gann, A., Levine, M., Losick, R. and Inglis, C. (2007) *Molecular Biology of the Gene*, 6th Edn. Pearson Education, San Francisco, CA.

More advanced reading

The following selected articles are recommended to readers who wish to know more about specific subjects. In many cases, they are too advanced for first-year students but are very useful sources of information for subjects that may be studied in later years.

Section A

Bergeron, J.J.M., Au, C.E., Desjardins, M., McPherson, P.S. and Nilsson, T. (2010) Cell biology through proteomics – ad astra per alia porci. *Trends Cell Biol.* 20: 337–345.

Brunet, S., Thibault, P., Gagnon, E., Kearney, P., Bergeron, J.J.M. and Desjardins, M. (2003) Organelle proteomics: looking at less to see more. *Trends Cell Biol.* 13: 629–638.

Egner, A. and Hell, S.W. (2005) Fluorescence microscopy with super-resolved optical sections. *Trends Cell Biol.* 15: 207–215.

Leis, A., Rockel, B., Andrees, L. and Baumeister, W. (2009) Visualizing cells at the nanoscale. *Trends Biochem. Sci.* 34: 60–70.

Lidke, D.S. and Wilson, B.S. (2009) Caught in the act: quantifying protein behaviour in living cells. *Trends Cell Biol.* 19: 566–574.

Lundin, V.F., Leroux, M.R. and Stirling, P.C. (2010) Quality control of cytoskeletal proteins and human disease. *Trends Biochem. Sci.* 35: 288–297.

McConnell, R.E. and Tyska, M.J. (2010) Leveraging the membrane – cytoskeleton interface with myosin-1. *Trends Cell Biol.* 20: 418–426.

Piston, D.W. and Kremers, G.-J. (2007) Fluorescent protein FRET: the good, the bad and the ugly. *Trends Biochem. Sci.* 32: 407–414.

Section B

Carugo, O. and Carugo, K.D. (2005) When X-rays modify the protein structure: radiation damage at work. *Trends Biochem. Sci.* 30: 213–219.

Feige, M.J., Hendershot, L.M. and Buchner, J. (2010) How antibodies fold. *Trends Biochem. Sci.* 35: 189–198.

Fitzkee, N.C., Fleming, P.J., Gong, H., Panasik Jr, N., Street, T.O. and Rose, G.D. (2005) Are proteins made from a limited parts list? *Trends Biochem. Sci.* 30: 73–80.

Harding, C.V. and Neefjes, J. (2005) Antigen processing and recognition. *Curr. Opin. Immunol.* 17: 55–57.

Hirokawa, N. and Takemura, R. (2003) Biochemical and molecular characterization of diseases linked to motor proteins. *Trends Biochem. Sci.* 28: 558–565.

Hogg, P.J. (2003) Disulphide bonds as switches for protein function. *Trends Biochem. Sci.* 28: 210–214.

Koonce, M.P. and Samsó, M. (2004) Of rings and levers: the dynein motor comes of age. *Trends Cell Biol.* 14: 612–619.

Livák, F. and Petrie, H.T. (2001) Somatic generation of antigen-receptor diversity: a reprise. *Trends Immunol.* 22: 608–612.

Manis, J.P., Ming Tian, M. and Frederick, W.A. (2002) Mechanism and control of class-switch recombination. *Trends Immunol.* 23: 31–39.

Royer Jr, W.E., Knapp, J.E., Strand, K. and Heaslet, H.A. (2001) Cooperative hemoglobins: conserved fold, diverse quaternary assemblies and allosteric mechanisms. *Trends Biochem. Sci.* 26: 297–304.

Yildiz, A. and Selvin, P.R. (2005) Kinesin: walking, crawling or sliding along? *Trends Cell Biol.* 15: 112–120.

Section C

Brunet, S., Thibault, P., Gagnon, E., Kearney, P,. Bergeron, J.J.M. and Desjardins, M. (2003) Organelle proteomics: looking at less to see more. *Trends Cell Biol.* 13: 629–638.

Goldman, R.D. (2000) Antibodies: indispensable tools for biomedical research. *Trends Biochem. Sci.* 25: 593–595.

Rappsilber, J. and Mann, M. (2002) What does it mean to identify a protein in proteomics? *Trends Biochem. Sci.* 27: 74–78.

Zhou, M. and Robinson, C.V. (2010) When proteomics meets structural biology. *Trends Biochem. Sci.* 35: 522–529.

Section D

Belenky, P., Bogan, K.L. and Brenner, C. (2007) NAD+ metabolism in health and disease. *Trends Biochem. Sci.* 32: 12–19.

Berger, F., Ramirez-Hernández, M.H. and Ziegler, M. (2004) The new life of a centenarian: signalling functions of NAD(P). *Trends Biochem. Sci.* 29: 111–118.

Daniel, R.M. and Danson, M.J. (2010) A new understanding of how temperature affects the catalytic activity of enzymes. *Trends Biochem. Sci.* 35: 584–591.

Krem, M.M. and Di Cera, E. (2002) Evolution of enzyme cascades from embryonic development to blood coagulation. *Trends Biochem. Sci.* 27: 67–74.

Section E

Bernards, A. and Settleman, J. (2004) GAP control: regulating the regulators of small GTPases. *Trends Cell Biol.* 14: 377–385.

Carafoli, E. (2004) Calcium-mediated cellular signals: a story of failures. *Trends Biochem. Sci.* 29: 371–379.

De Meyts, P. (2008) The insulin receptor: a prototype for dimeric, allosteric membrane receptors? *Trends Biochem. Sci.* 33: 376–384.

Gu, H. and Neel, B.G. (2003) The 'Gab' in signal transduction. *Trends Cell Biol.* 13: 122–130.

Higgins, M.K. and McMahon, H.T. (2002) Snap-shots of clathrin-mediated endocytosis. *Trends Biochem. Sci.* 27: 257–263.

Hofmann, K.P., Scheerer, P., Hildebrand, P.W., Choe, H.-W., Park, J.H., Heck, M. and Ernst, O.P. (2009) A G protein-coupled receptor at work: the rhodopsin model. *Trends Biochem. Sci.* 34: 540–552.

Merrill Jr, A.H., Wang, M.D., Park, M. and Sullards, M.C. (2007) (Glyco)sphingolipidology: an amazing challenge and opportunity for systems biology. *Trends Biochem. Sci.* 32: 457–468.

Mulkidjanian, A.Y., Galperin, M.Y. and Koonin, E.V. (2009) Co-evolution of primordial membranes and membrane proteins. *Trends Biochem. Sci.* 34: 206–215.

Neel, B.J., Gu, H. and Pao, L. (2003) The 'Shp'ing news: SH2 domain-containing tyrosine phosphates in cell signalling. *Trends Biochem. Sci.* 28: 284–293.

Parton, R. and Hancock, J.F. (2004) Lipid rafts and plasma membrane microorganization: insights from Ras. *Trends Cell Biol.* 14: 141–147.

Patel, S. and Docampo, R. (2010) Acidic calcium stores open for business: expanding the potential for intracellular Ca2+ signaling. *Trends Cell Biol.* 20: 277–286.

Taylor, C.W., da Fonseca, P.C.A. and Morris, E.P. (2004) IP_3 receptors: the search for structure. *Trends Biochem. Sci.* 29: 210–219.

Section F

Arezi, B. and Kuchta, R.D. (2000) Eukaryotic DNA primase. *Trends Biochem. Sci.* 25: 572–576.

Corpet, A. and Almouzni, G. (2009) Making copies of chromatin: the challenge of nucleosomal organization and epigenetic information. *Trends Cell Biol.* 19: 29–41.

Hübscher, U., Nasheuer, H.-P. and Syväoja, J.E. (2000) Eukaryotic DNA polymerases, a growing family. *Trends Biochem. Sci.* 25: 143–147.

Kelleher, C., Teixeira, M.T., Förstemann, K. and Lingner, J. (2002) Telomerase: biochemical considerations for enzyme and substrate. *Trends Biochem. Sci.* 27: 572–579.

Lansdorp, P.M. (2005) Major cutbacks at chromosome ends. *Trends Biochem. Sci.* 30: 388–395.

Section G

Bentley, D.L. (2005) Rules of engagement: co-transcriptional recruitment of pre-mRNA processing factors. *Curr. Opin. Cell Biol.* 17: 251–256.

Chadick, J.Z. and Asturias, F.J. (2005) Structure of eukaryotic Mediator complexes. *Trends Biochem. Sci.* 30: 264–271.

Conaway, R.C., Sato, S., Tomomori-Sato, C., Yao, T. and Conaway, J.W. (2005) The mammalian Mediator complex and its role in transcriptional regulation. *Trends Biochem. Sci.* 30: 250–255.

Granneman, S. and Baserga, S.J. (2005) Crosstalk in gene expression: coupling and co-regulation of rDNA transcription, pre-ribosome assembly and pre-rRNA processing. *Curr. Opin. Cell Biol.* 17: 281–286.

Guthrie, C. and Steitz, J. eds. (2005) Nucleus and gene expression. *Curr. Opin. Cell Biol.* 17, Issue 3. Whole issue devoted to relevant articles.

Hobert, O. (2005) Common logic of transcription factor and microRNA action. *Trends Biochem. Sci.* 29: 462–468.

Kornblihtt, A.R. (2005) Promoter usage and alternative splicing. *Curr. Opin. Cell Biol.* 17: 262–268.

Lilley, D.M.J. (2004) The origins of RNA catalysis in ribozymes. *Trends Biochem. Sci.* 28: 495–501.

Rino, J. and Carmo-Fonseca, M. (2009) The spliceosome: a self-organized macromolecular machine in the nucleus? *Trends Cell Biol.* 19: 375–384.

Russell, J. and Zomerdijk, J.C.B.M. (2005) RNA-polymerase-I-directed rDNA transcription, life and works. *Trends Biochem. Sci.* 30: 87–96.

Schellenberg, M.J., Ritchie, D.B. and MacMillan, A.M. (2008) Pre-mRNA splicing: a complex picture in higher definition. *Trends Biochem. Sci.* 33: 243–246.

Stuart,K.D., Schnaufer, A., Ernst, N.L. and Panigrahi, A.K. (2005) Complex management: RNA editing in trypanosomes. *Trends Biochem. Sci.* 30: 97–105.

Warf, M.B. and Berglund, J.A. (2010) Role of RNA structure in regulating pre-mRNA splicing. *Trends Biochem. Sci.* 35: 169–178.

Section H

Aebi, M., Bernasconi, R., Clerc, S. and Molinari, M. (2010) N-glycan structures: recognition and processing in the ER. *Trends Biochem. Sci.* 35: 74–82.

Connolly, K. and Culver, G. (2009) Deconstructing ribosome construction. *Trends Biochem. Sci.* 34: 256–263.

Giglione, C., Fieulaine, S. and Meinnel, T. (2009) Cotranslational processing mechanisms: towards a dynamic 3D model. *Trends Biochem. Sci.* 34: 417–426.

Hernández, G. (2009) On the origin of the cap-dependent initiation of translation in eukaryotes. *Trends Biochem. Sci.* 34: 166–175.

Hernández, G., Altmann, M. and Lasko, P. (2010) Origins and evolution of the mechanisms regulating translation initiation in eukaryotes. *Trends Biochem. Sci.* 35: 63–73.

Korostelev, A. and Noller, H.F. (2007) The ribosome in focus: new structures bring new insights. *Trends Biochem. Sci.* 32: 434–441.

Mitra, N., Sinha, S., Ramya, T.N.C. and Surolia, A. (2006) N-linked oligosaccharides as outfitters for glycoprotein folding, form and function. *Trends Biochem. Sci.* 31: 156–163.

Rodnina, M.V., Beringer, M. and Wintermeyer, W. (2007) How ribosomes make peptide bonds. *Trends Biochem. Sci.* 32: 20–26.

Rodnina, M.V. and Wintermeyer, W. (2010) The ribosome goes Nobel. *Trends Biochem. Sci.* 35: 1–5.

Section I

Blagoev, B. and Pandey, A. (2001) Microarrays go live – new prospects for proteomics. *Trends Biochem. Sci.* 26: 639–641.

Rommens, C.M. (2004) All-native DNA transformation: a new approach to plant genetic engineering. *Trends Plant Sci.* 9: 457–464.

Section J

Bolaños, J.P., Almeida, A. and Moncada, S. (2010) Glycolysis: a bioenergetic or a survival pathway? *Trends Biochem. Sci.* 35: 145–149.

Kim, J. and Dang, C.V. (2005) Multifaceted roles of glycolytic enzymes. *Trends Biochem. Sci.* 30: 142–150.

Papin, J.A., Price, N.D., Sharon J., Wiback, S.J., Fell, D.A. and Palsson, B.O. (2003) Metabolic pathways in the post-genome era. *Trends Biochem. Sci.* 28: 250–258.

Schmidt, S., Sunyaev, S., Bork, P. and Dandekar, T. (2003) Metabolites: a helping hand for pathway evolution? *Trends Biochem. Sci.* 28: 336–341.

Section K

Anderson, R.G.W. (2003) Joe Goldstein and Mike Brown: from cholesterol homeostasis to new paradigms in membrane biology. *Trends Cell Biol.* 13: 534–539.

Beglova, N. and Blacklow, S.C. (2005) The LDL receptor: how acid pulls the trigger. *Trends Biochem. Sci.* 30: 309–317.

Wallis, J.G., Watts, J.L. and Browse, J. (2002) Polyunsaturated fatty acid synthesis: what will they think of next? *Trends Biochem. Sci.* 27: 467–473.

Section L

Oster, G. and Wang, H. (2003) Rotary protein motors. *Trends Cell Biol.* 13: 114–121.

Tielens, A.G.M., Rotte, C., Hellemond, J.J. and Martin, W. (2002) Mitochondria as we don't know them. *Trends Biochem. Sci.* 27: 564–572.

Weber, J. (2007) ATP synthase – the structure of the stator stalk. *Trends Biochem. Sci.* 32: 53–56.

Section M

Braissant, O. (2010) Current concepts in the pathogenesis of urea cycle disorders. *Mol. Genet. Metab.* 100 Suppl. 1: S3–S12.

Downie, J.A. (2005) Legume haemoglobins: symbiotic nitrogen fixation needs bloody nodules. *Curr. Biol.* 15: R196–198.

Fontecave, M., Atta, M. and Mulliez, E. (2004) S-adenosylmethionine: nothing goes to waste. *Trends Biochem. Sci.* 29: 243–249.

Abbreviations

$\Delta E_0{}'$	change in redox potential
$\Delta G^{0\prime}$	standard free energy change of a reaction at pH 7
$\Delta G^{\ddagger}$	Gibbs free energy of activation
ACP	acyl carrier protein
ADP	adenosine diphosphate
AIDS	acquired immunodeficiency syndrome
Ala	alanine
ALA	aminolaevulinic acid
AMP	adenosine monophosphate
Arg	arginine
Asn	asparagine
Asp	aspartic acid
ATCase	aspartate carbamoyltransferase
ATP	adenosine triphosphate
ATPase	adenosine triphosphatase
bp	base pair
BPG	2,3-bisphosphoglycerate
BSE	bovine spongiform encephalopathy
bZIP	basic leucine zipper protein
Cα	α-carbon atom
cAMP	cyclic adenosine monophosphate
CAP	catabolite activator protein
cDNA	complementary DNA
CDP	cytidine diphosphate
CFP	cyan fluorescent protein
CFTR	cystic fibrosis transmembrane conductance regulator
cGMP	cyclic guanosine monophosphate
CJD	Creutzfeldt–Jakob disease
CM	carboxymethyl
CNBr	cyanogen bromide
CoA	acetyl coenzyme A
COPI	coat protein
CRP	cAMP response protein
CStF	cleavage stimulation factor F
CTD	C-terminal domain
CTP	cytosine triphosphate
Cys	cysteine
DAG	1,2-diacylglycerol
dATP	deoxyadenosine 5′-triphosphate
dCTP	deoxycytidine 5′-triphosphate
DEAE	diethylaminoethyl
dGTP	deoxyguanosine 5′-triphosphate
DIPF	diisopropylphosphofluoridate
DNA	deoxyribonucleic acid
dNDP	deoxyribonucleoside 5′-diphosphate
dNMP	deoxyribonucleoside 5′-monophosphate
DNP	2,4-dinitrophenol
dNTP	deoxyribonucleoside 5′-triphosphate
DPE	downstream promoter element
DRE	cAMP response element
dsRNA	double-stranded RNA
dTTP	deoxythymidine 5′-triphosphate
E	energy
$E_0{}'$	standard redox potential
EC	Enzyme Commission
ECG	electrocardiogram
eIF	eukaryotic initiation factor
ELISA	enzyme-linked immunosorbent assay
ER	endoplasmic reticulum
EtP	ethanolamine phosphate
ETS	external transcribed spacer
F-2,6-BP	fructose 2,6-bisphosphate
Fab	fragment antigen binding
FACS	fluorescence-activated cell sorter
FAD	flavin adenine dinucleotide
FBPase2	fructose bisphosphatase 2
FMN	flavin mononucleotide
FRAP	fluorescence recovery after photobleaching
FRET	fluorescence resonance energy transfer

G	Gibbs free energy
GDP	guanosine diphosphate
GEF	guanine nucleotide-exchange factor
GFP	green fluorescent protein
GlcNAc	N-acetylglucosamine
Gln	glutamine
Glu	glutamic acid
GLUT1	erythrocyte glucose transporter or uniporter
Gly	glycine
GPCR	G protein-coupled receptor
GPI	glycosyl phosphatidylinositol
GTP	guanosine triphosphate
H	enthalpy
HbA	adult hemoglobin
HbF	hemoglobin F
HbS	sickle-cell hemoglobin
HDL	high-density lipoprotein
His	histidine
HIV	human immunodeficiency virus
HLH	helix-loop-helix
HMG	3-hydroxy-3-methylglutaryl
HMM	heavy meromyosin
HPLC	high performance liquid chromatography
Hsp	heat shock protein
HTH	helix-turn-helix
Hyl	5-hydroxylysine
Hyp	4-hydroxyproline
IDL	intermediate-density lipoprotein
IF	initiation factor
IgA	immunoglobulin A
IgG	immunoglobulin G
IgM	immunoglobulin M
Ile	isoleucine
IP_3	inositol 1,4,5-trisphosphate
IPTG	isopropylthiogalactoside
IRES	internal ribosome entry site
ITS	internal transcribed spacer
K	equilibrium constant
k_b	reverse reaction rate
k_f	forward reaction rate
K_m	Michaelis constant
LCAT	lecithin–cholesterol acyl transferase
LDH	lactate dehydrogenase
LDL	low-density lipoprotein
Leu	leucine
LMM	light meromyosin
Lys	lysine
MALDI-TOF	matrix-assisted laser desorption ionization-time-of-flight
MAP kinase	mitogen-activated protein kinase
Met	methionine
miRISC	miRNA-induced silencing complex
miRNA	microRNA
miRNP	microribonucleoprotein
mRNA	messenger RNA
MS	mass spectrometry
MWC	Monod-Wyman-Changeaux model
NAD^+	nicotinamide adenine dinucleotide
NADH	reduced nicotinamide adenine dinucleotide
$NADP^+$	nicotinamide adenine dinucleotide phosphate
NADPH	reduced nicotinamide adenine dinucleotide phosphate
NAM	N-acetylmuramic acid
NHP	nonhistone protein
NMR	nuclear magnetic resonance
NOS	nitric oxide synthase
Nt	nucleotide
ORC	origin of replication complex
ORF	open reading frame
ORT	oral rehydration therapy
P	pressure
PAGE	polyacrylamide gel electrophoresis
PAP	poly(A) polymerase
Pc	plastocyanin
PEP	phosphoenolpyruvate
PFK	phosphofructokinase
PH	pleckstrin homology
Phe	phenylalanine
pI	isoelectric point

P_i	free inorganic phosphate
PI 3-kinase	phosphatidylinositol 3-kinase
pK	dissociation constant
PKA	protein kinase A
PP_i	inorganic pyrophosphate
Pro	proline
PSI	photosystem I
PSII	photosystem II
PTB	phosphotyrosine-binding
PTH	phenylthiohydantoin
RER	rough endoplasmic reticulum
RFLP	restriction fragment length polymorphism
RIP	regulated intramembrane proteolysis
RNA	ribonucleic acid
RNA Pol	RNA polymerase
RNAi	RNA interference
RNP	ribonucleoprotein
rRNA	ribosomal RNA
S	entropy
[S]	substrate concentration
SDS	sodium dodecyl sulfate
SECIS	SElenoCysteine Insertion Sequence
Se-Cys	selenocysteine
SER	smooth endoplasmic reticulum
Ser	serine
SH2	Src homology 2
SH3	Src homology 3
snoRNA	small nucleolar RNA
snRNA	small nuclear RNA

SREBP	sterol regulatory element binding protein
SRP	signal recognition particle
SSB	single-stranded DNA-binding
STR	short tandem repeat
T	temperature
TBP	TATA box binding protein
TCA	tricarboxylic acid
TFII	transcription factor for RNA polymerase II
TFIID	transcription factor IID
TFIIIA	RNA polymerase III transcription factor A
TFIIIC	transcription factor IIIC
TGF-β	transforming growth factor-β
Thr	threonine
Tris	tris(hydroxy methyl) aminomethane
tRNA	transfer RNA
Trp	tryptophan
trp	tryptophan operon
Tyr	tyrosine
U	enzyme unit
UBF	upstream binding factor
UCE	upstream control element
UDP	uridine disphosphate
UTP	uridine 5′-triphosphate
UTR	untranslated region
UV	ultraviolet
V	volume
Val	valine
VLDL	very-low-density lipoprotein
YFP	yellow fluorescent protein

Index